Environment, Development and Sustainability in India: Perspectives, Issues and Alternatives

Manish K. Verma
Editor

Environment, Development and Sustainability in India: Perspectives, Issues and Alternatives

 Springer

Editor
Manish K. Verma
Department of Sociology and Controller
of Examinations
Babasaheb Bhimrao Ambedkar University
(A Central University; Accredited 'A' Grade
by NAAC)
Lucknow, Uttar Pradesh, India

ISBN 978-981-33-6247-5   ISBN 978-981-33-6248-2   (eBook)
https://doi.org/10.1007/978-981-33-6248-2

This Springer imprint is published by the registered company Springer Nature Singapore Pte Ltd.
The registered company address is: 152 Beach Road, #21-01/04 Gateway East, Singapore 189721, Singapore

Dedicated to

Sir!

As a divine human being
your existence was a blessing
your reminiscence a fortune!
You are esteemed beyond words
and missed beyond measure!
Days will pass and turn into years
but you will be remembered forever
for your compassion, humility and
benevolence!

The lessons you taught
will forever remain treasured in my heart,
the novel concepts and perspectives you
propounded
will keep you eternal in the academia!

 November 11, 1932 - May 10, 2020

Prof. Yogendra Singh
Professor (Emeritus)
Professor of Sociology
Jawaharlal Nehru University
New Delhi

Preface

Due to upsurge in human population, development endeavours have increased manifold during the past centuries. To meet the overgrowing human development urge, human civilization has passed through different catalyst phases such as industrialization and globalization. These landmark phases of human development accelerated the process of mass production and consumption through the introduction of mega industries and machines, easy transportation and communication channels diluting the geographical barriers along with strengthening of trade and commerce. All this is witnessed at the cost of scarcely available natural resources. Due to incessant harnessing of natural resources beyond its carrying capacity, recent decades have shown disturbance in natural ecosystem which is not only causing manifold environmental problems but also looming as a threat to human existence and survival. Market failure coupled with institutional and policy shambles has aggravated the vulnerability of nature to a large extent. Consequently, fast depletion of natural resources, environmental catastrophe and most importantly widening of social and economic disparities among people and between nations are quite apparent today.

India, being the fastest growing economy and largest democratic country having second-largest population in the world, also could not remain untouched from the emergent development prerequisites and has got carried away in the wave to meet growing market demand of consumer products. Guha points out in a retrospective overview about Indian experience of development, 'in unanimously advocating rapid industrialization after independence, which further accelerated in the phase of globalisation, Indian economists have quarrelled merely over the means to achieve the ends. Not surprisingly, their discussions have ignored the environmental consequences of economic growth as well as finite nature of many natural resources'. Hence, the mad race to meet the development endeavours of a growing population not only marred the interests of the weaker section of the society in India but also created large-scale social and economic rifts, ecological imbalance and environmental destructions. Slowly and gradually, India has also witnessed similar backlashes which its western counterparts were experiencing. Under the circumstances, Pieterse has rightly put in, 'the crisis of developmentalism as a paradigm manifest itself as a crisis of modernism in the West and the crisis of development in South'. The figures of the most recent report released by Oxfam support the viewpoint. In signs of rising income inequality, India's richest 1% now holds a huge 42.5% of the country's total wealth as against 2.8% held by bottom 50%. Indian experience of inequality is in confirmation with the global scenario. World's richest 1% has more than twice the wealth of the humanity combined. Moreover, 46% increase in wealth of the top 1% is recorded, whereas only 3% increase in bottom 50%. The bigger worry now is utter inequalities in consumption and thus distribution of resources. In the background of rising inequality, the problem of global warming, climate change, depletion of ozone layer, decrease in forest cover, urban congestion, food scarcity, flood, tsunami, earthquake, issues of health, sanitation, drinking water, etc., is getting far more alarming than ever before.

India's speedy rise in population is impacting its environment in two major forms. The first includes consumption of resources comprising land, water, food, air, minerals and fossil fuels. The second is waste products involving pollutants (air and water), toxic materials and greenhouse gases. In the 1970s, many environmentalists warned of a possible crisis due to population explosion. In 1968, Garret Hardin wrote a paper 'The Tragedy of Commons', expressing concern about possible crisis humanity would face due to the exponential rise in population. The Stanford Professor Paul R Ehrlich and his wife Anne Ehrlich wrote 'Population Bomb' in 1968 which became overnight sensation. Proving the forecasts accurate in India, recent predictions warned that much of Mumbai and its suburbs could be under water due to climate change-related sea-level rise by 2050. Across India, 36 million people—equivalent to the population of state of Telangana—currently live below the elevation of an annual average flood. An ICIMOD (International Centre for Integrated Mountain Development) assessment of the Hindu Kush suggests that a 90% decline in glaciers through this century could place 86 million people in river basin at risk of flood and water insecurity and affect 10 times that number indirectly. The Intergovernmental Panel on Climate Change (IPCC) finds that warming already experienced between 1981 and 2009 has reduced wheat yields by 5%. Emergent 'attribution' science noted that a number of extreme weather events— floods, droughts and heatwaves—in South Asia are partially attributable to human-induced climate change. Moreover, the state of development-induced displacement is also not satisfactory in India due to being the densely populated country. At present, India hosts 16 per cent of the world's population with only 2.45% of global surface area and 4 per cent of the water resources. Enumerations confirm that during last four decades, over 20 million people involuntarily displaced due to development projects, but as much as 75% of them have not been rehabilitated, and more than 40% of them are tribal, rural and poorer section of the society. The statistics underlining that development-induced displacement is affecting rural and tribal people more stringently becomes much serious in the background that India is almost 81 per cent short of specialist in rural community health centres, which would have given health support to these aggrieved displaced and consoled them from the wrath of displacement. Vandana Shiva in the book 'Ecology and the Politics of Survival' examines how ecological movements have questioned the validity of dominant concepts of economic development in the world. The book is of invaluable help in understanding India's and the Third World's environmental predicament. In an analytical overview, she asserts that due to thoughtless acts, our environment has endured harsh treatment, and in many cases, irreversible damage. She holds modern economic development responsible for the conflicts and environmental degradation and proposes a new development theory which supports sustainable development and the people's rights to justice and peace. Murray Bookchin has rightly mentioned in this context, 'the assumption that what currently exists must necessarily exist is the acid that corrodes all visionary thinking'. Here, the emphasis is to alarm that the planet has limited natural resources and if harnessed recklessly will exhaust fast leaving us in a world of scarcity and chaos. Therefore, under these circumstances, in the 1972 book 'The Limits of Growth', the authors argued that either civilization or growth must end.

Most ironically, despite such a disturbing situation across globe, no one is willing to share the responsibility of environmental backlashes and the politics of allegation and blame-game is going on, wherein the developed nations are alleging developing and under-developed countries; and rich and dominant to poor and frail responsible for the present state of environmental catastrophe and similarly vice versa is also in vogue. Therefore, in a critical appraisal, Pieterse asserts that 'the present development discourse in its ahistorical and apolitical character is incapable of coming to terms with the realities of world power and global interests, as is evidenced in the question of Third World debt'. Beck, by indicating the contemporary state of Western countries, has rightly called them as 'Risk Society'. The same condition, by and large, has diffused to other parts of the world in the wave of globalization, industrialization, liberalization, free trade practices and the movement of goods, services, knowledge and most importantly 'people' across national and continental boundaries. Stretching the viewpoint to another level of discourse, scholars even go to the extent of considering imbalance and inequality in societies and nations due to myopic, short-sighted and transient development planning which neglected not only scant availability of natural resources but also human capital. Now, the psychologists are taking note of the gloom and despair in the backlash of earth's deteriorating health by confirming that it has been taking a toll on people's mental health. Swedish climate activist Greta Thunberg went into deep depression over the deteriorating state of the earth in a tender age of 11 years.

The inferences drawn from the present state of development and environment disembark readily comprehend, in the words of Bell, that 'environmental problems are not only problems of technology and industry, of ecology and biology, of pollution control and of pollution prevention, environmental problems are also social problems. Environmental problems are problems for society—problems that threaten our existing pattern of social organization and social thought. Environmental problems are as well problems of society—problems that challenge us to change those patterns of organization and thought'. Gradually, a clear understanding is filtering out from the discourse on environment and development that the people have created environmental problems and it is their responsibility to resolve them, and hence, 'alternative' path of development is urgently required.

Off late, the global world is getting awakened that the state of environment is under threat due to the looming risks, and hence, the planet earth is in deep crisis and on the threshold of extinction. Therefore, in the past couple of decades, many summits were held at global level to address such a burning issue. Most recent among them which extended mass appeal towards the spirit of environmental protection and sustainable development envisaged the Millennium Development Goals (MDGs) and Sustainable Development Goals (SDGs). The MDGs were eight international development goals for the year 2015 that had been established following the Millennium Summit of the United Nations in 2000, following the adoption of the United Nations Millennium Declaration. The goals seek reduction in poverty, illiteracy, sex inequality, malnutrition, child deaths, maternal mortality and major infections as well as creation of environmental stability and a global partnership for development. The SDGs succeeded MDGs. In 2015, the United Nations Member States agreed to adopt a shared blueprint for peace and prosperity for people and planet popularly called as 'The 2030 Agenda for Sustainable Development'. Set in 2015 by the United Nations General Assembly, the Sustainable Development Goals (SDGs) are intended to be achieved by the year 2030. The SDGs are extremely exhaustive in nature and include various aspects related to social, economic, health and environmental issues in the form of collection of 17 global goals conceived to be a 'blueprint to achieve a better and more sustainable future for all'. These goals acknowledge that ending poverty and other deprivations must go along with strategies that improve health and education, reduce inequality and stimulate economic growth. Moreover, the goals set in motion for a better future also promises to tackle climate change and to preserve oceans and forests. Therefore, by widening the scope of sustainable development, for the first time, SDGs not only considered environmental aspects but also social, economic, health and overall well-being of humans on the planet earth. Most remarkably, in order to make the whole process democratic, participative and inclusive, it envisages collective action of all UN member countries including developed and developing ones to realize the fruits of sustainable development. Nonetheless, despite such a meticulously planned social development goal and agreement among UN member countries to evolve a sustainable society, the desired result of SDGs is yet to figure out at global level. Since the objectives were set in motion in 2015, India being socially and environmentally sensitive welfare country has taken promising steps on meeting some of these goals but remains a long way off its target on many accounts, showing the 2019 SDG India Index compiled by Niti Aayog. The report underlines that of the 104 sub-indicators measured for the 17 Sustainable Development Goals, India has only met 6 targets so far which is quite dismal. On other environmental fronts also, India remains woefully behind the global world. In 2018's Environmental Performance Index, it ranked 177 out of 180. Similarly, it ranked 102 out of 117 countries on the Global Hunger Index in 2019. For a country with global leadership aspirations, India's woeful environmental ranking could be quite questionable.

However, as an outcome of summits and conferences, campaigns of NGOs and exponents of civil society along with large-scale media coverage on environment, the global world is united on approving that emphasis should be given to evolve mechanism of 'sustainability' and 'sustainable development'. The trends which are emerging question the inordinate development endeavours and long-term sustainability of contemporary industrial societies. Hence, Dunlap put forth a word of caution by affirming that 'moving toward a more sustainable society, for example, will require using natural resources far more efficiently in order to minimize both resources withdrawals and pollution resulting from resources extraction, use and disposal'. Therefore, the prime agenda should be to identify synergies and trade-off between development and environment. For that, to comprehend the environment and development paradoxes in its complexity and yield instant remedies, strategic move, improved governance, enhanced institutional ability and strong willpower are urgently required. The perennial situation essentially advocates long-term strategies for the integration of development endeavours, environmental policies and an effective blueprint for future growth based on the principle of 'MDGs' and 'SDGs'. The message is loud and clear that the quest of sustainability earnestly demands a perfect blend between social, economic and environmental needs so as to create ecological balance and a sustainable future. As a result, in recent decades, nature of developmental paradigm and issues of environmental sustainability have emerged as a focal concern for the whole world. Academicians, policymakers, planners and development practitioners are compelled to rethink the dominant developmental paradigm from an environmental perspective so that environmental social ills are remedied and environmental and social justice as the goal of sustainable development is accomplished. Now, the intelligentsia is vying for 'development that meets the needs of the present without compromising the ability of future generations to meet their own needs'.

Under this background, through the edited book, 'Environment, Development and Sustainability in India: Perspectives, Issues and Alternatives' attempt is made to examine the main ecological and developmental issues which are creating environmental threats in India and challenging the serene existence of human beings. Effort is made to construct a clear pattern of complex relation between development urges and ecological needs vis-à-vis environmental, social, cultural, economic, technological and political processes enveloping it in view of theorization of the issue to provide social scientists and policymakers a platform for present actions to be initiated and also direction for future course of action to save the planet earth. Moreover, by examining the recent trends of development in India and recording the development dilemmas which are creating ecological imbalances and environmental threats, the edited book explores certain alternative means of development in a view to engineer equitable, bearable and viable planet earth.

To fulfil the main objectives, the edited volume is divided into three parts, viz. 'perspectives on environment, development and sustainability', 'issues of environment' and the last section on 'alternatives for environmental sustainability'. Each section divulges into a very important aspect of the interlinkages between environment, development and sustainability. The volume is comprised of 18 selected papers, which includes introductory chapter, contributed by eminent subject experts. It begins with the introductory chapter which discusses the theoretical issues related to the title and uncovers the main discourse underpinning it while citing instances from Indian experience. Moreover, it provides a broader overview of the papers covered in the book with a critical appraisal. Part I, 'Perspectives on Environment, Development and Sustainability', is having five papers in all assessing various critical aspects which include culture, ecology and development; practice and ethics of development; appraisal of sustainability by revisiting enduring modern and identified tradition; paradoxical nature of rural development in contemporary India; and finally, man–forest interaction in a metropolis. Part II of the volume captioned 'Issues of Environment' underlines such aspects and dialogues which are creating ecological imbalances and eventually posing environmental threats. This section has seven articles penned by experts probing various alarming issues which have plagued planet earth from ecological sustainability standpoint. The issues covered in this part of the volume include environmental consequences of dams; impact of the River Ganga restoration and conservation efforts; agricultural distress and the problem of sustainability and nutritional security; environmental migration as a challenge to sustainable development; interface between tribes and eco-tourism; people's movement against ecological conservation policies in the Western Ghats; and lastly, assessment of balancing growth with livelihood sustainability through the lenses of land acquisition in India. The last section, Part III titled as 'Alternatives for Environmental Sustainability', is having five chapters which discuss the alternative approaches and means of development for bringing ecological balance and sustainability. This part comprises themes on constructed wetland for a sustainable approach to wastewater treatment; biofuels and environmental sustainability; alternative approaches to measure sustainability in a subsistence economy; traditional knowledge practice and medicinal plant for tribal health and sustainable development; and lastly, indigenous knowledge of women and water conservation.

The edited book is having a broader canvass covering the spectrum of both micro- and macro-issues. The papers included in the volume cover critical Indian incidences of environment, development and sustainability. Most importantly, all the chapters of the volume are written by renowned academicians and experts from India having contributed significantly through their valuable writings for the enrichment of the subject. The book is of immense significance to the academicians, researchers, post-graduate- and graduate-level students of social sciences and environmental studies; policymakers; development practitioners; and the NGOs working in the area of environment, development and sustainability.

Finally, it is time to extend gratitude and acknowledge all those who stood by me for the completion of the edited volume. At the outset, I owe a deep sense of gratitude to all the contributors of the volume. Without their novel contributions, dedicated hard work and support, the edited volume 'Environment, Development and Sustainability in India: Perspective, Issues and Alternatives' would have been a distant dream. They extended their full support, cooperation and forbearance at every stage of the completion of the book and show intense willingness to help me in the editing work to make the volume a concrete reality. I am thankful to Prof. Sanjay Singh, the Vice Chancellor of Babasaheb Bhimrao Ambedkar University, Lucknow, for sparing me from the university affairs which enabled me to pen down the volume in a focused way. The motivation and support of Prof. Kameshwar Choudhary, my senior colleague in the Department of Sociology, need a special mention here. He not only prepared me mentally to work for this volume but also provided all kinds of support and assistance in formulation, sketching and editing of the book. The meticulous discussions held with him to streamline the format of the book proved extremely helpful in shaping the volume in a presentable form. I also express my deep sense of gratitude to Prof. Vinay Shrivastava, Director, Anthropological Survey of India, Kolkata, whose support and guidance proved extremely valuable. He selflessly stood with me whenever I required any help from him. I take this opportunity to extend my gratitude to the most energetic, bright and promising academician, Dr. Venkatesh Dutta, who helped me in the completion of introductory chapter. I am equally thankful to the office assistant of the Department of Sociology, Mr. Ajay Kumar, for providing secretarial assistance. The contribution of my research scholar Mr. Pawan Kumar is quite appreciative in this regard who helped me immensely in collection of secondary literature required for the book.

It is high time to show my sincere gratitude towards my teachers. To begin with, I extend a deep sense of gratitude to the eminent sociologist, extremely nobleman and most fortunately my revered teacher, the legendary (late) Prof. Yogendra Singh, who always stood by me and showered his blessings during his lifetime. In a long list of virtuous teachers who played significant role in my carrier-building, the names of Prof. K. L. Sharma, Prof. T. K. Oommen, Prof. Nandu Ram, Prof. Anand Kumar and Prof. Ehshanul Haque are foremost notable. They merit applause at this special moment. It is their teaching and blessings which strengthened me at every moment of my life.

It will go without saying if I does not mention the contribution of my family members whose encouragement, cooperation and forbearance enormously helped me to complete the volume. It is with the inspiration and blessings of my late grandparents S. D. Verma and Kishori Devi that I am in this position to fulfil their dreams. Even from their heavenly abode, their blessings are encouraging and motivating me to work diligently to come up to their expectations. I take this opportunity to thank my parents, Sri U. K. Verma and Smt. Usha Kiran Verma, brother and sisters, in-laws and other relatives for their love and cooperation extended to me at every moment of my life. The cooperation, help and assistance extended by my wife Runu Verma and the patience of my beloved son Aradhya Dev are highly appreciable for the completion of this endeavour. After all, it is their encouragement, mental support and sacrifice which strengthened me to work on this volume. I must admit that my in-depth academic engagement for the timely completion of the volume undermined and took away their share of time and space which they would have certainly spent and enjoyed with me. Therefore, without their help, cooperation and forbearance, this work would not have taken a proper shape.

Finally, I owe a deep sense of gratitude to Ms. Satvinder Kaur, Mr. Ramesh Kumaran and their entire team at Springer Nature, New Delhi/Germany, for taking a personal interest in getting this volume out of press with a high level of efficiency and the least amount of aberrations and errors.

Finally, I must admit that I have benefitted greatly from the suggestions of many colleagues in the process of writing and editing this volume. However, the entire responsibility about the content and form of the edited volume is mine, and if there are shortcomings I am alone to be blamed.

Lucknow, India   Prof. Manish K. Verma

About This Book

The volume provides a comprehensive account of asymmetric linkage in the trilogy of environment, development and sustainability and its impact on society. It examines varied perspectives and issues of development related to environmental destruction and sustainability challenge reflecting on the Indian experience. By examining the recent trends of development and recording the dilemmas which are creating ecological imbalances, it explores some alternative ways of development to achieve sustainability. Divided into three parts, it has a broad canvass. The first section examines critically the 'perspectives' on ecology, practice and ethics, rural development and man–forest interaction in metropolis. 'Issues' of dams, river, agricultural distress, environmental migration, eco-tourism, ecological conservation and land acquisition are assessed in part second. 'Alternative' means of development is explored in unit third by incorporating chapters on constructed wetland, biofuels, subsistence economy, water and traditional knowledge practice. This interdisciplinary book is of immense significance to academicians, researchers, post-graduate- and graduate-level students of social sciences and environmental studies; policymakers; development practitioners; and NGOs working in the area of environment and development.

Contents

Editor and Contributors

About the Editor

Manish K. Verma is Professor, Head and Deputy Coordinator of the UGC–Special Assistance Programme in the Department of Sociology at Babasaheb Bhimrao Ambedkar (Central) University, Lucknow, India. Previously, he served in the Department of College Education, Rajasthan, as Lecturer, Senior Lecturer and Lecturer in Selection Grade for 10 years and briefly at NTPC, Corporate Office, New Delhi, as Sociologist. With a doctorate from Jawaharlal Nehru University, New Delhi, he has more than 22 years of teaching and research experience. He has published many books, *including Globalisation, Social Justice and Sustainable Development in India* (2017), *Peri-urban Environment* (2017), *Globalization and Environment: Discourse, Policies and Practices* (2015) and *Development, Displacement and Resettlement* (2004). Several of his research papers and chapters have been published in journals and edited volumes. He is a member of various professional bodies such as the International Sociological Association, Indian Sociological Society and Rajasthan Sociological Association. At present, he is a member of the Managing Committee of the Indian Sociological Society. His main research interests include environment and development, involuntary displacement, urban ecology, social justice and globalisation.

Contributors

Chittaranjan Das Adhikary is an Associate Editor of Sociology at Banaras Hindu University. His academic career spans over 17 years including few years of action sociology in close collaboration with civil society organizations. He holds a doctoral degree in Sociology and a degree in Rural Management from Xavier Institute of Management, Bhubaneswar. He specializes in sociology of democracy and development, religion and peasant economy. He has dozens of research papers in reputed journals in India and abroad and has authored three books including one on *Democracy, Religion and Exclusion: New Frontiers of Sociology*.

P. V. Basil is a research scholar from Western Ghats area currently pursuing research on Environmental Conservation.

Purba Chattopadhyay is Assistant Professor of Economics in the Department of Home Science, University of Calcutta. For the last fifteen years, she has been in teaching profession and bears interest in the areas of theoretical economics, Indian economic issues, food and nutritional security, health economics, human development and extension studies. She has published more than twelve papers in journals of repute and seven book chapters. She is Member of the Executive Committee of Bengal Economic Association and is Associate Editor of the peer-reviewed journal, '*Artha Beekshan*'.

Venkatesh Dutta is currently Program Director at DST-Centre for Policy Research and is Professor of Environmental Sciences at Babasaheb Bhimrao Ambedkar (Central) University, Lucknow, India. He is Fulbright Fellow and British Chevening Scholar. He also worked with the World Bank's Water and Sanitation Program (WSP) as a consultant during 2003–2004. He did his postdoctoral research at the School of Public Policy, University of Maryland, USA, and doctoral work in the area of Regulatory Policy at the Faculty of Policy and Planning, TERI School of Advanced Studies, New Delhi. He is Fellow of Society of Earth Sciences (SES), India, and Member of professional bodies such as International Association of Hydrological Sciences (IAHS), UK; Indian Water Resources Society (IWRS), IIT Roorkee; Indian Association of Hydrologists (IAH); Indian Science Congress Association and International Association of Ecological Economics, Maryland. He is the recipient of United Nations Institute of Training & Research (UNITAR) Fellowship for doing a training programme at EPFL, Lausanne; UGC Junior Research Fellowship (JRF/SRF), 2001–2005 in Environmental Sciences; Sir Ratan Tata Trust Award, 2006; GDN Research Medal Award, Global Development Network for best research paper in international development, St. Petersburg, Russia, 2005; GDN Research Medal Finalist, Kuwait, 2008, and Budapest, 2012; and Global Development Marketplace Finalist, The World Bank, Washington DC, 2008.

Ramanuj Ganguly is Professor of Sociology at West Bengal State University, Barasat, West Bengal, with 21 years of teaching experience. Apart from working as Registrar (officiating) in the University for one and half years, he has been engaged with four projects (funded by UGC and ICSSR). He is one of the founding members and executive office bearer of Sociological Association of West Bengal and Sociology Alumni Association of Calcutta University. He has worked with UGC-enlisted journals, as Associate Managing Editor of Bharatiya Samajik Chintan and as Assistant Editor of West Bengal Sociological Review. Presently, he is the Convener of the Research Committee on Sociology of Religion of Indian Sociological Society. He has published seven books and more than 37 papers in journals and/or book chapters.

Aritra Ghosh is Assistant Professor and Head, Department of Sociology, Serampore Girls' College, Hooghly, West Bengal. His working career started from 2011 at the Bara Bazar Bikram Tudu Memorial College, Purulia, West Bengal, and since 2012 to 2019, he has been engaged in teaching in the Department of Sociology, Chapra Bangaljhi Mahavidyalaya, West Bengal. He is also attached as an External Faculty Member in the Department of Rural Development Studies, University of Kalyani. He has been the author of more than 20 articles. He also worked as a UGC Project Fellow and ICSSR Research Associate in the Department of Sociology, University of Kalyani.

Namita Gupta is teaching human rights for the last 11 years at the Centre for Human Rights and Duties, Panjab University, Chandigarh, India. Presently, she is Chairperson of the Centre. She has participated in various events held globally such as UN Workshop on Climate Change Adaptation in Jeju, South Korea (2013); training programme on the Precautionary Principle by CEU, Budapest (Hungary) in 2015; World Congress of Sociology at Vienna, Austria (2016); World Congress of Sociology at Toronto, Canada (2018). She has authored a book titled *Environmental Administration in India: Issues and Concerns* and has edited another one on *Human Rights Issues in India*. Besides this, she has published more than 25 research articles and has presented 35 papers in national and international conferences. Her areas of specialization are environment and urban governance.

S. Gurusamy is currently serving as Professor and Director, Centre for Studies in Sociology in Gandhigram Rural Institute, Deemed to be University, Gandhigram, Tamil Nadu. He is also the Dean, School of Social Sciences in the University. He has also served as Director, Centre for Studies in Development and Resettlement in GRI. He has 160 articles in the journals and three sponsored research projects and conducted several seminars, workshops and symposiums and produced 20 Ph.D. so far. He has specialized in Development Sociology and Policy Studies. He has facilitated establishment of Centre for Study of Social Exclusion and Inclusive Policy in GRI and also served for many expert committees in various universities and also visited several foreign countries.

Amalendu Jyotishi is Professor at School of Development, Azim Premji University, Bengaluru. He is Ph.D. in Economics from Institute for Social and Economic Change, Bengaluru. Here he worked on the ecological economic issues of swidden agricultural systems for his Ph.D. thesis that received 'VKRV Rao Memorial best Ph.D. thesis award'. His research work covers issues relating to natural resources and institutions from institutional economics, commons and legal pluralism perspectives. He has published his research ideas in journals, books, edited volumes and working papers. He is in the Executive Committee of 'Asian

Initiative on Legal Pluralism' and Indian Society of Ecological Economics. He has collaborated in research projects supported by organizations like NWO, SIDA, IWMI, Oxfam (GB) Trust, AKRSP (India), SANDEE, ARC, SSHRC.

Ashok Kaul Professor of Sociology and IIAS Fellow (Baden-Baden) Germany, is author of four books that include the most illustrious fiction on Kashmir 'Nativity Reclaimed'. He was Postdoctoral Fellow at the University of Klagenfurt, Austria (1991–1992), and the Department of Sociology, University of Alberta, Edmonton (1993–1994), and has been invited Visiting Faculty in the Department of Sociology, Delhi School of Economics, University of Delhi (2002), at CSSS, Jawaharlal Nehru University, New Delhi (2005), Central University, Tezpur, and also at GND University, Amritsar (twice). He has been bestowed the prestigious KECCS Award, New Delhi (2016), and IIAS Fellowship at Baden-Baden, Germany, in 2018.

Er. Ravindra Kumar is Advisor, WWF-India, New Delhi; Advisor (Water), DST Centre for Policy Research, BBA Central University, Lucknow; Member (E-flow Group), Consortium of 7 IITs for National Ganga River Basin Management Plan (NGRBMP); Reviewer: International Journal on Management of Environmental Quality, UK; Convener: Indian Water Resources Society, Lucknow Centre; and Former Faculty at MN NIT, Allahabad, WALMI, U.P., SIRD and SDMA. He served Irrigation Department, GOUP in various capacities and superannuated as Drinking Water Expert, State Water Resources Agency. He was Coordinator: Climate Change Cell at SWARA for UP Water Mission, worked for World Bank Survey 2017 (Water Laws and Regulation in Uttar Pradesh State) and also for ongoing Survey 2019 Enabling Business for Agriculture.

Sambit Mallick is currently Associate Professor of Sociology in the Department of Humanities and Social Sciences, Indian Institute of Technology Guwahati. He specializes in the sociology of science and technology, which also includes historical sociology and philosophy of the social sciences among his research interests. His research and teaching are at the intersection of 'philosophy', social theory and science studies.

M. Manjula is Faculty at School of Development, Azim Premji University, Bengaluru. She holds an interdisciplinary Ph.D. in Agricultural Economics. Her research interests are in the field of community-based adaption to climate change, agro-biodiversity and natural resource management and rural livelihood issues. She has publications in the form of journal papers, book chapters and research reports. She has coordinated research projects supported by ENERGIA, AusAID, Indo-Swiss Collaboration on Biotechnology (ISCB), SANDEE and Ministries of Government of India including Women & Child Development, Ministry of Environment Forest and Climate Change, Ministry of Rural Development, Department of Science & Technology.

Samita Manna, Ph.D., D.Litt. is former Vice-Chancellor of Sidho-Kanho-Birsha University, Purulia (2012 to 2014), and presently Professor in the Department of Sociology, University of Kalyani, West Bengal. Her working career started in 1979 at the Indian Statistical Institute, Calcutta, and since 1980, she has been engaged in teaching and research in the Department of Sociology, University of Kalyani, West Bengal. She has been the author of more than 15 books and 80 articles.

Tapan R. Mohanty is Dean of Online and Distance Learning Programme and Chairperson of Centre for Socio-Legal Studies at National Law Institute University at Bhopal. He has two books and around 50 articles in reputed international and national journals. He has visited almost 20 countries of the world including conferences of Durban, Honolulu and Copenhagen. He is the member of State Supervisory Board of Implementation of PCPNDT Act in Madhya Pradesh and recipient of Erasmus Mundus fellowship of European Union. He is a life member of Indian Sociological Society, Indian Society of Criminology and member of International Sociological Association (Madrid, Spain).

Siddhartha Mukerji is Assistant Professor in the Department of Political Science, Babasaheb Bhimrao Ambedkar University, Lucknow. He has been a Baden-Wurttemberg fellow to South Asia Institute, Heidelberg University, Germany. His areas of interest are electoral politics in India, public administration and e-governance, and political economy. He has published research papers and book chapters with renowned publishers/journals like Sage, Routledge and Economic and Political Weekly. His recent book on *India's Software Industry: Politics, Institutions and Policy Shifts* (2018) is yet another scholarly contribution. He has been associated with EECURI network of the India Europe Project studying urban–rural patterns of electoral change in Indian states and presented papers in its international conferences held at JNU (New Delhi), King's College London and London School of Economics and Political Science.

B. K. Nagla has retired as Professor of Sociology from M. D. University, Rohtak, Haryana. After his retirement, he worked as consultant at Kota Open University and also joined Banaras Hindu University as Professor on Babu Jagjivan Ram Chair. He has been Visiting Professor in many universities in India and abroad. He has been awarded ISC Fellow for the Growth and Development of the Indian Society of Criminology by the Council of the Indian Society of Criminology, affiliated with the International Society of Criminology, Paris. He has been conferred lifetime achievement award from Rajasthan Sociological Society in 2013. In 2016, Sulabha International Social Organization awarded him 'Sulabha Swachhata Samman' for authoring an outstanding book on 'Sociology of Sanitation'. At present, he is editor of Indian Sociological Society (ISS) Hindi Journal *Bhartiya Samajshastra Samiksha*'.

R. R. Patil is presently serving as Professor in the Department of Social Work, Faculty of Social Sciences, Jamia Millia Islamia, New Delhi. He also served as 'Professor' and Head, Department of Social Work and Dean, School of Social

Sciences, Central University of Rajasthan, Ajmer, during 2012–2014. He is associated in different capacities with several universities such as University of Delhi, Indira Gandhi National Open University, Visva-Bharati, M. S. University of Baroda, Jiwaji University, Bharati Vidyapeeth Deemed University. He has published two books and several scholarly articles/research papers in the international and national journals. His areas of research interest include Development Studies, Education, Civil Society and Marginalization.

Biswajit Paul is Assistant Professor in the Department of Sociology in Sidho–Kanho–Birsha University at Purulia, having research interest in community study, caste, Indian sociology and development.

Karuna N. Pohekar is currently working as Sr. Lab Assistant in the Department of Biosciences, Jamia Millia Islamia, New Delhi, and pursuing Ph.D. in Environmental Sciences from Guru Gobind Singh Indraprastha University, Dwarka, New Delhi.

Dr. Neetu Rani is Assistant Professor, University School of Environment Management, Guru Gobind Singh Indraprastha University. She has obtained Ph.D. in Environmental Sciences from IIT Delhi.

Clare Lizamit Samling is Assistant Professor in the Department of Sociology, University of Calcutta, where she teaches courses on Social-Anthropological Perspectives, Rural Sociology, Sociology of Development and Research Methodology. Earlier she worked as Research Fellow in two projects, i.e. 'Tribal Rights and Impact of Panan HEP and Teesta IV hydel power projects in Sikkim', funded by National Human Rights Commission, from 2013 to 2014 and DECCMA (DEltas vulnerability and Climate Change: Migration and Adaption) project, funded by IDRC, Canada, and DFID, UK, from 2014 to 2016. Her area of interest lies in environmental sociology, displacement, resettlement and rehabilitation. She has published many articles in peer-reviewed journals along with selected working papers.

K. L. Sharma was Professor of Sociology at Jawaharlal Nehru University, New Delhi, India, and superannuated in 2004. Besides teaching and research, he has been engaged in academic administration as Head of the Centre (CSSS, JNU); Dean of Students, JNU; Rector (Pro-VC), JNU, Vice-Chancellor (Rajasthan University, Jaipur) and Pro-Chancellor (Jaipur National University, Jaipur). He has published more than two dozen books and nearly one hundred articles. His latest book is on Caste, Social Inequality and Mobility in Rural India (Sage, 2019). He also writes for newspapers on current social issues.

Ritu Sharma after attaining M.Phil. and Ph.D. from JNU, New Delhi, is working as Senior Faculty since over a decade at Kamala Nehru College, University of Delhi. She headed the Department of Sociology from 2014 to 2016 and briefly during December-April 2020. More than a dozen of papers in reputed journals and edited volumes are published by her. Her area of interest includes inequality and stratification, environmental sociology, gender and society, media studies, visual culture and cultural studies, etc.

Rahul Shukla is pursuing Ph.D. in Sociology in the Department of Humanities and Social Sciences, Indian Institute of Technology Guwahati, India, since July 2014. He holds a Bachelor of Engineering degree in Mechanical Engineering from Rajiv Gandhi Technical University, Bhopal, and a Master of Arts degree in Development Studies from IIT Guwahati. His research interests include the interface between science, technology and society, focusing on the Renewable energy and agriculture in India.

Vinay Kumar Srivastava (Late) was Director, Anthropological Survey of India. Before taking up this position, he was Professor of Social Anthropology and Head of the Department of Anthropology, University of Delhi. He also served as the Principal of Hindu College. He studied both anthropology and sociology at the University of Delhi; and later, earned his doctorate from the University of Cambridge. His area of interest was social theory and methodology, tribal India and comparative religion.

Manish K. Verma is Professor in the Department of Sociology and Controller of Examinations at Babasaheb Bhimrao Ambedkar (Central) University, Lucknow. Earlier, he served as Head and Deputy Coordinator UGC SAP DRS-I in the Department of Sociology, and Sports Coordinator of the University. Having obtained doctorate degree from Jawaharlal Nehru University, New Delhi, he has more than 22 years of teaching and research experience. During his two decades long stint in academic arena at various positions, he has published many books, which include, 'Globalisation, Environment and Social Justice: Perspectives, Issues and Concerns' (2018; Routledge, Taylor & Francis Group, UK), 'Globalization, Social Justice and Sustainable Development in India' (2017; Gyan Publishing House, New Delhi), 'Peri-Urban Environment' (2017; Winshield Press, Delhi), 'Globalization and Environment: Discourse, Policies and Practices' (2015; Rawat Publications, Jaipur & New Delhi), and 'Development, Displacement and Resettlement' (2004; Rawat Publications, Jaipur & New Delhi). Several research papers and chapters have been published by him in reputed journals and edited volumes apart from conducting research projects independently. He is member of various national and international professional bodies like International Sociological Association, Indian Sociological Society and Rajasthan Sociological Association. Moreover, at present, he is holding the position of Chief Coordinator (Programme), Research Committee WG 05 (Famine and Society), International Sociological Association (Madrid, Spain) and Member of the Managing Committee of Indian Sociological Society, New Delhi. At present, he is Chairman, U.P. State-level Committee to evaluate the Socio-Economic Studies Reports of various infrastructure development projects and suggests Rehabilitation Measures

(Poorvanchal Expressway, Lucknow Metro, etc.) and Member, 'Institutional Ethics Committee', Sanjay Gandhi Post Graduate Institute for Medical Sciences (SGPGI), Lucknow. His main research interests concern environment and development with micro-level studies in the areas of involuntary displacement, urban ecology, social justice and globalization.

List of Figures

List of Tables

Chapter 1
Introduction

Manish K. Verma and Venkatesh Dutta

Abstract The introductory chapter discusses the theoretical issues and the inter-linkages between the title 'environment, development and sustainability'. Further, it uncovers the main crisis and risks which are looming on planet earth due to incessant harnessing of natural resources and consequent degradation of environment by citing the Indian experiences. The steps taken by governance, intelligentsia and civil society to control the environmental pitfalls along with the analysis of their success and failure are the main content of this chapter. Moreover, the paper explores the possibility of alternative path of development for bringing sustainability. Lastly, it provides a broader canvass of the papers covered in the book with a critical appraisal.

Keywords Environment · Development · Sustainability · Environmental degradation · Development alternatives · Risks

Introduction

During the early 1960s, the term environment was 'marginal' and sparingly used in the context of development, with social construct of the nature as 'green ideas' over *business as usual*. It was portrayed as a corridor to all that is superior, organic and enviable in the society. The world witnessed a rapid industrial and urban growth. A minority concern and an armchair construct suddenly became important and the notion of sustainability was debated in the academic circle as traditional wisdom to be adhered to in a *precautionary* sense (Adams 2003). The concept of green development was still academically marginal. It became fashionable to talk about it among

M. K. Verma (✉)
Department of Sociology and Controller of Examinations, Babasaheb Bhimrao Ambedkar (Central) University, Lucknow, India
e-mail: mkvbbau@gmail.com

V. Dutta
Department of Environmental Sciences, DST-Centre for Policy Research, Babasaheb Bhimrao Ambedkar (Central) University, Lucknow, India
e-mail: dvenks@gmail.com

the economists and social scientists during the early 1990s. The concept of development ran parallelly with sustainability with dissipated connections and metaphorical elusiveness. Ecocentric demand of respecting nature's carrying capacity and humanistic demands of growth and production took different strands of ethical and moral arguments.

The concept of sustainable development linked two different worlds together—that of environment and development. It became the most commonly cited term with the publication of the Brundtland Report called *Our Common Future* as an outcome of the Rio Conference, which further shaped the ideas about taking an ecocentric approach in economic development (Brundtland et al. 1987). It gave a fresh look at the interconnected and interdependent issues of environment and development. It also gave a shared vision of long-standing environmental issues and aspirational development goals beyond the twenty-first century. Since the publication of this report, the world has witnessed rapid technological transformations with new and emerging threats to the humanity in the form of environmental disasters which are rooted in our erroneous and immoral policy choice and actions. Many scholars have seen poverty and inequality as the main cause of environmental degradation in the Global South. The development agencies have increasingly focused their strategies and funding to improve the well-being and livelihoods of poor people who are distanced from the mainstream development. The international programs and schemes brought greater possibilities of connecting environment and development together in a common thread with national and regional political commitments. However, it was later realized that such strategies and actions were based on attributes that are socially desirable, and are abound with local and project-level problems—they were not desirable from the ecological standpoint in true sense. There are three major pillars of achieving sustainability and they should be addressed from very beginning—*first of all*, the idea of long-termism of ecological resilience has to be safeguarded and not merely short-term cosmetic environmental improvements. *Secondly*, the basic needs of housing, food and security have to be fulfilled and *finally*, both intragenerational and intergenerational equity have to be ensured in resource use and consumption. This posed a huge challenge to the development professionals and policymakers as it was virtually impossible to meet all of these criteria. In fact, the concept became so complex and classically extraneous that many scholars started distancing from this term and preferred to use *development* in strict sense in guiding policymaking and project design. Categorically, the concept of sustainable development is a significant idea, which was noticeably pointed at the UN Conference on Sustainable Development (commonly called as *Rio+20*), held in Rio de Janeiro in June 2012. The concept of sustainability was shaped by various international debates and discourses, most of them focused on promoting intergenerational equity, quality of life, carrying capacity of the ecosystem and justice and fairness for all the people. These approaches attempted to bring together the concerns of varied constituencies on a single platform, with placing the development and environmental communities on the forefront.

Sustainability as the New Paradigm of Development

From a 'development catchphrase' to the 'new paradigm' of development, the concept of sustainable development has evolved over a long period with a lack of uniformity and consistency in its interpretation (Lélé 1991). Initially, their meanings remained vague, and the concept did not include poverty, community participation and social well-being, ultimately leading to contradictions in policymaking (Daly 1990; Costanza and Daly 1992). It also lacked clarity and rigour in its implementation. But this weakness provided a lead for academic research as well as the political opening of action projects by multilateral agencies and development banks (Harris 2000; Pearce 2014). Development practitioners acknowledged that the benefits of progress did not automatically 'trickle-down' to the base of the pyramid, to those who needed them the most. According to Harris (2000) (Fig. 1.1),

> …powerful paradigms of thought are usually relatively straightforward and easy to grasp. In the area of social science, ideas that affect millions of people and guide the policies of nations must be accessible to all, not just to an elite. Only thus can they permeate institutions from the local to the global level, and become a part of the human landscape, part of the fabric within which we define our lives.

While growth is mainly defined by the GDP by most of the countries globally, the concept of Human Development Index (HDI) introduced in the 1990s is recognized as a more sensible measure of progress. The HDI integrates social and sustainable goals in a much meaningful way, though it is not used as a standard measure of economic development. However, even HDI fails to capture broader environmental issues and challenges such as climate change and ecological deterioration (Hickel 2020). By focusing on income growth, one may underestimate its impacts on pollution and environmental disasters. This inherently violates sustainability principles as clear

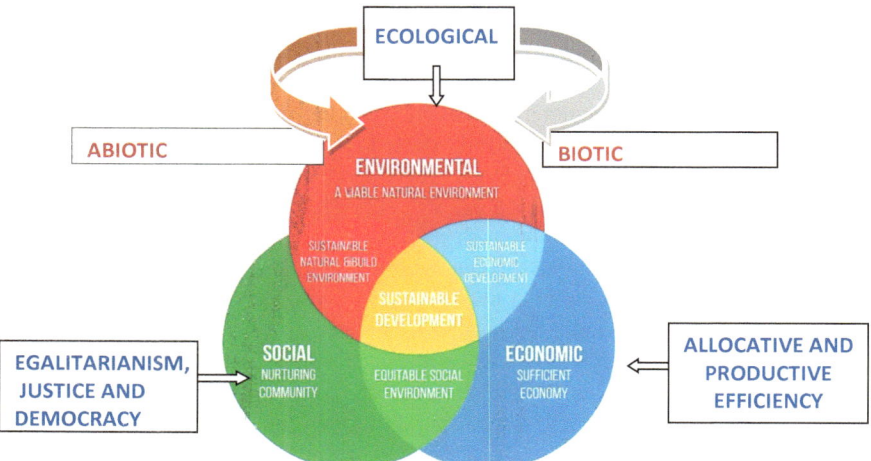

Fig. 1.1 Pillars of sustainable development and their interdependencies. *Source* Authors

from the fact that those countries that score high on HDI also account for higher per capita contribution of carbon emissions and pollution. So Sustainable Development Index (SDI) may be required as an integrative principle that incorporates social and ecological resilience and better measures nations' environmental efficiency in delivering social and human development goals.

Some scholars even tried to empirically measure main dimension of sustainability through indicators and thresholds—for example, sustainable development requires per capita ecological footprints less than 2.3 global hectares; more than 0.63 HDI; Gini coefficient less than 40 and share of renewable energy more than 27% (Holden et al. 2014). It is revealed that no countries in the world meet these four criteria, which calls for a massive transformative change in our economy, technology and behavioural pattern.

The Concept of 'Environmentalism' in the Development Context

The concept of 'environmentalism' in the development context only got enshrined during the early 1960s. The idea was not static and kept on changing meaning, evolving with economic realities and further influenced by the broader context of social and ecological resilience. This global scenario also applies to the Indian landscape wherein environmentalism developed and changed through various decadal cycles—from an ad hoc approach in the 1970s to a reactive framework during the 1980s. It was only during the early 1990s that anticipatory and integrative approaches were appreciated and discussed in the planning and policy literature. The post-1972 period (after Stockholm's conference) witnessed the framing of contemporary legislation on environmental laws in India. In 1972, the National Council for Environmental Policy and Planning was established. This institution evolved into a full-fledged Ministry of Environment and Forests (MoEF) in 1985. Later on, several judgements were passed by the Indian judiciary showing concerns for the protection and conservation of the environment. The proper integration of environmental concerns in the appraisal of projects only started after the framing of environmental impact assessment notification in 1994. The concept of allocative and productive efficiency in environmental resource use was still not a preferred subject to discuss in economic and social development discourses. Therefore, it is seen that various approaches have been adopted at different times as development policy has evolved following a long-term process of contestation and evolution (Fig. 1.2). The original prominence was given to rapid industrialization, urbanization and promoting more intensive agriculture through Green Revolution technologies.

The policies regarding environmental protection have changed very rapidly in India influenced mainly by the international diplomacy and through various legislations, creating of new institutions, setting up of National Green Tribunal as well as

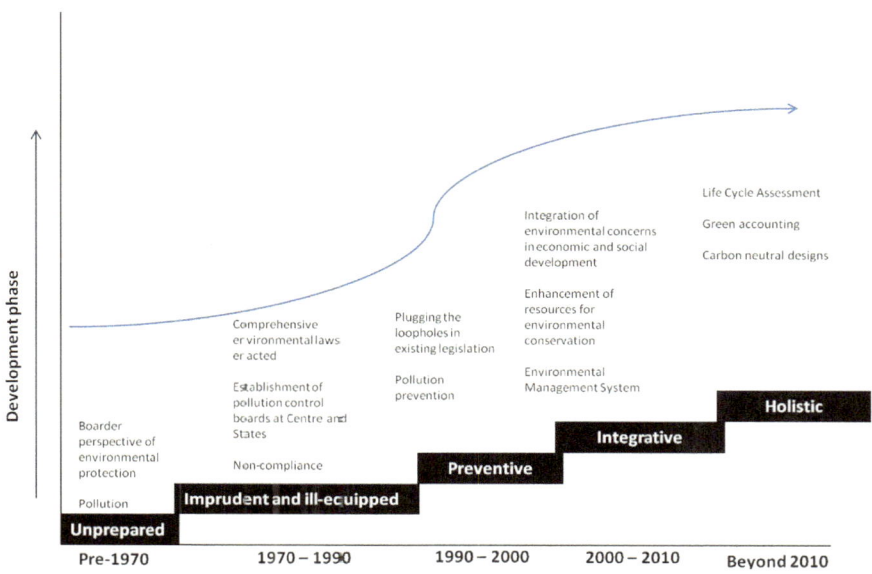

Fig. 1.2 Evolution of environmental framework in development policy. *Source* Authors

the critical judicial interpretations, but still, a holistic and integrative sustainability science approach is lacking, and there is need of further development in this regard.

During the 1980s, the focus shifted to the framing of general environmental laws and policies with the establishment of state and central pollution control boards. This phase also witnessed more comprehensive structural adjustments, beginning of the economic reform with liberalization of trade, eliminating government deficits through cess and taxation, and dismantling of inefficient parastatal organizations. The decade of 1990–2000 saw the preventive phase of policymaking wherein pollution prevention and polluter pay principle was given prominence. Various legislations were reviewed and amended, along with the framing of new ones. A flip side of the economic reform was that market-oriented policies failed to address the more significant issue of inequality and economic adversity of the poor even as productive and allocative efficiency improved. Therefore, it is argued that social inclusion is the primary driver of sustainability, which spurs resource efficiency and inclusive growth (Fig. 1.3).

Operationalizing Sustainable Development Goals (SDGs)

The Sustainable Development Goals (SDGs) set up by the United Nations in 2015 and the Paris Agreement on Climate Change are two important contemporary milestones in environmental policymaking that call for more profound transformative

Fig. 1.3 Wheels of sustainability—with social inclusion as the primary driver. *Source* Authors

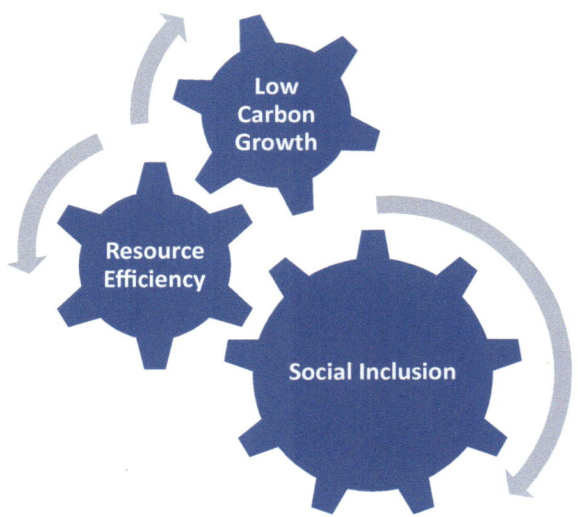

actions in every country. The SDGs have escalated considerable pressure on national and regional economies and business groups to achieve them by 2030. This will require harmonizing development with social equity and ecosystem resilience by governments, civil society, science and business groups (Sachs et al. 2019; Silvestre and Tîrca 2019). There are strong interdependencies across the 17 SDGs. However, there is a lack of clarity and rigour in a shared understanding of how these 17 SDGs can be operationalized across different regions. Sachs et al. (2019) have proposed six building blocks for achieving these SDGs: (i) education, gender and inequality; (ii) health, well-being and demography; (iii) energy decarbonization and sustainable industry; (iv) sustainable food, land, water and oceans; (v) sustainable cities and communities; and (iv) digital revolution for sustainable development. Each of these building blocks identifies the main factor of investments and regulatory environments that calls for significant transformative changes in governance and institutional design at local, regional and national levels.

Fukuda-Parr and Muchhala (2020) argue that the South played a crucial role in putting the sustainable development agenda on the global landscape—which questioned the consensus of the emerging powers of the North. The development actors and entrepreneurs from the Global South proposed an alternative vision of SDGs as a successor to the MDGs. It largely stemmed from contesting mainstream development views and advancing marginalized ideas.

Besides the three primary dimensions of sustainable development, there are secondary aspects which are often targeted by the development and policy professionals (Høyer 1999)—they are, quality of life concerns, nature's intrinsic value, public participation and overall protection of the environment (Fig. 1.4). Though, we may argue that the primary dimension of sustainable development as ingrained

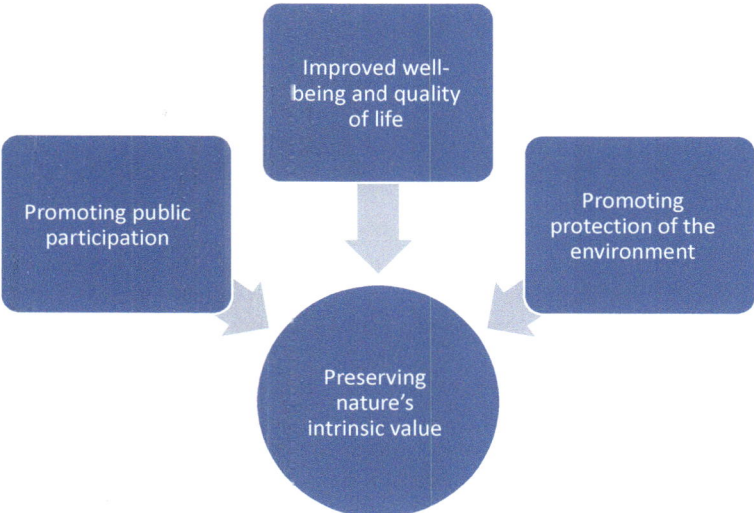

Fig. 1.4 Secondary dimension of the sustainable development based on Høyer (1999). *Source* Authors

in economic efficiency and social equity may be counterpart to an improved standard of living. But, in reality both economic efficiency and social equity may not be enough, and may be far away from what can be considered as truly sustainable. Public participation and access to the policy and decision-making system is very crucial to designing sustainable schemes and programs.

Structure of the Volume

The present volume is divided into three parts, viz. 'perspectives on environment, development and sustainability', 'issues of environment' and the last section on 'alternatives for environmental sustainability'. Each section divulges into a very important aspect of interlinkages between environment, development and sustainability. The volume comprises of eighteen selected papers, which includes introductory chapter, contributed by eminent subject experts.

The introductory chapter of the book is sketched by Manish K. Verma and Venkatesh Dutta. The chapter not only provides the conceptual understanding and theoretical debates but also offers contemporary discourses linking back the theoretical foundation. It uncovers the main crisis and risks which are looming on planet earth due to incessant harnessing of natural resources and consequent degradation of environment. It ponders into various perspectives and issues through the Indian lenses along with exploring the possibility of alternative means of development for bringing sustainability. Lastly, it provides a broader canvass of the papers covered in the book with a critical appraisal.

Thereafter, the book enters into the Part I captioned, 'Perspectives on Environment, Development and Sustainability'. It is having five papers in all. They assess various critical aspects including culture, ecology and development; practice and ethics of development; appraisal of sustainability by revisiting enduring modern and identified tradition; paradoxical nature of rural development in contemporary India; and finally, man-forest interaction in a metropolis.

The section begins with a chapter on 'culture, ecology and development' penned by B. K. Nagla. The paper examines the relationship between ecology, culture and development to understand human beings and their environment in contemporary society. The author delineates that ecology is the study of interrelation between organisms and their environment. By environment, he underlines, the natural environment including forests, rivers, lakes, seas, mountain, plants, etc. Thereafter, the author discusses in what way ecology is inextricably related with the processes of development and also highlights the problems which crops up when both the elements become dichotomous. By discussing so, the main focus is on to enunciate the role culture plays in the process of adaptation with natural surroundings on 'ecological niche' and also to highlight the cultural traits and complexes peculiar to particular society which develop in the process. Moreover, it also evaluates the role and significance of culture in the process of development with emphasis that, if overlooked, it fractures the very process of development. Afterwards, effort has been made to use diverse approaches of environmental and ecology to understand the connection between ecology and culture, culture and development, and ecology and development. Cultural ecological approach has been highly emphasized by the author while explaining relation between ecology, culture and development. Further, the paper also tries to explain the perspectives and imperatives as the perspectives provide vistas and visions of environment, and the imperatives call for the action that may bring true equilibrium, and thereby, improve the quality of life and environment.

Keeping tribes at focal point, Vinay Kumar Srivastava deliberates 'on development' by reflecting on 'practice and ethics' involved in it. He underlines three arguments which are in vogue with respect to the issue of development, which is described today as the main mantra of change. The author emphasizes in the sense that if society is not 'developing' (like some of the isolated tribal communities of the world), it is labelled as 'stagnant', 'changeless', 'backward and stuck' and 'primitive'. First, development is a Western concept, embedded in a cultural context qualitatively different from traditional societies of the east. It has been devised by the affluent among the west, being virtually imposed on rest of the world, particularly the traditional and poor societies, which had recently emerged free after the liquidation of the colonial rule, so that they come at par with the developed world. Second, the recipients of the 'development technology' that principally came from the west were apparently poor but 'squatted' on vast mineral and forest resources. As development proceeded, they became the most exploited, for they had to part with their precious natural wealth in exchange for the 'development benefits' they were getting. As development proceeded unabated, they became poorer and poorer, inhabiting ecologically unsustainable habitat. He stresses that development multiplied global inequality manifold. Within the 'developing nations', the benefits of change were not equitably distributed, so inequality sharpened at the local level as well. Third, the author further contends that overall impact of the western style of development has been villainous to the environment. It has lethally affected the biodiversity, accelerated global warming, rendering the earth uninhabitable. The changes, however, are not partisan, for they affect all parts of the world; therefore, the onslaught that the djinn of development is causing needs to be checked. Against this backdrop, Srivastava points out, the notions of 'alternatives to development' and 'alternatives in development' have come into being. Moreover, the contemporary discourse on development is intertwined with ethical issues. The article addresses these issues with insights from lifestyles that the Indian tribes had been leading for years.

In the article on 'sustainability', Ashok Kaul and Chittaranjan Das Adhikary is 'revisiting enduring modern and identified tradition'. The authors pronounce that it is an era of crisis; some manufactured and some caused by the rupture of modernity. The global world is multicultural, but the state boundaries are the packages of history that has made passports significantly essential. The disparities are across the globe, yet east is distinct than the west. The Western society did develop; when nature was to be conquered to sustain the humanity and the eastern societies, especially South Asian countries face double-edged problematic that nature is to be befriended for the sustainability. When the Western societies developed, world was compartmentalized in the hierarchical linearity and when the countries colonized and late to the process of development are to seek transformation in the quality of life, the world is globalized and the notions of privileging remain contested. The paper stresses that we have all become wanderers, some move with dignity with proper visas, passports and licences, while others transgress illegally internal or external boundaries of states. The referent is material. In its exploration, exploitation has become the canon. It is coming to saturation with the human interventions, where ecosystem and the conservation of energy are jeopardy. In this highly fluid state of interactions, the nature and the less privileged get trampled. There has to be rethinking so that the regenerative politics solves human problems with the consented privileges of the nature. In this background, the paper explores the subjectivity of history in tradition and modernity to understand the sustainability in holistic realm and offer the bridging with the identified traditions and modernity without Scienticism.

Rural development is not simply an offshoot of policies and programmes initiated by the Indian State and public-sector agencies and NGOs. It is far more inclusive and comprehensive as the rural people themselves aspire for their betterment and put efforts to that end writes K. L. Sharma while 'explaining rural development in contemporary India' and its 'paradoxical situation'. The author asserts, undoubtedly the constitutional provisions, policies, and programmes for rural upliftment have weakened some of the institutional bottlenecks, and granted a voice to the deprived and excluded sections of rural India. He cites urban bias as one such potent factor, obstructing egalitarian rural development. By drawing insight from his study of Six Villages in Rajasthan in the 1960s and again after half a century in 2015–2016 with a view to grasp the nature and direction of social change and development, he indicates that 'inclusive development' is the real concern of the people, implying access to assets, markets and opportunities on an equal basis, reduction in disparity of income, maximum benefits for the poor, increased focus on agriculture, employment, healthcare and education, and equitable sharing of public goods and services. Onslaught on traditional obstructive institutional social arrangements has paved a way for 'new actors', individuals and families to assert for their shares in the process of development. That led rural development to proceed in an inclusive form, both materially and socio-culturally, ensuring incorporation of all sections of society, particularly the poor and the deprived people. Thus, it is both a policy and a process, encompassing activities in public and private domains. The author concludes, rural development is change for a desired type of society and hence, has a human face as well.

Tapan R. Mohanty examines 'man—forest interaction in a metropolis' while taking the 'perspectives from hermeneutics'. He writes, the looming threat of climate change, terrorism and rising trend of conspicuous consumption have necessitated a process of global change that has not only brought death, destruction and damage but has also questioned the very ethics of civilization and human conduct. Our greed has long overridden our needs and hurtled us towards an economy and lifestyle that has little concern for ecology, environment or ethics. Indeed, the need to underline a social policy that would instil environmental consciousness would automatically make it imperative to take account of social diversities and human angularities. Hence, he argues, the need is to go beyond the instrumentalities of reason and competence of empiricism for that they do not enable us in depicting and comprehending the reality besides failing to take into account the variability of context. The WTO regime has seen plenty of conflict among the developed and developing nations over the issue of standardization of global environmental norms, for many developing nations it is not merely a choice between ethics or economics but the very survival in a competitive market, needless to add that it has also its corollary in local context as well. The dominance of developed states and communities in exploiting natural resources is a case in point. In this context, the paper attempts to use the hermeneutic framework to interpret man-forest interaction in metropolis and develop a framework for understanding current environmental concern.

The Part II of the volume, 'Issues of Environment', underlines such aspects and dialogues which are creating ecological imbalances and eventually posing environmental threats. This section has seven articles inscribed by experts probing various alarming subjects which have plagued planet earth from ecological sustainability standpoint. The issues covered in this part of the volume includes environmental consequences of dams; impact of river Ganga restoration and conservation efforts; agricultural distress and the problem of sustainability and nutritional security; environmental migration as a challenge to sustainable development; interface between tribes and eco-tourism; people's movement against ecological conservation policies in the Western Ghats; and lastly, assessment of balancing growth with livelihood sustainability through the lenses of land acquisition in India.

The first paper of the section by Namita Gupta on 'environmental consequences of dams' is 'a study of select hydroelectric projects in India'. Her paper discusses the long-term environmental consequences of some select hydroelectric projects in India by highlighting the issue of inadequate compensation and improper reha-bilitation for those displaced. She begins with the argument that the fundamental assumption behind the emergence of development studies after the Second World War was overall well-being for the mass of people. Pushed by this development strategy, the construction of large dams was justified on economic, social and polit-ical grounds. Few studies have been carried out to determine the ecological, economic and cultural importance of rivers before they are dammed. Similarly, it has been the case of long-term environmental and social consequences of areas after the construc-tion of a dam. However, the author further contends, the ongoing struggle against the large dams across the globe challenges this dominant model of development that holds out the promise of material wealth through modernization but perpetuates an unequal distribution of resources and wrecks social and environmental havoc. Once classified by Jawaharlal Nehru, the first Prime Minister of independent India, as the 'temples of modern India'; the large multi-purpose dams and river valley projects have today become the focus of widespread agitation in India. In support of her argument, she cites the instances of the Tehri and Pong dams in the north; the Kosi, Ganghak, Bodhghat projects in east; the Narmada valley project in central India; Bedthi, Bhopalpatnam and Ichampatti in the west; the Tungbhadra, Malprabha and Ghatprabha projects in south, which faced tremendous resistance from the local community. She concludes by underlining that the core issue of discontent against these major projects is location and design of the sites along with the social and ecological consequences of the projects on local community.

'Impact of efforts on Ganga restoration and conservation' is assessed by Ravindra Kumar in his most researched piece of work. He writes, it is a well-known and recognized fact that water impacts everyone. During recent years, India's water usage has increased and diversified, creating both increased water shortages and water quality degradation in rivers, aquifers and thereby threatening broader envi-ronmental sustainability. River healthcare gaps blight the quality of life of such average Indian who have been relegated to the periphery. He further argues, whereas governments are expected to restore and maintain the wholesomeness of the rivers ensuring environmental flows and preventing the pollution ingress into the water bodies, however, it is also noteworthy that the protection of water and environ-ment infrastructure is a social responsibility of every citizen—both individually and collectively. In this background, the paper examines the Ganga conservation and rejuvenation strategies and its impact on environment and drinking water. Author underlines that government is committed for conserving and rejuvenating national river Ganga, along with addressing inter-related issues of sustainable agriculture, basin protection against floodplains disasters, river hazard management, urban river management, wastewater management and revival of water bodies for providing environmentally safe sanitation. Through a well-designated scheme of afforestation in riparian zones to purify base flows and run-off draining into the river, they are also working on to enhance the ecosystem services of rivers and water bodies so that it remains healthy for downstream users. In this regard, institutional network of inte-grated water resource management plan, policy and regulatory governance provides synergy and help to other key stakeholders, experts, investors and well-wishers.

On micro-level, the author argues, key points of river restoration include aspects of flows *(aviralta, nirmalta)*, functions of river as geologic entity and ecological entity. One may track river science, engineering and operations including afforestation and biodiversity to suggest ways of improving overall efficacy of aquatic ecology, ecological restoration (lateral, longitudinal and vertical connectivity) and geological safeguarding (sediment transport, assessing quantity, quality and nutrient value). Critical success of recovering wastewater and restoration of drains are components of urban river management. Decentralized infrastructure can greatly enhance the speed of water treatment leading to one city-one operator through reuse of treated sewage/trade effluents. In order to have a successful water economics, it essentially requires creating supporting environment for sustained infrastructure management through water valuation, pricing and effective implementation of the urban river management strategies and well-functioning water markets. The paper concludes that if Ganga restoration and conservation has to succeed then the governance must establish their own technology innovation and commercialization financing mechanisms. The paper highlights that Environment Technology Verification (ETV) is one such process being implemented by the National Mission for Clean Ganga.

'Locating agricultural distress in India' is the theme of contribution of Purba Chattopadhyay and her exercise is intended to 'realigning for sustainability and nutritional security'. She asserts, contemporary India is a major agricultural producer and exporter. The Green Revolution in 70s ensured food self-sufficiency. There has been a fourfold increase in grain yields per hectare, leading to 3.7 times increase in output since then. Presently, we are contemplating second generation of Green Revolution. Towards this end, specialists of sustainable agriculture seek to assimilate three main goals into their work: a healthy environment, economic profitability and socio-economic equity. However, the author argues, a close scrutiny of facts reveals that entire discussion of sustainable agriculture is relying on the farmers, a large percentage of whom are undernourished, uneducated, barely clothed, mentally broken or in short, whose own existence is unsustainable. Added to this, climate change is tightening its grip and threatening food productivity. The present paper attempts to analyse the existing situation in two parts. In the first part, the consumption and production trends of agricultural produce are examined on the basis of available secondary data, whereas the second part engages with primary data gathered from a cross sectional survey of aspirations of the next generation of farmers in West Bengal. The paper engrosses with the issues like household dependency on agriculture, operable land possession, income from farming, productive investment, consumption expenditure, household debt condition, outcome versus satisfaction among farmers and occupational continuance of next generation from farmer households. She concludes by pointing out the gaps in understanding of sustainable agriculture where a cybernetic negotiation exists between the growers, food processors, distributors, retailers, consumers and waste managers.

Clare Lizamit Samling examines 'environmental migration' as 'a challenge for sustainable development'. The article underlines that anthropogenic factors are the cause of Climate Change and Global Warming is now being increasingly ascertained.

The Intergovernmental Panel on Climate Change (IPCC) has reported alarming estimates and predictions that are expected to have greater impact on the poorest and most vulnerable people, especially those residing in the low lying island regions like the Sundarbans. Migration (voluntary or forced) of people has been identified as one of the major impact of global environmental changes. In this background, the paper attempts to focus on the facet of human mobility due to environmental changes and the complexity that the subject arouses in the global as well as Indian scenario. She asserts, migration of people, voluntary or forced, has garnered attention in the Sustainable Development Goals (SDGs). It is one such area other than environmental change that have received major attention. However, the author contends, India in its commitment towards Sustainable Development has far more demanding issues to tackle with; therefore, 'climate migration' is not featured high on its policy agenda. To conclude, the chapter emphasizes that environmental migration is a burgeoning challenge for most of the developing countries, especially India, in their path towards achieving the SDGs and require solutions that are tangible as well as sustainable.

In their paper, Biswajit Pal and Ramanuj Ganguly interrogate 'interface between Tribes and Eco-Tourism' in a view to 'study sustainability and development in Purulia, West Bengal'. The article elaborates, Purulia is located in the western most part of West Bengal, where social backwardness, poor economic conditions, exotic environment and various ethnic groups prevail. Thus, in view of authors, the prevailing condition makes the locale significant and ripe for the scholars of social sciences to understand the changing Indian scenario in present times. Despite several infrastructural problems, geographical specificity and social backwardness, the State Government, in this district, has been promoting eco-tourism steadily as a potential method and tool to exploit the exotic environment for attracting tourists who would pump in resources to the local economy. The major challenge that the government wishes to counter is that of the Maoist activities, often threatening the democratic fabric of the state. Here does the paradox lies. Since the motive and design of the state is to promote development of the people, whereas, eco-tourism is based on the principles of 'environment over development'. That means, they argue, environment should be given more priority while taking up any developmental policy as the natural resources and environmental conditions are limited and non-renewable. If sustainability of the environment is ignored while taking up developmental programmes then the development will jeopardize both environment and culture. In this context, the paper attempts to study: the eco-tourism project in the Baranti area, a small tribal village of Santuri block under Raghunathpur subdivision within Purulia District to understand what these projects have achieved in tangible terms, like employment, housing, assets and skills; cultural impact of the project in terms of life style practices; landholding pattern of the place; nature of shifting identity; changing nature of aspirations of the local people; and possible social mobility among the people of the area. On the basis of the primary findings, the authors examine the paradoxical debate between 'environment over development' and 'development over environment', and attempt to draw conclusions for policy framing.

The chapter captioned 'people's movement against ecological conservation policies in the Western Ghats' enunciates the need for ensuring environmental sustainability by reflecting on people's resistance over ecological conservation in Idukki District in the state of Kerala, India. The authors, S. Gurusamy and Basil P. V., elaborate by mentioning that Western Ghats has been conferred by UNESCO with 'World Heritage Status' and it is one among the eight 'biodiversity hotspots' in the world. Idukki District, a part of Western Ghats, having major population residing is of migrant farmers from lower lands of Kerala settled through state promoted schemes for poverty eradication. The diverse moves initiated in the Western Ghats with regard to conservation of biodiversity created insecurity among people and led to Farmers Movements in Idukki. Misunderstandings and disinformation campaigns on Madhav Gadgil Report and Kasthuri Rangan Report regarding conservation of Western Ghats added fuel to fire and converted the movements into Social Action form. Moreover, locals received information regarding Environment Sustainability and Conservation Policies mainly from the discussions of local tea shops, Church, Election Campaigns of Political Parties, controversial news and media discussions which act as catalyst to the movement. The authors stress that land rights and environmental sustainability are complimentary to each other which deserve proper address from stakeholder's point of view. The paper concludes that proper environment education and participatory environment conservation is need of the hour to tackle the barriers in conservation and ensuring sustainability of Western Ghats.

Siddhartha Mukerji assesses 'social and political action over land acquisition in India' in a monumental work on 'balancing growth with livelihood sustainability' which is last contribution of the unit. The paper articulates the issue of land acquisition in India with respect to growth and livelihood sustainability. In the present times, the author pens by taking critical account of historical narratives of development in India, the agenda of growth informed by the precepts of market has given rise to livelihood insecurities to millions of those who stay in the margin. Land acquisition is one such phenomenon that has caused tremendous displacement and unsettlement without providing adequate compensation. Additionally, it has led to exploitation of natural resources like forests, water and other resources that remained preserved under a sustainable agrarian economy. The necessity of land acquisition emerges from the relentless capitalist expansion process that is underway especially since the emergence of SEZs and the booming real estate sector in India. Legislations in recent times have attempted at smoothening the process of land acquisition to give a boost to the commercial drive. However, it has faced resistance at both societal and institutional levels. The dispossessed have resisted big land grabs that have taken place after independence. However, the social and political agitations have intensified with greater civil society activism and consolidation of opposition forces. Democratic politics has provided both the strength and motivation to not only resist land acquisition but also express public dissatisfaction over the unregulated drive towards corporatization that has led millions of people dispossessed, displaced, unemployed and unsettled.

The last section of the volume, Part III titled 'Alternatives for Environmental Sustainability', is having five chapters which discuss the alternative approaches and means of development for bringing ecological balance and sustainability. Various alternative means are discussed in this section which include constructed wetland for a sustainable approach to waste water treatment; biofuels and environmental sustainability; alternative approaches to measure sustainability in a subsistence economy; traditional knowledge practice and medicinal plant for tribal health and sustainable development; and lastly, indigenous knowledge of women and water conservation.

The section begins with the paper of R. R. Patil, Karuna N. Pohekar and Neetu Rani who deliberate on the possibility of 'constructed wetland' in a view to explore 'a sustainable approach for wastewater treatment'. They write, disposal of wastewater in urban area relegate to sewage treatment plant located at remote places. Contrary, in rural area wastewater disposed off into rivers and lakes through open drainage without pre-treatment which ultimately causes public health issues in the urban and rural areas. As the volume of wastewater is high and the number and capacity of treatment plant are low, it leads to water pollution and reduces its quality for daily use. Further conventional method for the treatment of wastewater is expensive and requires trained personnel for the maintenance and operation. Hence, other alternate easy to build, inexpensive method for the treatment of wastewater are essentially required. Similarly, as there is increase in scarcity of freshwater, the benefits of wastewater reuse are equally necessary to reduce the pressure on the use of water from the natural resources. It is in this context, other than conventional method, new emerging constructed wetland technologies using various locally available macrophytes for the treatment of wastewater is receiving greater attention due to their cost effective and environment sustainable approach. In this background, the paper explores alternative means by focusing on the types of wetland, its use for the treatment of wastewater and its reuse for gardening, irrigation and pond reclamation purpose, so as to reduce the pressure on the use of freshwater from the natural resources. Moreover, for a comprehensive understanding, the paper highlights brief history of constructed wetland, its types, uses and importance for the treatment of different types of wastewaters by using various macrophytes and model/design.

Rahul Shukla and Sambit Mallik examine the utility of 'environmental sustainability and discursive flexibility of Jatropha in India' in their most illustrated piece of work on, 'biofuels for what'? The authors argue that environment discourses typically represent biofuels as a sustainable and clean source of energy vis-à-vis prevalent fossil fuels. Biofuels are also portrayed as panacea to multiple challenges, viz. climate change mitigation, international trade deficit, energy crisis, unemployment and rural development—especially in developing countries like India. India has been promoting biofuels under the larger ambit of renewable energy sources since late twentieth century. However, various concerns over the adoption of biofuels and its diverse social and political implications have been raised, both in India and the world over. Against such backdrop, this paper attempts to understand the contested meanings arising out of the debates among the scientific communities engaged in adoption, promotion, research and development of biofuels in India. Furthermore, the study examines the legitimizing discourses associated around Jatropha (the preferred biofuel crop as per the biofuel policies of the Government of India), and, how different conflicting agencies of biofuels embody the crop with several meanings, and in turn draw flexibility to strategically switch over the multiple objectives and challenges which are publicized in the biofuel policies. The study from Science, Technology and Society (STS) perspectives argues that Jatropha credited with various objectives including improving environmental conditions has not been stabilized yet. Indeed, discursive nature of scientific-claim making in the promotion of Jatropha only creates a space to compensate the consequences emanating from the Jatropha cultivation in India.

Following in the section, Amalendu Jyotishi and Manjula M. propose 'alternative approaches to measure sustainability in a subsistence economy' by citing 'empirical insights from shifting cultivation in Odisha'. The chapter attempts to critique the existing and popular criteria of measurement of sustainability especially in the context of subsistence economy. They write, shifting cultivation or swidden has been a widespread form of land use since Neolithic period. Globally, it has been viewed as an unsustainable form of agriculture in the mainstream policy. Therefore, it is an appropriate context to test the alternative/complimentary criteria of measuring sustainability. In process of doing so, the paper suggests three alternative/complimentary ways to look into the factors related to sustainability applicable for subsistence economy like shifting cultivation, viz. carrying capacity of the land, employment and consumption needs. The authors empirically verify the alternatives in the context of shifting cultivation in the villages of Odisha by conducting a research in the year 2000, and a revisit to the same region after a gap of 17 years to verify how these three criteria are useful in understanding sustainability of one of the earliest forms of agricultural practices. Based on the findings, the paper shows how policies, awareness and collective action play critical role in reshaping the alternative sustainability goals. Moreover, in this direction, how new policies may play a critical role in manifesting the three criteria.

The paper on use of 'traditional knowledge practice and medicinal plant' is contributed by Samita Mann and Aritra Ghosh for ensuring 'tribal health and sustainable development'. They assert that sustainable livelihood ensures proper human development providing with the basic necessities of everyday life. India is a country of multi-ethnic groups having more than 500 tribal communities along with different religious groups. Among these tribal groups, more than 80% live in different forest environment and fulfil most of their basic needs from the surrounds. In general, these tribal people preserve good notions of health among them based on their perceptions about nature. Overtime they are facing crisis due to non-sustainability of production, consumption and uses of forest goods. Bestowing to the welfare approach adopted since independence for the tribal communities, a special human value loaded attitude is shown to them for shielding their human rights; however, protection of their rights in nature has continuously being ignored, whereas sustainable development helps to preserve the natural resources from the environment for its economic growth and social viability from present generation to future generation. To be more precise, they argue, the tribals are the main source of indigenous knowledge about the medicinal plants used by them for healing and curing ailment of day to day life. Therefore, medicinal plants should be protected and be used by them for sustaining their rights as humans. In this article, by emphasizing on the significance of the sustainability of medicinal plants and its importance in tribal life, the authors focus on an alternative development method which can improve the social environment in general and the tribal development in particular.

The concluding paper of the section and the volume is contemplating 'indigenous Knowledge of women' by Ritu Sharma for exploring 'alternative methods of water conservation in Rajasthan'. The author asserts that this is integral in locating gender dynamics in natural resource intervention to value water as a resource. Women are neglected in water policy, when in fact, they mitigate acute shortage by collecting water from farthest places to restore its multiple use. Their training in water conservation evolves around fetching from water bodies to be passed on to generations via folklores and local narratives. The social fact of patriarchy exerts constraints on women and water-burden falls upon them as 'cultural labour' reiterating inequal right (collection and storage) as compared to male counterpart. Women's role has been recognized globally, yet their inclusion in governmental planning and water resource management is dismal. This raises questions of women's dispossession of power in equity and governance, despite their active role in accordance with low budget alternative methods of water conservation. This highlights the feminist paradigm of 'knowledge as power' reviving alternative methods disguised in cultural and religious discourses of water conservation. Hence, based on vast in-depth literature of identifying women as 'sustainable-community' becomes central to ecological consciousness of preserving water. Thus, the aim is to encourage gender mainstreaming by accommodating indigenous knowledge of water conservation as an alternative in Rajasthan. By drawing a correlation between sustainable water practices of dry-arid (Jaisalmer) to tribal semi-arid areas (Dungarpur) to elevate the status of women as 'primary leaders cum handlers' in water conservation, the paper addresses disproportionate relationship with water both in scarcity and abundance due to its run-offs, wastage and problems subject to gendered control and inequality. The work is unique in two ways: firstly, it discusses the 'indigenous as alternative methods' to conserve water by being resilient to scarcity, and secondly, to have all-inclusive approach of gender mainstreaming to natural resource equity and governance.

Policy Implications and Way Forward

As we have seen through various timescales and academic landscapes, that there is still no agreement or consensus—both politically and scientifically, on the construct and definition of sustainability and sustainable development. It still remains a concept that is idealistic with optimistic *target-driven* agreements and protocols rooted in SDGs. Scholars have linked the concept with related idealistic terms such as liberty, democracy and social justice which are globally desirable but enormously difficult to accomplish (Meadowcroft 2007; Lafferty 2004). Improving the local-level situation through project-centric interventions is considered good from the funding agencies perspective that are mainly focused on dismantling *status-quo*, often changing an unsustainable circumstance to a less unsustainable condition, but the outcome cannot be considered as sustainable. It can be conveniently said that sustainability is a *process* and sustainable development is the *product* of socio-economic development where economic efficiency, social equity and ecological resilience are given equal importance as fundamental objective values, entailing the same widths and breadths and the same policy implications. These are non-negotiable and are not subjective preferences of a market or an individual (Daly 2005). This is a *majority construct* that presently dominates the academic and to some extent political debate surrounding sustainable development. Following this common sense, it is argued that economic growth cannot be the primary measurement of sustainable development. This discrepancy runs in opposition to the well-liked 'triple bottom line' model focusing on striking the *stability* between environmental, social and economic goals (Holden 2012).

It is clear that the meaning and frames of sustainability vary according to the specific context, even though the primary and secondary dimensions have been identified with measurable indicators. Sometimes, researchers argue that the meaning of sustainability is relatively stable, if the difference in our understanding of the context that changes its frame. Therefore, contradictions arise, as the problems are perceived by the different actors with different interests with differing set of contextual interpretations. It is also advisable that the development professionals should be more worried about the implications of the concept affecting the *status quo*, rather than the meaning itself. The concept and meaning, though logically appealing, in reality fails to address how to improve the quality of life of people without disturbing the ecological resilience, improving the well-being of natural and human-built system. A sustainable practice or state is one that can be maintained for the foreseeable future without progressive shrinking of the *shared values*. There are many questions that need to be answered while interpreting the broader meaning across the scales, namely, what kinds of practices and circumstances need to be continued in the larger interest of maintaining and improving the ecological and human well-being? What are the foundations and elements of the foremost threats to achieving the sustainability? How development professionals should approach the context and what places should be examined from a micro-perspective and what should be deliberated to find out the potential problem areas? Can sustainability be designed as friendly sub-set of other development goals or policy choices? Therefore, it is desirable to seek answers to these questions through empirical case studies. The present volume attempts to present scholarly work that testifies the complexity of the problems and, at the same time, emphasizes the need for stronger policy actions at local, regional and global levels for advancing the quality of life of people through sustainability and innovation.

References

Adams, W. M. (2003). *Green development: Environment and sustainability in the Third World*. Routledge.

Brundtland, G. H., Khalid, M., Agnelli, S., Al-Athel, S., & Chidzero, B. J. N. Y. (1987). Our common future. *New York, 8*.

Costanza, R., & Daly, H. E. (1992). Natural capital and sustainable development. *Conservation Biology, 6*(1), 37–46.

Daly, H. E. (1990). Toward some operational principles of sustainable development. *Ecological Economics, 2*(1), 1–6.

Daly, H. E. (2005). Economics in a full world. *Scientific American, 293*(3), 100–107.

Fukuda-Parr, S., & Muchhala, B. (2020). The Southern origins of sustainable development goals: Ideas, actors, aspirations. *World Development, 126*, 104706.

Harris, J. M. (2000). Basic principles of sustainable development. *Dimensions of Sustainable Development*, 21–41.

Hickel, J. (2020). The sustainable development index: Measuring the ecological efficiency of human development in the anthropocene. *Ecological Economics, 167*, 106331.

Holden, M. E. (2012). *Achieving sustainable mobility: Everyday and leisure-time travel in the EU*. Ashgate Publishing, Ltd.

Holden, E., Linnerud, K., & Banister, D. (2014). Sustainable development: Our common future revisited. *Global Environmental Change, 26*, 130–139.

Høyer, K. G. (1999). *Sustainable mobility: The concept and its implications* (Doctoral dissertation, Institute of Environment, Technology and Society, Roskilde University Centre).

Lafferty, W. M. (2004). Introduction: Form and function in governance for sustainable development. In *Governance for sustainable development: The challenge of adapting form to function* (pp. 1–31). Edward Elgar Publishing Limited.

Lélé, S. M. (1991). Sustainable development: A critical review. *World Development, 19*(6), 607–621.

Meadowcroft, J. (2007). Who is in charge here? Governance for sustainable development in a complex world. *Journal of Environmental Policy & Planning, 9*(3–4), 299–314.

Pearce, D. (2014). *Blueprint 3: Measuring sustainable development*. Routledge.

Sachs, J. D., Schmidt-Traub, G., Mazzucato, M., Messner, D., Nakicenovic, N., & Rockström, J. (2019). Six transformations to achieve the sustainable development goals. *Nature Sustainability, 2*(9), 805–814.

Silvestre, B. S., & Țîrcă, D. M. (2019). Innovations for sustainable development: Moving toward a sustainable future. *Journal of Cleaner Production, 208*, 325–332.

Part I
Perspectives on Environment, Development and Sustainability

Chapter 2
Ecology, Culture and Development

B. K. Nagla

Abstract Ecology is the study of interrelation between organisms and their environment. By environment, here it means the natural environment including forests, rivers, lakes, seas, mountain, plants, etc. The paper discusses how ecology is inextricably related with the processes of development and also highlights the problems when both the elements are considered dichotomous. The main focus is on what role culture plays in the process of adaptation with natural surroundings on 'ecological niche' and in the process what cultural traits and complexes peculiar to particular society develop. Moreover, it also analyses the role of culture in the process of development and if overlooked fractures the very process of development. Therefore, an effort has been made here to know the different approaches to environmental and ecological issues related to ecology and culture, culture and development, ecology and development. Emphasis has more been placed on cultural ecology approach while explaining between ecology, culture and development. Further, it also tries to explain the perspectives and imperatives as the perspectives provide vistas and visions of environment, and the imperatives call for the action that may bring true equilibrium, and thereby, improve the quality of life and the quality of environment. Finally, the paper examines the relationship between ecology, culture and development to understand human beings and their environment in contemporary society.

Keywords Ecology · Culture · Development · Environment · Ecosystem

Concept of Ecology

Ecology signifies the way we perceive at the world and how we feel at ourselves. Each geographical region in the world comprises a particular ecosystem which is an interrelated habitat for plants and animals shaped by climate and terrain. These ecological factors have a strong effect a change upon culture as well. Ecology is the study of

B. K. Nagla (✉)
Department of Sociology, M.D. University, Rohtak, Haryana, India
e-mail: bnagla@yahoo.com

© The Author(s), under exclusive license to Springer Nature Singapore Pte Ltd. 2021   23
M. K. Verma (ed.), *Environment, Development and Sustainability in India: Perspectives,
Issues and Alternatives*, https //doi.org/10.1007/978-981-33-6248-2_2

multidisciplinary or an interdisciplinary discipline like genetics, anthropology, sociology, etc. which is viewed as the study of interrelation between organisms and their environment. For example, sociologists try to understand the interrelationship between human beings and their environment particularly the natural environment including forests, rivers, lakes, seas, mountain, plants plains and oceans, forests and plants, flora and fauna, etc. These are support parts of ecology.

The concept of ecology was coined by combining two Greek words, namely *Oikos* (meaning household or living place) and *logos* (meaning the study of) to denote such relationship between the organisms 'at home'. There is some controversy associated with the author who coined the term ecology and first used it in the literature. For instance, the Kosmondy gave credit to Henry David Thoreau in 1858 for the first use of the term ecology. There are, however, references in the literature attributing German biologist H. Reter also who is said to have used this term for the first time in 1868. Although there is uncertainty about the original coinage of the term, however, there is consensus that the German biologist Ernst Haeckel first gave substance to this term.

The concept ecology is first used by Haeckel in 1886 and he viewed the ecology of an organism as… 'the knowledge of sum of the relations of organisms to the surrounding outer world, to organic and inorganic condition of existence'. There are two branches of ecology: one *autecology*, the study of the individual organism's interaction with environment and the other *synecology*, the study of the correlation between the organisms engaged with the given unit of environment. The latter study continued to exist and determined the basic indicator of ecology. Concepts and techniques of investigation of bio-ecologists that are used by sociologists in their work. In brief, the concept ecology refers the web of physical and biological systems and processes which employ humans are one element.

In order to comprehend the etymological and conceptual issues involved with the terms ecology, culture and development along with their correlation to unmask the contemporary state of environment and society, the present paper highlights three important aspects which are as follows:

I   Ecological approaches;
II   Concepts of Culture and Development;
III Ecology, Culture and Development: interface.

Our interest is here to understand first the ecological approaches which explain human ecology. This would help in the analysis of ecology, culture and development.

I Ecological approaches

In the field of sociology, human ecology has been used as the ecological approach which was included in the curricula of American universities in 1920s. There are different approaches to the study of the interrelationship between organisms and their environment. Among them most important are as follows which are discussed by Rambo (1983):

1. Environmental Determinism and Environmental Possibilism;
2. Cultural Ecology;
3. Ecosystem Model;
4. System Model.

1. **Environmental Determinism and Environmental Possibilism**

   The early generations of sociologists and anthropologists around the turn of this century believed that environmental setting in which the group was located caused the cultural features of a group. Environmental determinism (or environmentalism) held that the reason of the presence of a particular society was traced to the habitat in which it occurred.

   The ecology factors affect both plant, animal and human being's life of a particular place in context to the geography and hydrology. They adapt and shape according to the ecological conditions whether it is the desert or the area of scarce rainfall, rocky or sandy soils, and extreme temperatures.

   On the basis of comparative cultural data accumulated early in this century, the theory of environmental possibilism claimed that environmental factors limit the adaptive possibilities for a given culture but do not determine which kind of adaptations or choices a society makes. Kroeber (1939) acknowledged, of course, that the environment did play a limiting role. But within these limits, societies chose freely from a number of possibilities.

   Possibilism advanced but did not resolve the environment-culture question. By the middle of the twentieth century, anthropological thought moved towards a more balanced concept by describing it as the interplay between the physical environment and cultural system. The later approach known as cultural ecology related to the ways in which particular environmental features are interwoven with specific aspects of culture and emphasized on selecting only those environmental characteristics and cultural arrangements which can be functionally related.

2. **Cultural Ecology**

   Cultural ecology focuses on the relationships of specific cultural features of a group's adaptation to its total habitat. Steward (1955), the primary advocate of cultural ecology, was interested in understanding the patterns of activity adopted by social groups in order to exploit a particular niche. Kroeber (1939) emphasized that various cultural adaptation technique is possible within a given environment. This also refers to cultural ecology, as defined by Bennett (1976: 310)—'This is the heart of cultural ecology: the way man-man relations modify man-nature relations in particular representative cases, and how the results affect the future of both'.

   The idea of cultural ecology is to know the origin of specific cultural patterns which describe specific cultural zones than arriving on general principles applicable to any cultural-environmental situation (Steward 1955: 36, 1968: 337). This emphasizes to the study of the particulars of local ecological conditions instead of distinct cultural histories (Vayda and Roy 1968: 483). Identification of exploitative and demographic patterns which shape kinship organization is done through Steward's method (Helm 1962: 631).

3. **Ecosystem Model**

The holistic concept of ecosystem—a type of general system capable of including the activities of man—has recently gained wide acceptance. The ecosystem conceptually unites the biology, organization and behaviour of man and the other animals, plants and inorganic concepts within a single framework in which the interaction of the components may be studied (Anderson 1973: 183). This is particularly appealing to anthropology since it allows for the study of the mutually dependent interactions of organic, inorganic and socio-cultural components. However, the ecosystem model points out that human beings play a powerful role in maintaining the balance between organism and their environment. Human beings can preserve or destroy nature. The main advantage of this ecological model is that it is dynamic. The model points out that nature cannot destroy entire towns and cities now in a normal circumstance. It may be possible in a calamity situation like a major earthquake. But this is a factor not to be easily ignored. It has been seen that deserts infiltrate into villages and floods inundate many fertile fields with silt. Thus, the ecosystem model indicates that there will be a severe backlash from natural disasters if human beings do not mend their careless apt of doing. This is relevant for human beings and their environments including forests, urban dwelling, transport system and even developmental paradigms or models.

4. **System Model**

This is useful model as it puts human beings at the centre of the strategy of ecological plans. However, it continues its interactive view. This is much helpful as it is interrelated between human beings and nature. This view is applicable to stop ecological devastation which is possible through a new, creative and regenerative attitude. Thus, the environment cannot be held responsible. Human beings cannot expect the forest to re-generate itself. Therefore, human society should take initiatives to reverse the damage of all ecological areas in which acute deforestation is a real problem. Therefore, we cannot make responsible to environment and also expect the forest to re-generate itself. Human beings have to do efforts to reverse the loss of ecological devastation. We would be able to save the acute deforestation if we do the efforts in this direction.

As we know that the members of society, as a whole, selectively exploits the ecological resources. At the same time, the ecological bears to society and suffers its technological needs. The social system takes assistance from the ecological system for the consumption of fuel, fodder, petroleum, etc. Even then, human beings generally do not repay to nature for what they have got.

Effort has been made in the paper to use all the approaches for the study of ecology and culture, and ecology and development. However, emphasis has more been placed on cultural ecology approach while explaining relationships between ecology and culture. Along the line, several other necessary approaches have also been looked into as part of the member of the planet to think clearly, act consciously to avoid disaster.

II Concepts of Culture and Development

In the preceding section, we have discussed the concept of ecology and its approaches. In this part, we would like to explain two very significant concepts which have been used repeatedly in this paper. These are:

(i) Culture and its Pattern Theory;
(ii) Development and the Phases of Development.

(i) Culture

In sociological literature, the term culture is often used in a sense which is more restricted than its anthropological meaning. Alfred Weber has distinguished between culture and civilization. Civilization according to him includes scientific and technical knowledge and the control that they give over natural resources, while by culture he means the artistic, religions, philosophical and other such products of a society. R. M. MacIver has made similar distinction, he said that 'our culture is what we are; our civilization is what we use'. Such use of the term culture obviously limits its meaning, and this has significant implications to know the relationship between culture and development. There are two ways for using the concept of culture: first, as a pattern in relation to ecology and second, dichotomy between culture and civilization which is more sociological meaning of culture, when culture is used as a heuristic device and a standard for the process of development.

The Pattern Theory of Culture

Kroeber and Kluckhohn and critical reviewed historically hundreds of the concept of culture to explain the pattern theory of culture which would be acceptable to most of the scholars of social sciences.

'Culture consists of patterns, explicit and implicit of and for behaviour acquired and transmitted by symbols, constituting the distinctive achievement of human groups, including their embodiments in artifacts; the essential core of the culture consists of traditional (i.e. historically derived and selected) ideas and especially their attached values; culture systems may, on the other hand, be considered as products of action on the other as conditioning elements of future action' (Kroeber and Kluckhohn 1952).

According to pattern theory, culture is created by individuals and groups and interacts with them as well as with the environment. Such a theory assumes the process of cultural growth as a historical process of a particular group of its past choices, conscious and unconscious, at any given time, is related to a set of patterns. This perspective of culture precipitates 'present in persons, shaping their percepts of event... Culture is a blending variable between human "organism" and "environment"' (Kroeber and Kluckhohn 1952: 186).

(ii) **Development**

There are varied connotations of 'development' put forth by various scholars. This is explicitly a problem to arrive at consensus, but it can be done through turning the pages of history of development. For Myrdal (1970), 'Development means the process of moving away from 'underdevelopment'; or arising or rising out of poverty; it is sought and perhaps actually achieved by means of planning for development. However, United Nations defines development as: 'the ultimate purpose of development is to provide increasing opportunities to all people for better life' (UNESCO 1975).

Schumpeter (1960) has made distinction between development, economic development and social development. By development he means that such changes occurred in economic life by its initiate within itself rather forced upon it.

For him, this is the simple aim of historical economic development which is not only associated with the rest for purposes of exposition. It is not possible to explain economic changes merely on the basis of previous economic conditions alone due to fundamental dependence of the economic aspect on everything but by considering the comprehensive situation of social development.

Schumpeter observed that historical transformation establishes neither a circular process nor pendulum movements. This is the concept of social development is explained by the two circumstances along with other facts: for example, we are not able to explain adequately the historical state of things for further actions. We accept the existence of a controlled one but do not solve the problems.

Phases of Development

During the last five decades development theory has taken several sharp turns. Dube (1988) has deliberated the concept of development into four phases based on the growth and diversification or specification reflecting planning and development in India.

In the first phase, development refers basically economic development which focused exclusively on economic growth. Realizing the limitation of economic growth as a measure of development, they put attention in terms of 'redistribution with growth on a trinity of economic growth, equity and self-reliance' (Weissokf 1975: 775–794).

In the second phase, it was realized that social development is also essential need of the society. Therefore, it is emphasized the relationship between economic and social development for social change. Hence, institutional framework of society was modified to change the attitudes and values for the process of economic development. In the period of this phase, the economic approach to development continued to maintain its autonomy and did not forge links with the revolution in the behavioural social sciences. The revolution gave birth to the modernization paradigms.

The third phase was reactive and responsive to the development. It was consequences of serious reaction to the unequal paradigms of development and modernization. This translated an idea into an action for more successful development. New concepts of centred development emerged and called for greater access for the common man to planning process, to cater the basic needs human resource mobilization/and capital formation. It requires rethinking of new environment and energy consciousness for development.

The fourth phase was a reflexive one for the development. This aims to understand the world order and also the national orders. There is a need to change both the orders for human existence. A modern global economic order is necessary for the North and for the South nations; for the developed, the developing and the underdeveloped. The national framework of development for underdeveloped societies as well as of the developed areas has to undergo modification. The culture of development thus has to be changed. A new style and new idioms are needed.

Can we do something about global armament? Can adequate resources be transferred to programs of human harmony and survival? Can we do something about the alienating character of technology and industry? What kind of social formations will offer greater satisfactions and reduce tensions as well as unrest with the new technologies? Can education, public health and other social services be reorganized to offer greater distribution benefits to the general mass of people? How does one handle problems in the area of management of change? Many of these problems are beyond those of development. Therefore, development planning should be done in such a manner which can influence the man for the proper use of nature. In this context, sociologists are interested to understand the interrelationship between human beings and their environment.

III Ecology, Culture and Development: Interface

In this section, the main focus would be on what the role culture plays in the process of adaptation with natural surroundings, i.e. 'ecological niche' and in the process what cultural traits and complexes peculiar to a particular society develop. Another important aspect brought out in this section is the study of cultural role in context to development, how culture stands as sine quo non-development and if overlooked fractures the very process of development. Furthermore, the section deliberates upon how ecology is inextricably related with the process of development and also highlights the problem when both the elements are considered dichotomous. The whole debate which revolves around the trilogy of ecology, culture and development is based on the following issues:

  (i)   Ecology and Culture;
 (ii)   Culture and development;
(iii)   Cultural ecology;
 (iv)   Ecology and development.

(i) Ecology and Culture

Implicit in all ecological studies in anthropology and sociology is the notion of adaptation. That is, a human population adapts to its habitat, it is able to survive and perpetuate itself. Culture is the means by which humans make this adaptation possible. Some populations are better adapted than others are. They are in Darwinian terms, 'winners' in the struggle for existence.

Culture and environment studies have generally made an attempt to deal with two different kinds of problems. First, they seek to explain why particular elements or sets of elements exist at specific times and in particular places. Second, they attempt to interpret how these particular elements function within a given culture and natural habitat.

Geertz's comparison of irrigation in Bali and Morocco, with example, shows that each culture is so much part of its ecological niche that it is difficult, if not possible, to think of the culture independent of their ecological settings. For Geertz, ecology plays a central role in the joining of all elements of culture. As he puts it, nature is not just a stage upon which culture performs. 'A society turns itself to its landscape, mountainside, river ban, or foot hill oasis, until it seems to an outside observer that it could not possibly be anywhere else than it is so, it could not be otherwise than what it is' (Geertz 1963). On the same strand, Mann has explained the role played by ecology in the formation of cultural traits and complexes in the process of adaptation in 'culture and ecology in Ladakh' (Mann 1990).

Looking to the above context, we may view that ecology is the interrelationships between human being and their environment. They influence to each other. Society shapes nature and nature also shapes society.

This is a two-way process. Just as nature shapes society, society shapes nature. For example, Indo-Gangetic floodplain has the fertile soil for intensive agriculture which allows dense inhabitants. This generates enough surpluses to support other, non-agricultural activities causing to complex hierarchical societies and differences in the states. Contrary to this, pastoralists of desert Rajasthan move from place to place for their proper livestock supplied with fodder. We have also instance of ecology shaping the structure of human life and culture. For example, the environmental conditions of Leh and Ladakh influence the human life irrespective of the caste and religion.

(ii) Culture and Development

Max Weber made a serious attempt to understand the elective affinity between irrational (religion) and the rational (capitalism). He provided a thesis wherein Calvinistic Ethic provided cultural stimuli for the rise of modern capitalism. Similarly, Parsons and Hoslitz argue that values or achievement helped in economic development. The absence of achievement motivation is associated with a kind of education in which less emphasis is placed on technical skills and on contribution to production, with more emphasis on acquiring culture and civilization. To put it in more general terms, cultural value preferences emerge from the matrix of a society's historical experiences, which include religious experiences. And these value preferences or culture are crucial for the development.

Capitalism has created the commodification of nature for buying and selling things in the market for profit. A river has multiple cultural meaning, for example, spiritual and aesthetic values on the one hand and ecological and utilitarian for the profit and loss from the sale of water for an entrepreneur on the other. Socialist values of equality and justice have led to the seizure of lands from large landlords and their redistribution among landless peasants in a number of countries. Religious values have led some social groups to protect and conserve sacred groves and species and others to believe that they have divine sanction to change the environment to suit their needs.

The main cultural problems of development faced by the Indian society can be attributed to the peculiar nature of the social organization, which is formed on the basis of Hindu religion and caste. The Indian social organization has its existence for the past several centuries. Despite the ongoing process of modernization, it has not undergone any radical change. The values associated with religion and caste find their expression in the day-to-day activities like mode of production, selection of occupation, relation between members of the family, consumption of food, eating habits, sexual relations, fertility rate, etc., which ultimately arrest rationalization of socio-economic activities.

(iii) **Cultural Ecology**

Cultural ecology is a scientific approach reflecting the interaction between human society and natural environment. Scholars of Faculty of Arts at Charles University in Prague, Czech Republic used this approach. They emphasized the interrelationships between cultural ecology and landscape ecology which indicates interdisciplinary study of biophysical as well as societal processes and patterns in landscapes. In this context social scientists focused on the relationship of humans and environment who used the concept of cultural ecology. Cultural ecology refers to the relationship between culture, man and environment which was also introduced as human ecology.

The first human ecology appears in the 1920s in Chicago. Human ecology of Robert E. Park and his colleagues was mostly aimed at urban sociology. A totally different approach is represented by cultural ecology of Julian Steward, who created it as an anthropological sub-discipline stressing the adaptive function of culture. Social ecology of Murray Bookchin brought more philosophy and social activism into the discussion.

In the 1970s, Gerald Young and environmental sociology of William Catton and Riley Dunlap introduced human ecology who used environmental factors back to the studies of modern complex societies. Besides the USA, we also would like to mention here the Czech disciplines and scholars investigating the human-nature relationship, for example, Bohuslav Blazek applied the concept of social ecology as the sociological approach mostly concentrating on rural areas towards environmental problems of Jan Keller.

Mukerjee (1942) studies social ecology which is unparalleled. The discipline of social ecology requires the co-operation of sciences and social sciences. This substantiate our earlier view that social/human ecology has reciprocal relationship between human beings and their environment and they shape to each other in particular environment conditions related to the geological, geographical and biological factors constitute ecological zone.

Mukerjee (1926) explicates the area of human ecology 'as a synoptic study of the balance of plant, animal and human communities, which are systems of correlated working parts in the organization of the region'. American pioneers do not consider cultural factors in ecological relations. According to them, such relations are similar to plants and animals. But, Mukerjee viewed that human beings and their environment are largely similar in ecological relations. Cultural norms have a very important role in case of human beings. Human ecology uses this fact in the formation of an ecologic unit like 'region', social habits, values and traditions. Social ecology emphasizes the ever-complex reciprocal relations between man and the region which reflect in ecology and society.

The development of ecological zones is the outcome of a dynamic process that is the challenge of the environment and the response of the people who establish a settlement. Ecological balance is not possible without a symbiotic relationship with the ecology or environment of the area and actions of human beings which weakens or destroys the social fabric. For example, natives of the concerned locations very often are moved to new settlements due to building industrial plants or constructing irrigation plants or constricting irrigation dams in India. It badly affected community's life of the locals. In the new environment, it may not associate that kind of relationship with the surrounding.

Industrial development senselessly exploits natural resources reflects its 'security threatened due to the exhaustion of coal and petroleum', consequently, it falls the supply of minerals and vitamins which cannot be artificially manufactured. Of course, there is a need to recognize the importance of ecological values in the industrial society keeping in view of sustainable development for future generation. But it is essential that these values should 'have reached the level of standards of moral behaviour' (quoted from Nagla 2015: 88).

Technology and use of tools are the means by which human populations at all times and at all places; exploit the environment in order to satisfy their basic needs for food and shelter. Two thinkers who have placed great stress on tools and technology in the growth and development of culture are V. Gordon Childe and Leslie White.

Childe argued that tools are principal means whereby any society adapts to its natural setting. White (1949), on the other hand, saw technology in even more deterministic terms. He divided culture into three subsystems—a technological system, a sociological system and an ideological system. He saw the three systems hierarchically layered, so that the technological layer determined the nature of the two, while the reverse was not true. Marvin Haris took White's belief in the primacy of technology and energy, the material aspect of cultural systems. His analysis (1974) of the multiple role of cow in India uncovers and makes sense of the cultural importance of cattle to the energy and technological needs of the country.

Mukerjee was worried about senselessly spread of urbanization. He supports and favours the idea and process of urbanization from the ecological point of view. There should be a check for urban at the expense of the countryside. This is possible when agriculture should be diversified, and industries should be decentralized. Therefore, Mukerjee pleaded for regional development keeping in view the ideas of social ecology. He believed a balance between economic growth and natural resources. He stood for a balance between economic growth and ecological fitness. Even traditional crafts and skills like weaving should be restored for economic growth of a region without any great damage to its ecology. Mukerjee warned to the local countrymen for deforestation. He seriously pleaded for conservation of forests and protection of ecological balance.

Mukerjee points out his concern to the ecology as follows:

(i)   over grazing;
(ii) shortsighted devastation of trees and scrubs; and
(iii) defective techniques of cultivation brings about a serious imbalance in the biophysical constitution of the entire region. It severely impairs nature's cycle (cited from Bhattacharya et al. 2003: 98).

Removal of vegetation brings about a chain of unfavourable consequences as follows:

(i)   denudation of the topsoil;
(ii) fall in the underground water level;
(iii) decline of rainfall;
(iv) increase of aridity; and
(v)   acceleration of 'river', sheet or gully and 'wind erosion'.

These have led to serious and continuous agricultural deterioration (cited from Bhattacharya et al. 2003: 99).

Ecology and Development

Sociology has one of the areas to study ecology and development is still younger within the discipline. It has been differently used to study as 'ecological anthropology', 'social ecology' and 'environmental sociology' (Baviskar 1997). Social historian, Guha (1994), refers it the field 'social ecology' aptly following a well-known sociologist, Radhakamal Mukerjee.

In the 1920s, Radhakamal Mukerjee introduced the concept of region as a concept as a synthesis of ecology and sociology. He considered: 'The region is at once an ecological aggregation of persons, an economic framework and a cultural order' (quoted in Guha 1992: 62). From the region, Mukerjee (1942) elaborated a theory of 'social ecology' from the region which he further examined through empirical studies such as those of the Indo-Gangetic Plain and of agricultural productivity in the princely state of Gwalior.

The theory of 'social ecology' and 'human ecology' is used interchangeably and refrained model of 'ecology' in the natural sciences. Concepts such as community of human beings are interwoven through the 'webs of life' balancing the natural resources which are taken from ecology. Social ecology systematically includes the human species within its scope than ecology. Baviskar (1997) prefers to use 'environmental sociology' which emphasized social relations. These are also affected

by the biophysical world in which social beings live. In this context, we discuss here
ecological perspectives and imperatives.

Perspectives and Imperatives

The discussion of ecology and development has received some considerable weigh-
tages during the last few decades owing to the facts that (1) human interference
with environment has resulted to an ecological imbalance that threatens not only
the contemporary human beings but also the future citizens of the world, and (2) the
placement of greatest emphasis on economic development through the highest degree
of exploitation of science and technology has resulted to a socio-cultural imbalance
that affects the mechanism of smooth maintaining and continually evolving the nexus
of social structure. The socio-cultural situation has become ineffective to efficiently
make use of the best advantage of super-organic development of human civilization
rather than organic development that brings forth temporary satisfaction to material
living of human beings (Bhowmick 1988: 11) (The term super-organic development
is introduced by Herbert Spencer and his choice of this term reflects his view that
evolution must be viewed as transformation that has taken place in three realms: the
inorganic, organic and super-organic.).

The very situation has induced human actions to take appropriate measures for
understanding the nature and extent of the implications of ecological imbalance
on the one hand and the question of development on the other. Hence, this paper
aptly indicates 'perspectives and imperatives' as the perspectives provides a vistas
and visions of environment, and the imperatives call for action that may bring true
equilibrium for better the quality of environment and life of human beings'.

Perspectives

There are different perspectives of ecology and its interaction to culture of human
beings. These differences based on individual characteristics whether these are innate
or are influenced by environmental factors. For example, people are poor as they do
not get proper opportunity not because they are innately less talented or hard working.
Theories and data related to the environment and society are generated by the social
conditions in which empirical studies are done.

'The primary motivation for development is human welfare, which is also the
primary motivation for environmental quality' (Kayastha 1992: 40). These two moti-
vations do not come into conflict but disapprove the mistaken dichotomy between
environment and development. Hence, Brundtland Commission Report (1987) indicated the basic issues about poverty in the developing countries, international
inequality and sustainability of development are taken in the same subject.

Development Dichotomies

To begin with, there is the overwhelming priority accorded to industry over agriculture. The origins of the industrialize-or-perish model may be traced to visionaries like Visvesvaraya and Meghanad Saha. It was formally institutionalized in the second five-year plan, for which underdevelopment was 'essentially a consequence of insufficient technological progress' (Government of India 1956). Here rapid industrialization on the western model was the key to progress. As the Lok Sabha observed in 1954, the objectives of economic policy and planning were that 'the tempo of economic activity in general and industrial development in particular should be stepped up to the maximum possible extent' (Mahalanbois 1955).

In the Second Citizen Report on the state of India's Environment 1984–85, Agarwal and Narain (1985) highlighted the implications of this strategy for natural resource utilization. The planners have directed the processes of natural resources utilization in a way as to primarily benefit large industry. Thus, forest policies have consistently favoured the promotion of commercially valuable (and often ecological inappropriate) species of trees. Moreover, timber has been sold at throwaway prices to the processing industry. Again, the massive infra-structural investment, by the state in mining and the generation of electric power, has been dictated by the requirements of industry both public and private. These hidden subsidies are not the only cost borne by the public—thus industries have been allowed to dispose of solid, liquid and gaseous wastes with scant regard for the surrounding population.

The development of industrialization and technology affected seriously ecosystems in unprecedented expansion of population and industry. This requires proper equilibrium management systems of human organism and the natural resources. We have seen the consequences of the dangers inherent in industrial environment due to the occurrence of nuclear disasters like Chernobyl, industrial accidents like Bhopal, and Mad Cow disease in Europe. We will be always in the risk if we do not give proper attention for sustainable development.

Where state policy has looked at the countryside, it has placed its bets squarely on the commercialization of agriculture. The Green Revolution Strategy, by handsomely subsidizing modern irrigation and chemical inputs, not only increased the dependence of agriculture on industry, but also encouraged the process of social differentiation and the extinction of family forms. From an ecological standpoint, the most significant consequences have been the breakdown of commercial institutions regulating the utilization of ponds, grazing lands and forests. In replicating the classic problems of the development of agrarian capitalism, this privatization of common property resources deprived the poor of an important means of subsistence (Guha 1986). This leads inevitably to the division between town and country.

The real function of Green Revolution is to make self-sufficient in food gains and to supply food to a growing population; it is hardly bothered to mitigate the maldistribution of nutritive sources within the village. Other instances of the diversions of resources towards the city include the takeover of agricultural land and consumption of firewood, e.g. both Bengaluru and Delhi consume several lakh tones of firewood annually and consequently give way to deforestation. The firewood trade directly related to the proliferation of slums, in symptomatic of a deepening urban environmental crisis.

Most important question regarding development dichotomies is issue relating to big dams. During 1950s and 1960s the main concern of planning was economic development. And it is in this context big dams were considered indispensable for economic development. Heavy reliance was place on these kinds of multipurpose projects for the development of irrigation, control of floods and generation of electricity. It is for their multiple activity and high potential for economic development that Nehru called them 'modern temples' of India.

Benefits from these (Mega projects) began to be realized when National Agricultural and Industrial growth rates increased substantially. These benefits have not occurred without paying costs in terms of loss of land and forests, risk of dam failure causing loss of life and property, spread of waterborne diseases, problems of rehabilitation, etc. Per capita availability of food grains went up to a level of 478 grams per day in 1986 as compared to that 395 grams increased from 510 lakh tones in 1950 to 1700 lakh tones in 1989–90. Net sown area increased from 11.0 Crore hectare to 14.1 Crore hectare during pre-plan period was 225-lakh hectare. It increased to 675 lakh hectares by the end of 1984–85. Fertilizer consumption increased from 0.62 lakh tones in 1950–51 to 124.33 lakh tones in 1989–90. Installed power generating capacity has increased from meagre 1400 MW in 1947 to 64,000 MW in 1991. In 1956 only 7294 villages are electrified, this number increased to 460,536 in 1990 (India 1991).

The main problem has been the issue of displacement (Verma 2004a, b, 2015: 245–275, 2016: 23–48). The government's performance regarding rehabilitation and resettlement of outsees in the past has been quite dismal. This has been one of the main reasons of the protest against the big dams. To cite an example, 36,000 people were displaced in case of Bhakra Nangal Project, which was completed in 1959. Land was acquired from 2180 families in the districts of Una and Bilaspur in Himachal Pradesh only. They were promised rehabilitation in the districts of Sirsa and Hissar. But till 2001 only 730 families had been resettled. The few were given property rights; as a result, they cannot take loan against this land if they so desire. One major objection raised by environmentalists is that the constructing of big dams results in involuntary migrations.

Another issue relating to problems arising out of big dams is the issue of deforestation. Any development project that causes deforestation becomes a target of the environmentalists. In India, the forest cover is diminishing at a rate of about 1.5 lakh hectare per year. A dam effects forest cover in two ways: (1) by submergence of forest under the reservoir on the upstream side, which is around 10% of the irrigation potential created by a big dam. (2) Forest cover is also reduced because of the dam building activities such as construction of project roads and housing colonies. There are other problems apart from these, which are the serious fall out of the big projects like land degradation, sedimentation of reservoir, problem of seismicity, waterborne diseases, etc.

Hence, the development which was aimed in the national interests for the betterment of the million, and for placing country on the path of progress have, in fact, fractured the very premise of assumptions relating to human progress and of development and created environmental degradation which threatens, if cumulated, the very existence of the people. If this stands as a dilemma of development in dichotomy with ecology, is there an alternative or imperatives for following a holistic path of development, which ushers in happiness and equity among the people.

Imperatives

The concept of ecology has been changing in human action as time passes in context to the natural features of the environment. There are the instances of the widespread impact of human activity on nature like deforestation, climate change, global warming. Over a time, ecology and human action often influence each other in ecological transformation of biophysical properties and processes, e.g. flow of the river water, species composition of forest and also other human activities for agriculture, industry and urban development, etc.

Martell (1994) enumerates four existing proposals put forward to overcome developmental problems and environmental degradation. These are: the first issue is taken for the global slowing or halting of growth in line with natural finitude. The second proposal is to continue growth in less developed countries (LDCS) while restricting of no growth strategies to the developed world (Barkenbus 1977). The third alternative is a combination of 'sustainable development' with self-reliance. This proposal is to accommodate development while protecting the environment. The fourth proposal is for the development strategies, which needs appropriate technology.

Appropriate technology is sensitive to the environment in two ways: (i) it accelerates indigenous self-reliance. Local people become independent without outside help for development. (ii) Appropriate technology should fit to local culture, circumstances and abilities.

The issue of 'environment and development' is crucial in context to political and the intellectual necessity for modifying the object of development for the welfare of human beings to make them self-reliance having knowledge of actual conditions. This would allow autonomy and self-reliance to use firsthand knowledge of actual conditions in local contexts.

There are number of pathbreaking studies, however, we have to learn more for basic insight which is possible through the more creative minds that are exercised on the issue combines for the concerns as follows: (i) structure of equity to ensure autonomy and self-reliance, (ii) local participation (Oommen 1990). This requires specific empirical studies for sustainable development strategies for the quality of life in all regions which would be equitable international order.

Policy Implications

The study of ecology, culture and development is to understand the relationship between organisms and their environment. The present paper discusses about how ecology is inextricably related with the processes of development and also highlights the problems when both the elements are considered dichotomous. The main focus is on what role culture plays in the process of adaptation with natural surroundings on 'ecological niche' and in the process what cultural traits and complexes peculiar to particular society develop. An effort has been made here to know the different approaches to environmental and ecological issues related to ecology and culture, culture and development, ecology and development. Therefore, it is important to understand ecological approaches which are relevant to policymakers for the development of the society.

Ecology contributes to 'identification of social factors, such as norms affecting fertility and living arrangements, that must be taken into account when predictions are made regarding future patterns of exploitation; study of possibilities and mechanics of social control to limit population and preserve the environment; and an analysis of perspectives that suggest revised social goals and policies—for example, new meanings of conservation, new life-styles, new conceptions of property and responsibility. To overcome the ecological crisis, there is a need to understand the principles of ecosystems and the various forces that threaten them. We may view that eco-development is an action-oriented construct based on cultural ecology. We must cope with situations as they arise, atypical as they may be. In a sense, each case is unique, as it represents a specific combination of natural and cultural factors, and as it occurs in a particular historical and sociopolitical setting.

Conclusion

The rapid growth of technology, the spread of industrialization and the increasing world population have produced ecological crisis in two forms: pollution of the environment and depletion of natural resources. The energy crisis depletion or destruction of natural resources and pollution of environment, on the philosophical side, 'reflect limitations in man's thinking. To save man from his own destruction, he should concern himself with his own inner realm, and that younger generation should be taught facts, but should be encouraged to question and understand facts, as well as to look for answers themselves' (New York Times 1980). The reconciling development requires harmonious relationship between human beings and their environment.

Ecology contributes to identification of social factors, such as norms affecting fertility and living arrangements, that must be taken into account when predictions are made regarding future patterns of exploitation; study of possibilities and mechanics of social control to limit population and preserve the environment; and an analysis of perspectives that suggest revised social goals and policies—for example, new meanings of conservation, new life-styles, new conceptions of property and responsibility. To overcome the ecological crisis, there is a need to understand the principles of ecosystems and the various forces that threaten them. We may view that eco-development is an action-oriented construct based on cultural ecology. We must cope with situations as they arise, atypical as they may be. In a sense, each case is unique, as it represents a specific combination of natural and cultural factors, and as it occurs in a particular historical and sociopolitical setting.

References

Agarwal, A., & Narain, S. (Eds.). (1985). *The state of India's environment, 1984–85: The second citizen's report*. Delhi: CSE.

Anderson, J. N. (1973). Ecological anthropology and anthropological ecology. In J. J. Honigmann (Ed.), *Handbook of social and cultural anthropology* (pp. 179–239). Chicago: Rand-McNally.

Barkenbus, J. (1977). Slowed growth and third world welfare. In D. Pirages (Ed.), *The sustainable society: Implications for limited growth*. New York: Praeger Publications.

Baviskar, A. (1997). Ecology and development in India: A field and its future. *Sociological Bulletin, 46*(2), 193–207.

Bennett, J. (1976). *The ecological transition: Cultural anthropology and human adaptation*. Oxford: Pergamon.

Bhattacharya, S. K., Gupta, S. K., & Bhadra, R. K. (2003). *Understanding society* (pp. 98–99). New Delhi: NCERT.

Bhowmick, P. K. (Ed.). (1988). *Ecology and human development*. New Delhi: Inter India Publications.

Brundtland, Commission Report (1987)

Dube, S. C. (1988). *Modernization and development: The search for alternative paradigms* (p. 1988). N.J., USA: Zed Books.

Geertz, C. (1963). *Agricultural innovation: The process of ecological change in Indonesia*. Berkeley: University of California.

Government of India. (1956). *Second five year plan*. New Delhi: Planning Commission of India.

Guha, R. (1986). Ecological roots of development crisis. *Economic and Political Weekly, XXI*(15).

Guha, R. (1992). Prehistory of Indian environmentalism: Intellectual traditions. *Economic and Political Weekly, 27*(1&2), 57–64.

Guha, R. (1994). *Social ecology*: Delhi: Oxford University press.

Helm, J. (1962). The ecological approach in anthropology. *American Journal of Sociology, 67*(6), 630–639.

Kayastha, S. L. (1992). Environment and development. In M. Raza (Ed.), *Development and ecology: Essays in honour of Prof. Mohammad Shafi* (p. 40). Jaippur: Rawat Publications.

Kroeber, A. L. (1939). *Cultural and natural areas of nature North America, American archeology and ethnology* (Vol. 38). Berkeley: University of California Press.

Kroeber, A. L., & Kluckhohn, C. (1952). Culture: A critical concepts and definitions. *Papers of the Peabody Museum of American Archeology and Ethnology, 47*, Harvard University Press.

Mahalanbois, P. C. (1955). *The approach of operational research to planning*. Calcutta: Reprint.

Mann, R. S. (1990). Culture and ecology in Ladakh. *Man in India, 70*, pp. 217–227.

Martell, L. (1994). *Ecology and society: An introduction*. Cambridge: Polity Press.

Mukerjee, R. (1926). *Regional sociology*. Century: New York.

Mukerjee, R. (1942). *Social ecology*. London: Longmans, Green and Co.

Myrdal, G. (1970). *Approaches to Asian drama: Methodological and theoretical selection from Asian drama*. New York: Vintage.

Nagla, B. K. (2015). *Indian sociological thought* (pp. 86–88). Jaipur: Rawat Publications.

Oommen, T. K. (1990). *State and society in India: Studies in nation building.* New Delhi: Sage Publications.

Rambo, A. T. (1983). *Conceptual approaches to human ecology.* Honolulu, Hawaii: East-West Center.

Schumpeter, J. A. (1960). *Theory of economic development.* New York: Oxford University Press.

Steward, J. (1955). *Theory of culture change: The methodology of multi-linear evolution.* Urban: University of Illinois press.

Steward, J. (1968). Cultural ecology. *International Encyclopedia of the Social Sciences, 4,* 337–344.

UNESCO. (1975). *Goals of development.*

Vayda, A. P., & Roy, A. R. (1968). Ecology: Cultural and non-cultural. In J. A. Clifton (Ed.), *Introduction to cultural anthropology* (pp. 467–498). Boston: Houghton Mifflin.

Verma, M. K. (2004a). *Development, displacement and resettlement.* Jaipur: Rawat Publications.

Verma, M. K. (2004b). Development induced displacement: A socio-economic study of thermal power projects. *Man in India Journal, Ranchi.* July–Dec 2004.

Verma, M. K. (2015). Globalization and environment: Discourse, policies and practices. Rawat Publications: Jaipur, New Delhi.

Verma, M. K. (2016). Development induced displacement, SEZs and the state of farmers in India: Some insights from the recent experiences. *Journal of the Human Rights Commission, 15,* 23–48.

Weisskopf, T. E. (1975). China and India: A comparative study of performance in economic development. *Economic and Political Weekly,* No. 5–6, Annual No. 87, pp. 775–794.

White, L. (1949). *The science of culture: A study of man and civilization.* Farran, New York: Strauss.

Chapter 3
On Development: Practice and Ethics

Vinay Kumar Srivastava

> *Our worry should not be development; it should be inequality. Development that reduces inequality is the best.*
> —D. K. Bhattacharya (Personal Communication, 1 December 2017)
> *The trouble with life is not that there is no answer; it is that there are many answers. The purpose of anthropology is to make the world safe for human differences.*
> —Attributed to Ruth Benedict

Abstract Three arguments are in vogue with respect to the issue of development, which is described today as the main mantra of change, in the sense that if society is not 'developing' (like some of the isolated tribal communities of the world), it is labelled as 'stagnant', 'changeless', 'backward and stuck' and 'primitive'. First, development is a Western concept, embedded in a cultural context qualitatively different from traditional societies of the east. It has been devised by the affluent among the West, being virtually imposed on rest of the world, particularly the traditional and poor societies, which had recently emerged free after the liquidation of the colonial rule, so that they come at par with the developed world. Second, the recipients of the 'development technology' that principally came from the West were apparently poor but 'squatted' on vast mineral and forest resources. As development proceeded, they became the most exploited, for they had to part with their precious natural wealth in exchange for the 'development benefits' they were getting. As development proceeded unabated, they became poorer and poorer, inhabiting ecologically unsustainable habitat. Development multiplied global inequality manifold. Within the 'developing nations', the benefits of change were not equitably distributed, so inequality sharpened at the local level as well. Third, the overall impact of the Western style of development has been villainous to the environment. It has lethally affected the biodiversity, accelerated global warming, rendering the earth uninhabitable. The changes, however, are not partisan, for they affect all parts of the world; therefore, the onslaught that the djinn of development is causing needs to be checked. Against this backdrop, the notions of 'alternatives to development' and 'alternatives in development' have come into being. The contemporary discourse on development

V. K. Srivastava (Deceased)

© The Author(s), under exclusive license to Springer Nature Singapore Pte Ltd. 2021   41
M. K. Verma (ed.), *Environment, Development and Sustainability in India: Perspectives, Issues and Alternatives*, https://doi.org/10.1007/978-981-33-6248-2_3

is intertwined with ethical issues. The article addresses these issues with insights from lifestyles that the Indian tribes had been leading for years.

Keywords Sustainable development · Post-development · Nature reverence · Tribal knowledge · Polarization · Endogenous development · Cultural rootedness

Development is one of the most contested and controversial subjects (and concepts) in contemporary discourse, both academic and popular.[1]

Development: An Angel, an Ideal

It was not so when Harry S. Truman, the Thirty-third President of the United States of America, in his famous inaugural speech of 20 January 1949, described development as if it were the Aladdin's Lamp, the gentle rubbing of which would be miraculous, yielding whatever human communities need for their survival and future growth.[2] Truman was philanthropic in his approach. His agenda was to make available to the 'peace-loving' people of the world the benefits of the technical and scientific knowledge, the great strides of which had taken place in the United States. In his approach, the exploitation of people for profit had no place as was the sole objective of the colonial and imperialist governments in different parts of the world, which transferred technology to new states with the primary goal of maximizing their own profit, or they siphoned off the local resources (most of them, not-so-easily-renewable) for building up the empire at home. Truman proposed a 'programme of development' which purportedly rested on the principle of 'fair-dealing' in a democratic manner. The emphasis in Truman's vision was on technical and economic features of development (growth indices, industries, market hubs, infrastructures), for he believed, as did the later economists, that the key to 'prosperity and peace' lay in 'greater production'.

Eventually, all this was going to transform the world into a 'market society', on which would be dependent the life-chances of people. How would this be possible? Production would be enhanced by a wider and more intensive application of modern scientific and technical knowledge; thus, attention should be given to the disciplines of engineering and financial planning. The so-called undeveloped, underdeveloped and traditional nations should come forward with an enterprising and a 'path-breaking' spirit—to work vigorously towards achieving the standards that the developed nations of the world had already attained. It is well known that the present-day developed world was very different in its pre-development days; it was neither undeveloped nor underdeveloped, because these states of 'un-development' and 'under-development' were a by-product (or a 'concomitant product') of development of the West (Escobar 1995). The development of the West created its opposite everywhere. The present-day developed world took a long time to reach the stage at which it now is. Its results of accumulated knowledge in technology, science, economics and administration, and the manuals of its governance and laws, could be handed over to the developing world. The upshot of this in the non-Western world, the recipient, would be faster, without the severity of the so-called 'teething' and 'gestational' problems that the West experienced, and if at all they surfaced, they could easily be handled against the background of the sophisticated technical and scientific apparatus.

Eventually, the world would scale towards convergence. Prosperity of the developed world would be visible all over. Economic progress would alleviate poverty; newer institutions would come into existence, making the life easier; general health standards would improve, longevity would enhance, with people ageing gracefully; upward mobility of people would become uncomplicated, with the sequel that their children would do better in life, surpassing the accomplishments of their parents.

The world that development was expected to create was the ideal, straight from the happy-ending fictions. It was a state of freedom from the scourges of poverty, destitution, inequality, and injustice. A swathe of optimism followed from this notion of development. Unsurprisingly, development was deified—it was regarded as the 'best contraceptive', to recapitulate Dr. Karan Singh's words of 1974, or as a Kamadhenu Cow, as said a tribal respondent of mine in Udaipur (Rajasthan) in January 2017.[3] The first thought was given in the context of population control, based on the idea that the 'rich get richer and richer, with fewer children', whereas the 'poor get children, and lead a life of penury, depressing further in its scale'. The second was an analogy—as the milk of the celestial cow (Kamadhenu) does not dry, so do the gifts of development never decimate. While inaugurating the construction of the Bhakra Nangal Dam on 8 July 1954, Pandit Jawaharlal Nehru said that the structures like it were the 'temples of modern India'.[4] The then Prime Minister's statement could be interpreted to mean two things: first, these miracles of modern technology would solve the lingering problems of poverty, scarcity, and unemployment in India; and second, it implied the replacement of 'traditional religion' by 'modern religion', which was nothing but secularism. These 'temples' of resurgent India were for all people; they were not like village wells where caste monopoly in drawing water could be exercised. Modern institutions and infrastructural developments did not discriminate between people and communities; hence, they were not only the symbols but also the powerful agents of Indian unity. That is why these institutions were dedicated to the entire nation.

Development: A Polysemic Concept

One of the multi-meaning concepts in social science is of development.

In 1986 came Roy and Srivastava's book, aptly titled *Dialogues on Development*, which captured the views, opinions and feelings of the common people in Bihar on and about development. Conducted through open-ended and unstructured interviews—'pleasant chitchats', as they are sometimes called in research methods—the book was a break from the conventional approaches in development studies, where the notions and theories of the experts and advisors were given precedence over any other set of ideas. It was an approach 'from above', to enquire from them (the experts and advisors) what they meant by development, and what would be most suitable for the local people for whom a directed change had been planned, since in this bundle of thinking, the people were supposed to have a 'muddled mind', not knowing what would be good for them in the long run. Moreover, their internal differentiation, the fact they comprised different communities and classes, would rarely allow the emergence of a consensual view on what they needed, because each segment of society would think from its point of view and interests.

Against this backdrop, the best approach was that the outside professionals and specialists decided, after deliberations among them and on the basis of their comparative knowledge of the development programmes in different contexts, what would be the best for the people under question. The latter should be advised (in some cases, even cajoled and forced) to accept the programmes as they were and adapt to them. The passivity of people was well accepted in this thinking, which, it may be noted, continues unabated even today in several cases and many development experts tend to think so, and talk about it informally, although they may not voice this opinion of theirs in writing. Even when the so-called paradigm shift seems to have occurred from 'experts to people' in designing development, the belief is still held that people generally are not able to think clearly about what they want and what is good for them, for all. That is why they need the outsiders' assistance. Therefore, the change should come in the 'attitude and behaviour' of the experts.

Incidentally, Robert Chambers titled one of his early booklets on PRA (Participatory Rural Appraisal. Participatory Research and Action) as ABC of PRA. By the acronym ABC, he meant 'Attitude, Behaviour Change'.[5] The onus of changing one's mode of thinking and perspective should fall on the experts; they should 'learn from people', and their transformed demeanour would have a lasting impression on people. The expert should be a prime mover of change not only through his actions but also behaviour—'how he conducts himself while carrying out the action plan'.

The central method Roy and Srivastava adopted was to reverse the paradigm—'go to the below', to the people who were the 'targets' of development, more often the 'guinea-pigs', who in a patronizing language were termed the 'beneficiaries', and ask them in a relaxed atmosphere, without the operation of any 'reactivity syndrome' or fear, what they thought about the terms (and the range of ideas connected to them) which had saturated their environment, namely *unnāti*, *vikās*, *pragati*, and expressions like *āgebadhte rahnā*, *taraqui karnā*, and several others of the ilk, all implying the meaning attributed to the English term 'development'. Their method was to gather the local knowledge and understanding on development (and not to initiate any programme themselves, for they were not development workers) which might be diametrically opposed to the views of the experts, most of them outsiders to the community where they were entrusted with the task of introducing the planned changes.

Roy and Srivastava's eminently readable text will impress anyone with the polyvocality of views on development that the people hold. Needless to say, the variables of gender, age, strata and disability greatly affect the understanding of development. Since the publication of this book, it has become a common practice in the sociological studies of development that the native perspectives are given as much importance as those of the outside experts.[6] This grasping of the local meaning (the 'native point of view', as it is sometimes called) is essential for the success of the development plan, its modification and reception, for the agency is now supposed to be transferred to the 'recipients' of development, rather than its 'implementers'. But as said earlier, the old thinking that people are 'passive' and 'unclear' still continues to hold the grip of many of us.

Meanings of Development

A survey of the vast literature on development, across the disciplines, acquaints us with its several meanings. Its first meaning is mediated by the class to which the people belong. For those who have been affluent in the traditional economic systems, those who have been living comfortably for a couple of generations, those who have never experienced the culture of poverty and scarcity, development means having as much access as is possible to goods of diverse variety. For them, a ceaseless change in goods and fashions is a mark of development. Access to more goods at all times is an indicator of good life, and may be of psychic satisfaction.

By contrast are millions of people who have lived in abject poverty for generations (like the present-day de-notified and nomadic communities, and many tribal and rural people) and have no avenues of upward mobility (like the construction workers, whose children will be stepping into their parental occupations, since they do not get an opportunity to learn new skills or acquire schooling). They seem to have reached their highest and the farthest point, where unfortunately they would remain stuck. In the language of sociology, they are the 'blocked spiralists'.[7] For such people, development means getting a regular opportunity to work; to secure a modest supply of food, water, and shelter; having an access to the basic healthcare services; and protection from exploitation by the powerful actors of the state and civil society.

Thus, two contrasting concepts of development crystallize here. For those who occupy the upper, privileged, strata of society, development is the 'modern' way of living—thus, development is used interchangeably with terms like industrialization and modernization, and epithets like 'technical progress', 'market and specialized institutions and an untrammeled access to them'. For those at the lower strata of society, development is procurement of the 'basic living with dignity', and when this is denied, the proclivity of people lapsing into anti-social activities increases, for this is the only option left to them for eking out their bread.

Against these meanings of development, one may look at the thoughts of the Latin American authors (Escobar 1992, 1995). They have interpreted development in non-material terms, in comparison with its Western conception where what matters is 'goods economy'. Interpreted as liberation, development in Latin America means the empowerment of poor people—the transfer of political power to them from the traditional elites and from those who control modern technology. Development would be defined here as having occurred when the powerless people are able to transform their lot with the agency being transferred to them. Thus, development is not to be confused with a multiplication of material goods and accessibility of people to them, but with a transformation of society which liberates the hitherto oppressed masses. The essence of this idea is that 'true development' is anti-inequality and anti-oppression. The Western conception is concerned with the production and circulation of goods, empowering people to buy them, without abolishing inequality. In fact, the Western notion of development has exacerbated inequality, helping in the concentration of wealth in fewer individuals, financially strengthening middle classes, but impoverishing the lower strata. It is thus not surprising that one per cent of India's population holds 73% of her wealth. India has witnessed 34% increase in the number of people having a net worth of Rs. 1000 crore or more.[8] These individuals collectively own $719 billion, which is alone a quarter of the nations's GDP.[9] And, in a nation of 1.2 billion people, it is appalling that more than 800 million souls live on less than Rs. 20 per day![10]

Development: Ensuring Good Life

Both these meanings differ. Economy dominates the first, the Western; politics, the second. The first promotes more production, more visibility of the new material culture, whereas the second is for a structural transformation of society, the elimination of the exploiters and a change in the power relations.

At this point, an important question is: What is common to all the different notions of development, irrespective of the social and political contexts in which they are embedded? To answer this question, we shall look at what all human beings, irrespective of the social and cultural milieus they belong to, want. What do they long for? What is their craving? Anthropologists are particularly advantageously placed to answer this question, because they have painstakingly collected accounts of the life ways of people, spending years with local communities, endeavouring to know what people hanker for. What kind of life they want?

All humans have the survival instinct. They all want to live. They all want good life, free from lean periods of scarcity and hunger, free from debilitating illnesses, free from natural furies, free from servitude and exploitation. The aspiration that one's progeny should do better than them may not be universal, for the hunting and food gathering societies hope their children to be like them, or like some of their well-remembered ancestors who excelled in catching games, or weaving baskets, or sharpening arrows, or making a highly effective boomerang. Different cultures may add different notions to the common denominator of 'good life'.

Development, thus, may be defined as the directed and planned process of material and non-material change which ensures good life to people, or promises to make their lot better, trying to eliminate the hurdles they tend to encounter in leading a life of dignity and satisfaction. The execution of the development process, since it entails the use of new and sophisticated technology, necessitates the involvement of the outside experts, but it should not be viewed as an exogenously introduced change, because as experience has shown us time and again, that its chances of local acceptance and involvement are slender if it does not have an active involvement of the people, the receivers of change. The ultimate aim of development should be to empower people, so that they freely take decisions about their lives. In other words, the final objective should be the self-mobilization of people. In this sense, development is not any kind of change, for the latter may be progressive or retrogressive. It is a change oriented towards the good life, where it is a healthy partnership of the 'receiving people' and the 'outside experts', working in a dialogic fashion, which charts out the course of change, ultimately aiming towards, as Chambers (1993) says, 'handing over the stick to them [the people]', where the people become the masters of their destiny.

What Is Good Life?

The concept of 'good life' has evoked a lot of interest among the philosophers. When a man is able to do his normal duties well and derives happiness and satisfaction from his work and activities, he is said to lead a good life. This idea, largely influenced by the writings of Aristotle, is generally accepted in social science and philosophy. However, the anthropologists believe that in addition to 'some universal standards of good life', which arguably may be indisputable, each community has its 'own ideas of what constitutes a good life'. Here, the notion of cultural relativism comes into play, which submits that things are meaningful in their respective social and cultural contexts, and once they are taken out of that, they are rendered meaningless. This thesis, which American anthropology espoused, particularly one of its leaders, Boas (1940), has been at the forefront of anthropological investigations and methodology.[11] Thus, when we are trying to build up the concept of good life, which development should ensure, it is expected of us that in addition to the universal parameters, we shall also try to discover the local images and thoughts of the good life.

It is erroneous to believe that all the customs and practices of a community are the best, subscribing to the universal values. Some patriarchal communities may practice genital mutilation on girls, thus aiming to deny them the coital pleasure (Sulkin 2009). Among the Dani of New Guinea, the mourners, especially the women, are expected to chop off the tips of their fingers when a loved one of theirs dies (Heider 1970). The belief behind this practice rests on twin ideas: first, to physically experience the pain of a permanent loss, and second, to keep the spirits of the dead away, for they can meddle in the affairs of the living, causing mental and physical agony. Some communities, like in village India, may deny school education to girls, for they believe in the redundancy of this experience for women, who in any case are supposed to look after the household chores. In some cases, primary health care may also be denied to women; pregnancy complications may not be referred to the hospital, since they may not be regarded as important members of the household. True, differences do not imply inequality, but it is on these differences that inequality may be socially constructed. Thus, by implication, the local design of inequality may be considered by people as an index of 'good life'—thus, a situation where girls do not attend the school, learn household duties or remain veiled, do not challenge the oppressive authority of males, may be called by people as an instance of 'good life'.

Here come the questions of value. Cultural relativism should be distinguished from what is called 'ethical relativism', which means that morality is to be defined in terms of the normative system of the society under study. In that sense, the witch accusations, which are used to punish a family or deprive a woman of her property, are as 'ethically good' as are the conventions of food sharing. Anthropologists disagree with such conclusions, because for them, cultural relativism is a methodological devise, the aim of which is to understand the meaning of the local practice or custom, and certainly, it is not to render its justification or promotion.

On the contrary, the action anthropologists initiate animated debates with people regarding some of the local practices, continuing from the days of the yore, for which they may have a just sense of pride, but people may not know that these may be thwarting their progress or infringing upon the rights of human beings to live with dignity, which may in fact be slowing down or increasing many of their vital indices, like infant mortality rate. The anthropologists are beholden to people for the help they provide in making their fieldwork successful, but it does not imply that they approve of each aspect of their living. One of the earnest duties of the anthropologists is to have a critical look, a look of balanced appraisal, at the local culture and the lifestyles of people, and in case some of them appear to be contributing to their plight, this observation should be shared with them.

One may suspect that people will not take kindly the critical comments on them. They may be ethnocentric, and consider any of the evaluations of their ways of living as an attack on their pride, and that too, from an outsider who happens to be with them because of their compassion. Furthermore, it is thought that any such act on the anthropologist's part would slacken his rapport with the people, and may arouse their indifference, even fury.

However, the experience shows the opposite to be true. When an anthropologist lives with the members of a community in their natural habitat for a long time, often for more than a year, he becomes an 'ineluctable part of their self'. With great humility, he learns from them. With the passage of time, he becomes their friend, where, with an element of authority, he can advise them on several matters and also accept their suggestions on many counts. Often he becomes their representative, their spokesperson, who, on behalf of the people, takes up the issues of their interest with the authorities, presenting the local perspectives and their relevance. In a nutshell, the result of such an empathetic and lengthy living and interaction on an equitable plane (often called 'rapport' in anthropological literature) is that the people listen to the comments (critical as well) of the anthropologist (the 'well-wisher') on their life style. The field anthropologist initiates a dialogue with the people, to which the latter would respond, sometimes offering strong rejoinders to the ideas put forth. In summary, it is the commencement of a reciprocal exercise; the anthropologist will place his point before the people about the changes they should collectively bring about in them to make their future brighter, and would wait for the people to respond to that. It is this persisting dialogue that would help in bringing about an endogenous change, and also in systematizing a vision of good life. In other words, the concept of good life is built up over time; it is not a given, unchanging, entity.

Development: A Conundrum, an Ominous Djinn

Today, development has become a cliché, an exhortation, a yardstick of measurement (of individuals, societies and nations), triggering national and international migrations, positing reference groups to be emulated, a political slogan, an electioneering strategy, the agenda of new nations, and for social scientists, the process of progressive change, which in ideal terms, is destined to bring about a metamorphosis in the lifestyles of people. The aspects of development are so engulfing, and equally worrying, that an academic discipline can only afford to ignore it at its peril. Not only are the courses on development an integral part of every discipline under the head of social science, but also separate interdisciplinary degree programmes titled 'development studies' have been floated to cater to the requirements of planned and directed changes. So infectious is the study of development that one cannot think of a school or college subject which does not make reference to it. It will not be an exaggeration to say that we live in, and are shaped by, the repertoire of development. What Redfield and Singer (1954) meant by 'future-oriented' societies, is what we mean today by 'development-oriented' societies.

Even when the term 'development' is being subjected to rigorous interrogation, with several scholars delineating its evils and curses, some in hard-hitting expressions while some politely, the echo of 'more development', 'lack of development', 'development being a distant goal', or 'sidelined in development plans and activism', are heard from remote corners of the world, including almost inaccessible tribal hamlets (Pieterse 2000). People crave for development, for they see it as offering a remedy to their emaciated economy and lack of opportunities in their area. The cause of ethnic violence, fissiparous tendencies in the country, separatist movements, and the general social malaise is sought in lop-sided development or its absence. The common argument is that people rise in revolt when they encounter gross inequalities that the development process creates all over the world. Whereas some become richer, because development outputs reach them, the others who get scanty or no benefits at all, depress into poverty. As the riches heap up, with time and over generations, in a similar way, poverty accumulates, from which are born several other interrelated maladies, like poor health, criminality, exploitation of women and children, low social esteem and cessation of the desire to improve one's station in life or even, the yearning to live. The images of inequality are all over—from media to the concrete living—which are brought to the doorsteps of people, making them ask why they have been left out? Why can't they have the same kind of living as the privileged lot has? Would they and their children continue to dwell the same state of despicability and deplorability as their genitors lived? Would there ever be a silver lining in their future?

The paradox is that the members of these communities know well that development has caused them misery, spoiled their lifeline of resources and has dispersed its gains unequally. The economically and politically powerful people have monopolized over the gains, and thus their posterity's future is secure, whereas those at the bottom of the hierarchy have remained alienated from whatever little prosperity could have come to them from the developmental benefits, and thus, they and their offspring live in the midst of gloom and uncertainly. The talks on these issues, as well as the inside thoughts, the 'soul-thoughts', so to say, often lapse into the shenanigans of upper classes vis-à-vis the gullibility, honesty, and simplicity of the lower and depressed people. Although they know that fatalism does not provide a succour to their problems, they often sink into moods of depression, having a feeling where they think that their alley-ways are all sealed.

If on the one hand development is seen as exacerbating the crises of the already precariously-destined people, on the other, it is seen as the only panacea of improving their collective life. It is outlandish to think that the communities can return to their pre-development phases. The damage to the communities has already been caused. Changes are irrevocable. One of the biggest questions before the governments and development experts is to recognize this paradox as real, and thus instead of swaying with the current of the anti-development crescendo, they must think of improvising a kind of development in which the ethical and moral principles are fully integrated. Instead of being a subject of technicality, where the objective principles of engineering and economics operate, it should become humanistic, a matter of catering to the subjectivity of happiness, satisfaction, and self-esteem and dignity.

Development: A Pursuit of Economics and Engineering

In this section, I intend to explain why development came to be equated with Westernization and modernization, and became in straightforward terms an economic issue, with the effect that economics was regarded as the main discipline that would take up the entire calculus of development (Mair 1984; Dube 1994). However, since development projects dealt with technology, infrastructures and material appurtenances—in a nutshell, with visible, concrete outputs—the discipline of engineering was coupled with economics. Development was thus an economics-cum-engineering venture, and from this followed the battery of policy prescriptions and decision making (Sen 1982).

As I proceed, I shall argue that as we move from simple societies (that the tribal societies are supposed to be) to the present ones, called complex, in an evolutionary and a comparative matter, we find the separation of the realm of economy from that of society, and with this, a separation of ethicality from economics. This de-linking of economy from society, and ethics from economy, will remind us of Weber's rationalization, which included, beside the primacy of scientific and technical order, an exit of traditional morality, beliefs, customs and values. When Weber spoke of the 'disenchantment of the world', he not only meant the decline of the colourful folklore and mysticism (the stories of fairies, magical feats, taming of demons) but also of the relations of affectivity, surrendering one's intellect to sentimental factors.[12] Wb's world, bereft of love and affection, and any sort of a preferential treatment of some, was what the later urban sociologists called the world of anonymity, segmental relations, secondary groups and superficiality of ties.[13]

In this skein of thought, global inequality was justified in terms of the idea of 'equality of opportunities'. People were given equal opportunities to compete for the positions and resources, and if they could not come up, the problems of inefficaciousness and not living 'up-to-the-mark' lay with them, and not with the society, which all through acted justly. In other words, the 'rationalized state' was concerned with the 'all, rather than with some', notwithstanding the conditions in which the 'some' live, because for it 'all are same'. Thus, for the welfare of all, if 'some' had to sacrifice their interests or were made to sacrifice, it was legitimate, rationally justified, for rationalization was the antithesis of any modicum of sentimentality. The theory of development, thus, was concerned with the overall impact, the overall visibility, rather than with those who were left behind. It seemed to be quite Darwinian in its spirit, in the process of evolution (read, development), the fit will survive, the rest are fated to elimination.

Simple and Complex Societies

Travelling through the roads of Kolkata, you come across colossal posters on which is written 'Bengal means Business'. You are perplexed, because for you Bengal is synonymous with art and aesthetic institutions, with the poetry of the Gurudev and the films of Satyajit Ray, the scholarly pursuits of a large number of intellectuals, with a diversity of arguments and salubrious, and leisurely, discussions on a variety of subjects in small congregations, which gradually contribute to an expansion of the outlook of people, and a simple, ascetic and contented life, which is a sharp contrast to the acquisitive and consummeristic living that some other communities (of economically successful migrants) lead in Bengal or in other parts of the country. When 'Bengal becomes Business', it is the economic-centric view of development that takes precedence over all the other ways in which you might like to think of Bengal. In fact, the pictures that accompanied this slogan, and many others of the same type, are of the work stations and production units, with young men and women, healthy, dressed in Western-styled clothes, manning these places, and appear to be totally devoted to the tasks that have been assigned to them. These pictures could have been from any part of the developed world; so, when globalization-enthusiasts speak of homogenization crystallizing in the contemporary world, these are the sort of images they have in mind.

There were several new nations, which reeled under the grip of neo-colonialism, oppressive political states, under the domination of traditional elite, and were poor in resources, for these had depleted over time. They had a low quality of life, suffering from intermittent periods of hunger and starvation. They craved for industries, business enterprises, international economic collaborations, loans for setting up the local units of production. In a nutshell, the emphasis was on the generation of wealth. For this pursuit, the cultural heritage was a non-issue. If its sacrifice was imminent for economic goals, for it was thought to be thwarting the quest for wealth, it must be done without moral scruples. This would explain how traditional culture of people, which integrated them, was reduced to the backseat, and a new culture, some kind of a poor 'carbon copy' of the Western culture, started taking its place.

In traditional societies, economy was embedded in society. Social principles, which essentially were moral, regulated economy, the processes of production, exchange, and consumption; but once the path of development was chosen, which was essentially the 'path of modernization as the West had experienced', economy got itself separated from society, in the process, was also gradually liberated from social/moral norms. This was the 'non-moralization of economy'; one may also call this process the 'non-humanization of economy', since the social/moral ensemble defined 'what is human in a given context'.

Economy was fully human in traditional societies. Concerned principally with the satisfaction of the basic needs of people, of survival, what characterized it was the 'smallness of its scale'. For instance, only that many fruits, leaves, and nuts were gathered which could be consumed by members of the household. Food was eaten only when people felt hungry, unlike the modern world, where people eat even when they are not hungry, because plenty of food is always available; and it seems to me that the mere sight of food triggers our reflexes and we reach out for food even when we may be full. Simple societies valued scarcity. They knew that the forests had plenty but not at every point in time the plants and trees were able to yield the same quantum of resources. People knew the seasons, the moments in the ecological cycle, when certain foods would be available, and they must be ready for their collection. Forests (trees and plants, or animals) were also like 'human', growing slowly, with their pace, and would yield their products not just for humans but also for all living beings. Forests were non-discriminatory. Ethnographers of small communities have carefully noted and documented that the universe of people comprises all forms of 'lives'. Their world consists of plants and animals, but it also contains things like lakes, mountains, glaciers, rainbow, which are not as living as are the plants and animals, but have their own 'lives'. It was in this context that the idea of animatism was espoused, which meant that the things that the Westerners regarded as non-living were considered by the members of a small community as living. The idea that all have 'souls' (thus, feelings and sensations) gave a new meaning to the environment these people inhabited. That is why the forests were 'living beings', and like the guardians, they cared for their dwellers and dependents. So are the mountains and lakes, rivers and ponds, boulders and sand, stars and sky.

Let us look at another fact. Modern men and women take delight in buying not because they need the things, but because the markets are brimming with products and people have money to buy these, for their personal future use, as investment for their posterity, or as gifts of love.[14] The simple societies think differently. For them, the entire forest is theirs—whenever they need food, they can always visit it, collect in as much of its products, which are obtainable at that time, as they require, hunt as many animals as they want. The idea of ravaging the forest, terrorizing the animals, hoarding the resources is alien to them. Whichever food has been collected is consumed that day. The task of storing it for the following days is non-existent, because people know that the following day, when they would need it; they would go out on a food-collection round. If the local area is unable to fulfil their needs, they would fold up their flimsy huts and the other belongings, and move to another location in their habitat, where they expect to find an adequate supply of food and other resources. What has been presented here is a generalized account of the life ways of simple societies, particularly of foragers, based on a number of empirical accounts (Sahlins 1972).

When we move on the continuum from 'simple societies' to the 'complex', we find the occurrence of at least four processes. First, a 'communitarian ownership' of resources gives way to an 'individual ownership'. Second, the 'non-acquisitive' behaviour is replaced by the 'acquisitive', which in pejorative terms is called 'hoarding'. Communities become more and more surplus-producing. Third, to social differentiation, which is a universal feature of all societies, is added the division of people into strata; thus, an 'egalitarian society' becomes stratified, unequal, with tendencies of internal exploitation. Finally, institutions become differentiated from a 'common mass', which is society. Economy which earlier was rooted in society now has its own defined existence, almost autonomous, with tenuous links with society. Thus, it is independent from the moral and ethical norms of society. The same happened with other institutions (like politics, kinship). Each of these institutions developed its own codes of conduct, procedures of working, and rationality. As said previously, the separation of the institutions from society also meant their liberation from the norms of morality and ethics, which essentially were the creation of society. When the introduction of planned change came to be termed 'development', by then the economy had become more 'economic' than 'social'. To put it in other words, the 'economic context of social behaviour' evolved into a separate 'economic institution'. Thus, the process of development became a matter of engineering and economics than of morality.

Think Ethically: From Some to All, from Pain to Gain

Ethics was not out of our radar. In fact, much before the term 'development ethics', which Gasper (2009) defines as an 'agenda of questions about major value choices involved in the processes of social and economic development', came into vogue, nineteenth-century political economists, such as John Stuart Mill and Karl Marx, raised the questions of ethics in economy, and worked towards devising concrete steps for eliminating hunger and oppression from the globe. Not only that, the ethical ways of living, extending a care to all, so that all are happy and contented, can be traced to classical traditions, which culminated in the thoughts of Mahatma Gandhi (Barua 2015); it is therefore unsurprising that a collection of articles on post-development includes a short article from Gandhi where he talks of the capacity of the earth to satisfy the needs of all, but not the people's greed.[15] The message delivered to all was that they should make a clear distinction between 'need' and 'greed'; it is the latter that has to be consciously purged. However, as I have argued earlier, the separation of economy from society (the moral fabric) also meant its 'de-ethicization'—ethics and economy became two different streets.

The title of 'development ethics' seems to have come from the work of Louis-Joseph Lebret (1897–1966), who advanced the notion of 'human economy'; in other words, he sought to 'put the economy at the service of man' (Cosmao 1970). A major writer in this direction was Goulet (1931–2006), whose definition of 'development ethics' is oft-quoted: 'It [development ethics] is that field which examines the ethical and value questions related to development theory, planning and practice'.[16] He emphasized three basic components, which must ideally be met by every development programme: life-sustenance, self-esteem, and freedom. His work became a precursor to the writings of Sen (1999) and those who contributed to the evolution of the Human Development Index (Gasper 2004).

The experience that the tribal communities in India had of development was devastating. More than 40% of people who have been displaced because of one or the other development project requiring their land were tribal, and the exact number of those who were affected by these projects (the so-called project affected people) is not known, but certainly it is not less (Mathur 2013). The abysmal state in which these people find themselves had its genesis in the highly skewed development projects, where the benefits went to the already-privileged sections, whose hold on the wealth strengthened. In other words, the already marginalized people (that the tribespersons are) were further pushed to the brink of starvation and a never-ending move in search of jobs and shelter.

Against this backdrop, the critical and reflexive approach of development ethics submits: (1) Economic growth should not be equated with societal improvement (or human development); (2) Costs incurred on development and their distribution should not be ignored; (3) The ethics of benefits, costs and risks should be worked out; (4) Criteria for good social and global development should be delineated, and also periodically evaluated; (5) The idea that goods do not ensure good life should be reinforced and its veracity examined empirically; (6) The rights of the individuals that have to be respected and protected should be closely evaluated; and (7) The notion that free choice is a marker of development should be critically scrutinized.

Ethical parameters need to be examined case-wise, keeping in view, on the one hand, the universal parameters of good life, and on the other, people's understanding of the kind of life they would like to lead. For this work, the best approach is of ethnography (the 'ethnography of development'), for it submits that whatever is to be planned for the people should involve them as much as is possible. For both the experts and the people, it should be a mutually conducted learning process, where both learn not only the points of view of each other, but also the interests of the entire society, because often some communities have to sacrifice their own interests for the sake of others, for instance, evacuating their traditional habitat for building a dam, which may not eventually benefit them, if they are destined to move far away. People may argue that such a dam would benefit the rich peasantry rather than the foragers or part-time food-growers. If on one end exists the 'mathematics of gain', on the other is an 'intense experience of pain'. Berger put forth this point in his *Pyramids of Sacrifice* (1974); Cernea (2011: 104) also writes: 'Some get the gains, while others get the pains'. The paradigm needs to be reversed, for creating 'human economics', 'human development' and a 'humanistic discipline'. Keeping ethical standards in mind, as a method, development ethics requires an assessment of a situation before initiating any programme of change. In addition, as the process of development goes on, the need of a 'side-by-side' evaluation is necessary to see whether the ethical considerations remain intact. We may assign this role to field anthropologists (sometimes called the 'resident anthropologists'), those who happen to be residing with the people and examining the process in situ, to keep vigil so that the ethical principles are not undermined or flouted.

To think of human actions as bereft of morals and ethics is utopian. Each act of ours benefits a few, may remain neutral to others and may harm some. One of the central contributions of development ethics is to rehabilitate the battery of values and endeavour our best to make the world worth living. I can do no better here than to reproduce the following words of Henry Dicks, a psychiatrist, quoted in Kluckhohn's article on values (1955: 131):

> It is ludicrous for any of us to pretend that we are neutral, desiccated scientists to whom all ideas are equal, and to deceive ourselves that our sole value goal is the disinterested search for truth.

Policy Implications

Let us now take up the implication of these ideas to the issues of planning and delineation of policies for people. Two approaches are discernible. The first is the positive approach, also known as 'scientific', or after Amartya Sen, 'engineering approach'. It is where the technical aspects overrule all others. Development is a matter of equipping people with material goods, infrastructural progress and technical expertise. The decision about the anatomy of development lies within the competence of the technical and financial experts; people's voices are muted before the shrills of the outsiders.

Material appurtenances, as a result, may be seen all around, but people are unable to connect with them, for these do not augment their way of living. On the contrary, they may be a liability. By contrast to the positive approach, the humanistic view is 'people first', meaning that all kinds of impositions where the technocrats and bureaucrats are the decision-makers should be discouraged. Their competence should be tailored according to the needs and demands of people. Irrespective of who we are, our first commitment should be to the people, for development should give agency to people by empowering them.

The ethical issues submit that development is not value-neutral. It is Janus-faced. It is an instrument. It may ameliorate the lot of people or may lead to their destitution. Therefore, the question that all policy makers should ask is: 'who is going to be the beneficiary of the development programmes?' If the benefits seem to be finally going to the affluent and the actors in power structure, then such ventures should be critically re-examined. Such evaluations require the people's concerted and consolidated view, and for this task, the social scientists can play an indispensable role. Every team of planners and policy makers must have social scientists who have a first-hand experience of working with people at the grass-roots.

End Notes

1. For some important contributions to this theme, see Collier et al. (2006), Connelly (2007).
2. Harry S. Truman, Inaugural Address, 20 January 1949, Inaugural Address of the Presidents of the United States. Bartleby.com (https://www.bartleby.com/124/press53.html). [Accessed 12 May 2019].
3. Quoted in Mathai (2008).
4. Temples of Modern India. Financial Express, 16 August 2003. https://www.financialexpress.com [Accessed 23 May 2019].
5. See Chambers (1994), Kumar (1996).
6. See the ethnographic accounts of Chaudhury (1993), Patnaik (1996).
7. Frankenberg, building upon the work of W. Watson, gave this concept in 1966. Also see Payne (1973).
8. Oxfam Report: Which Sectors are Creating so many Billionaires in India? https://www.bsinesstoday.in/current/economy-politics/oxfam-report-which-sectors-are-creating-so-many-billionaires-in-india/story/312067.html [Accessed 23 May 2019].
9. Wealth of richest 831 Indians equal to quarter of GDP: Report. https://timesofindia.com/business/india-business/richest-831-indians-collectively-own-a-quarter-of-gdp-report/articleshow/65955083.cons [Accessed 22 May 2019].
10. Nearly 80 pct of India lives on half dollar a day. https://www.reuters.com/article/idUUSDEL218894 [Accessed 22 May 2019).
11. Also see Wrong (1997) for cultural relativism as an ideology, and how it could be distinguished from a method.
12. For Weber's ideas, see Weber (1919), Bendix (1960).
13. Many ideas that developed in Wirth's paper of 1938 were examined by other authors in their respective writings.
14. On this point, see Miller (2013).
15. See Gandhi's short article titled 'The quest for simplicity: My idea of Swaraj' (pp. 306–8) in Rahnema and Bawtree (eds.), 1997.
16. See Goulet (1996: 1); also see his 1971 book. He drew inspiration from the French religious intellectuals, 'worker priests', and from the 'hunger and thirst for justice' of the Gospel of Matthew.

References

Barua, A. (2015). Towards a philosophy of sustainability: The Gandhian way. *Sociology and Anthropology, 3*(2), 136–143.

Bendix, R. (1960). *Max Weber. An intellectual portrait.* Garden City, New York: Doubleday & Co., Inc.

Boas, F. (1940). *Race, language and culture.* New York: The Macmillan Co.

Cernea, M. M. (2011). Broadening the definition of 'population displacement': Geography and economies in conservation policy. In H. M. Mathur (Ed.), *Resettling displaced people, policy and practice in India* (pp. 85–119). London: Routledge; and Delhi: Council for Social Development.

Chambers, R. (1993). *Rural development: Putting the last first.* London: Routledge.

Chambers, R. (1994). The origin and practice of participatory rural appraisal. *World Development, 22*(7), 953–969.

Chaudhury, S. K. (1993). *Myopic development and cultural lens: An evaluative study of tribal development among Konds of Orissa.* New Delhi: Inter-India.

Collier, D., Hidalgo, F. D., & Maciuceanu, A. O. (2006). Essentially contested concepts: Debates and applications. *Journal of Political Ideologies, 11*(3), 211–246.

Connelly, S. (2007). Mapping sustainable development as a contested concept. *Local Environment, 12*(3), 259–278.

Cosmao, V. (1970). Louis-Joseph Lebret, O.P. 1897–1966: From social action to the struggle for development. *New Blackfriars, 51*(597), 62–68.

Dube, S. C. (1994). *Tradition and development.* Delhi: Vikas.

Escobar, A. (1992). Imagining a post-development era? Critical thought, development and social movement. *Social Text, 31*(32), 20–56.

Escobar, A. (1995). *Encountering development. The making and unmaking of the third world.* Princeton & Oxford: Princeton University Press.

Gasper, D. (2004). *The ethics of development—From economism to human development.* Edinburgh: Edinburgh University Press.

Gasper, D. (2009). Development ethics and human development. *Human development report networks, United Nations development programme,* Issue 24, 1–3.

Goulet, D. (1996). *A new discipline: Development ethics.* Working Paper 231. Notre Dame, NI: The Helen Kellogg Institute for International Studies.

Heider, K. G. (1970). *The Dugun Dani, A Papuan culture in the highlands of West New Guinea.* New York: Routledge.

Kluckhohn, C. (1955). Implicit and explicit values in the social sciences related to human growth and development. *Merrill-Palmer Quarterly, 1,* 131–140.

Kumar, S. (1996). ABC of PRA: Attitude and behaviour change. *PLA Notes,* Issue 27, 70–73. London: IIED.

Mair, L. (1984). *Anthropology and development.* London: Macmillan.

Mathai, M. (2008). The global family planning revolution: Three decades of population policies and programmes. *Bulletin of the World Health Organization, 86*(3), 161–240. https://www.who.int. Accessed May 23, 2019.

Mathur, H. M. (2013). *Displacement and resettlement in India. The human cost of development.* London: Routledge.

Miller, D. (2013). *A theory of shopping.* London: Wiley.

Patnaik, S. M. (1996). *Displacement, rehabilitation and social change.* New Delhi: Inter-India.

Payne, G. (1973). Typologies of middle class mobility. *Sociology, 7*(3), 417–428.

Pieterse, J. N. (2000). After post-development. *Third World Quarterly, 21*(2), 175–191.

Redfield, R., & Singer, M. B. (1954). The cultural role of cities. *Economic Development and Cultural Change, 3,* 53–73.

Sahlins, M. (1972). *Stone age economics.* Chicago & New York: Aldine Atherton, Inc.

Sen, A. (1982). *Choice, welfare and measurement.* Oxford: Blackwell.

Sen, A. (1999). *Development as freedom.* Oxford: Oxford University Press.

Sulkin, C. D. L. (2009). Anthropology, liberalism and female genital cutting. *Anthropology Today, 25*(6), 17–19.

Weber, M. (1919). Science as a vocation. In H. H. Gerth & C. Wright Mills (Eds.), *From Max Weber: Essays in sociology* (pp. 129–156). London: Routledge.

Wrong, D. H. (1997). Cultural relativism as ideology. *Critical Review, A Journal of Politics and Society, 11*(2), 291–300.

Chapter 4
Sustainability: Revisiting Enduring Modern and Identified Tradition

Ashok Kaul and Chittaranjan Das Adhikary

Abstract It is an era of crisis; some manufactured and some caused by the rupture of modernity. The global world is multicultural, but the state boundaries are the packages of history that has made passports significantly essential. The disparities are across the globe, yet east is distinct than the west. The western society did develop, when nature was to be conquered to sustain the humanity and the eastern societies, especially South Asian countries face double edged problematic that nature is to be befriended for the sustainability. When the western societies developed, world was compartmentalized in the hierarchical linearity and when the countries colonized and late to the process of development are to seek transformation in the quality of life, the world is globalized and the notions of privileging remain contested. We have all become wanderers, some move with dignity with proper visas, passports and licences, while transgress illegally internal or external boundaries of states. The referent is material. In its exploration, exploitation has become the canon. It is coming to saturation with the human interventions, where ecosystem and the conservation of energy are jeopardy. In this highly fluid state of interactions, the nature and the less privileged get trampled. There has to be rethinking so that the regenerative politics solves human problems with the consented privileges of the nature. The present paper explores the subjectivity of history in tradition and modernity to understand the sustainability in holistic realm and offer the bridging with the identified traditions and modernity without Scienticism.

Keywords Identified tradition · Reflective modernity · Sustainability · Holistic episteme · Ecosystem · Conservation of energy

A. Kaul (✉) · C. D. Adhikary
Department of Sociology, Banaras Hindu University, Varanasi, India
e-mail: ashokkaulbhu@gmail.com

C. D. Adhikary
e-mail: adhikary@bhu.ac.in

M. K. Verma (ed.), *Environment, Development and Sustainability in India: Perspectives,
Issues and Alternatives*, https://doi.org/10.1007/978-981-33-6248-2_4

Introduction

It is an era of crisis; some manufactured and some caused by the rupture of modernity. The global world is multicultural, but the state boundaries are the packages of history that has made passports significantly essential. The disparities are across the globe, yet east is distinct than the west. The western society did develop, when nature was to be conquered to sustain the humanity and the eastern societies, especially South Asian countries face double edged problematic that nature is to be befriended for the sustainability. When the western societies developed, world was compartmentalized in the hierarchical linearity and when the countries colonized and late to the process of development are to seek transformation in the quality of life, the world is globalized and the notions of privileging remain contested. We have all become wanderers, some move with dignity with proper Visas and passports and licences, while transgress illegally internal or external boundaries of states. The referent is material. In its exploration, exploitation has become the canon. It is coming to saturation with the human interventions, where ecosystem and the conservation of energy are jeopardy. In this highly fluid state of interactions, the nature and the less privileged get trampled. There has to be rethinking so that the regenerative politics solves human problems with the consented privileges of the nature. The present chapter will explore the subjectivity of history in tradition and modernity to understand the sustainability in holistic realm and offer the bridging with the identified traditions and modernity without Scienticism.

The Age of European Centrality, Colonialism and Objectification

The success of the European adventures in the sixteenth century was instrumental in the rise of capitalism, making the beginning of a new epoch. Its roots were in Protestant ethics. Subsequently, the reformation movement paved the way for the prerequisite elements deemed necessary for the ascendancy in the new epoch. It started with the objectification of the earth, its management and its race for subjugation of non-European territories. Despite contesting each other in the colonial pursuit, the common ground was the formation of the European centrality. This was possible because Europe set peace in the process of development and scientific inventions to incur that realm of power and ideas. The seventeenth century was the intellectual response to the collapse of the old structure. The powerful single unified revolution in the natural sciences supplemented the new intellectual writings in social sciences and philosophy. It was thought that there was intrinsic relationship between scientism and human values. Culture specifics were ignored. Judgement was given on the culture, compared with the European monoculture. Its final articulation was thought to be the modern project that gave a define direction to the history (Gupta 2014; Smith 2016; Wagner 2012). The social change was thought to be holistic and its direction linear.

It emphasized that the new modern society will stand as a contrast to the old society. The wheel of Progress was the pivotal to the agenda of modernity. The consensus framed was that the entire world had to be come under the dictates of the European monoculture. This was justified by the process of colonization, the civilizing mission of the other. The conquest of the people and to break them from tradition and their environment was a mission that had redefined the nature and environment, subservient to the humans, who hold power. Lands were subjugated and the natural resources were transferred from one country to the European heartland that bestowed it the centrality, the superiority among the comity of nations. If the Enlightenment was a break with the past to pave way for the triumph of reason and ensure the path of linearity towards equilibrium and parity, the process of colonization and notion of race superiority raised an instrumental rationality to alter the course of modernity to journey of Scienticism. The universal values of the Enlightenment were subdued by the agenda of modernity (Adorno and Horkheimer 1997). It undermined the strength of the other cultures. It established the European centricism and its hegemony. The colonized world was given the referent that the direction of history was straight its plank to be modernization and the agencies the nation states. Thus the process of colonization clubbed with the Enlightenment project produced the inner contradictions of logic in the path of capitalism. The triumph of reason in the Enlightenment project had come from the scientific ideas of the seventeenth century, well supported by materialist philosophies of Hobbes, Descartes Leibniz. Its rationale was positivism that produced to mega-theories of Functionalism and Conflict School. While Functionalism treated the dependency model of parts to frame the whole, a perspective led to the specialization of fields and its internal autonomy and dynamics. The collective common interests would be guided by rational calculations of common interest; the result would be 'an ordered society'; it denotes the terrific surge of the ideas, which blew apart the ancient regime in France, initiated revolution over there and in North America, and launched the classical doctrines of radical liberalism. The framing of the Enlightenment project was thought to be the universal solution to the global order. The maps were formulated in such a way that showed the distribution of power and resources from the centre that was Europe, as if it was given. Any notion other than European was inferior and comparable to the European central ideas. The European self was the Centre and all affluence and resources were accumulated in these countries across from the oceans and seas. The subjugated eastern world was the other, the Oriental (Said 1979). This site of power marked clear-cut categorization, the notions of 'superiority/inferiority, domination/subordination, rationality/irrationality, Enlightenment/disposition and occident/orient'. This helped to manage history and frame the distinction of the East and the West. At the same time, it was an empirical knowledge to understand the unknown colonized world. The colonial social engineering done for the consolidation of power and plunder of the raw materials were helped by this knowledge. During the process of the empire building and the colonization, the orient experienced structural transformation of the entire society. Not only social engineering of colonial rule created new classed and distribute poverty through the drain of resources and plunder of raw materials, but it had its regenerative impact also over the new western education (Chaudhury et al. 2015). Transport and communication, telegraph and railways knitted India together and subsequently, a new educated middle class was created that imbibed the European monoculture as its referent and this class proved to be instrumental in the implementation of the agenda of the Enlightenment project. Even after the de-colonization, the European mainstream culture remained the basic plank for the mode of the development and modernization, the result was that the diversities were either to get submerged with the plural culture or had to vanish, giving way only to mainstream culture. The referent for which was the European monoculture evolved and presented through the Enlightenment known as the modern project. The modern project had a mission to promote a certain set of ideas as universal ideas in order to encompass the diversity of the world and to impose practical rationality through the agency of the nation state. The developmental programs were so designed and made as to make this culture flourish and emphasis was laid for the better quality of life of the new rich and the new middle class in the colonized countries. Resources were thought to be unlimited and expansionism a way of understanding to capture the resources. The territorial expansion innovation, state formation and miniaturization received prime attention and subsequently provided the basis for the European triumph and hegemony. Woods were cut, indigenous people were forced to accept the new referent culture or else had to perish as 'landless poor peasantry'; the mission for the civilizing was given the loud tune. The entire world was brought under this agenda. The agenda was the Enlightenment programme (Leidman 1997). The project has a promise, a dream to realize. The intellectual mooring for the agenda were provided by the mega-theories of positivism. Functionalism and the conflict theories became the foundational theories to understand the nature of social change. The social change was thought to be holistic and the society that would come after the change would stand as a contrast to the one that existed previously. Functionalism tried to establish that 'the societies generate practices, which were the function of system-maintenance and historians would explain it in terms of its society-supporting function'. On the other hand, Marxism offered itself 'as the antithetical body of praxis and ideas capable of fighting bourgeois domination'. The conflict will result in a new historical synthesis in which the free development of all was the condition of the free development of each. But the promise made by Marxism about the vision of the new society and its realization failed. The ideological divide and the Cold war perpetuating the rationale of bipolarity never allowed the world to go beyond the Cold war dynamics. The issues of sustaining humanity were either ignored or neglected. These were never allowed to emerge. The perception was that 'one world begins to have a unitary consciousness of itself, it also sensed but could not count the entire world's particularity' (Inglis 1993: 23). This led to the failure of the human sciences to develop a just world order, devoid of domination and power, where nature and humanity could live together in consensus of levity human project.

Enlightenment Crisis, Materiality and Strategic Silence on Ecology

The dialectic of modernity had its tacit ramification, which not only created disparities but also questioned the very notions of positivism that had given European culture the power to judge the communities and traditions. It did run its course to exclude people on race and culture and on the plank of moderation. Modernity which is supposed to lead towards establishment of civil society actually led to a social order marked by gross inequality and violence. We have been witness to genocides, world wars and consolidation of nation states as an expression of bipolarity in the modern era. The fair intentions of the project were demeaned by the new race for the expansionism promoted and prompted by scientism with its goal-rationality; and its subjectivity was the domination and power.

The result was that the 'protective cover and the resource base'; both gradually were diminishing. The militarization for hegemony and expansion through high technologies in selective hands made natural resources also to vanish on selective basis. It derailed the ecosystem leading to enormous wastage of energy through overuse of natural resources, for retaining hegemony and domination of western centrality (Fulekar et al. 2014; Park and Labys 1998). The consequences were increase in the levels of carbon dioxide emission of certain greenhouse gases, global warming, toxic pollution, deforestation and erosion in the biodiversity. Earth to hold the human and technological wastage did come to brim. The new century opened this human pandora box, which previously remained hidden under the cover of strategic silence (Sachs 2008, 2015; Goldin and Winters 1995). Development at the cost of natural resources was deemed to be transformation to materiality. The mode of development was thought to be either to command economy, wherein the state was the main transforming agency. Policies were formulated at the highest level of state apparatus. The other was the trickle-down economy, presuming the cultural diversities and differences would melt down in a common culture. It could ensure the benefits of the development to the lowest rung of the society. In both the cases, the hegemonic centre with its monoculture was the referent culture (Blewitt 2008). It became a power that destroyed the very essences of the Enlightenment agenda. The simmering discontent was first experienced in the 60s, when the Functionalism lost its claim and credibility as a grand theory due to the political upheavals and the student unrest all over the world. And with the advent of the 80s, the Marxian orthodoxy has been shattered. The period from the late 60s to the advent of the new century witnessed different movements all over the world (Pieterse 2007). Outbreaks and revolts at the level of masses all over the world questioned the legitimacy of the established states. This caused a disruption in understanding social change and viability of social transformation. The new century created a new consciousness. Market in the absence of ideologies did become new rationale to make human to move from one place to another more swiftly and more wilfully. The global village was not a world with parity but it had its centres that created conditions for unequal bargain. And multicultural world lived like imagined communities without any belongingness with the present (Anderson 1983). The romantic past was explored and God came back to fill the space of ideologies, even in the public realm. It has not only knit the whole world together through the electronic media, but also has brought serious rupture to modernism—a breakdown of the process. The enormous post-Cold war migrations have made the world more multicultural than before (Watson 2000; Bhargava et al. 2007). With the weakening control of the nation states, loyalties are shifting quickly and the boundaries, national as well as others, have become fluid. Society is finding its reference beyond the boundaries of a nation state. This has resulted in the segregation of the communities. Nationalism once thought to be perennial identities of history and hold the citizen together is losing its relevance. Sub-national and the ethno-national are the new identities of history and culture which thrives on imagined communities and long distant nationalism, aligning with new primordialism which traces connectivity to genealogy. Late capitalism has robbed off social and has made a person rootless nomad, moving from one corner to another (Jameson 1991).

Uneven Development and Exclusion as a Problematic

The arrival of the new century places its emphasis on the conditionality of life chances and market mechanisms in a highly digitally connected world. The process of exclusion does not need interventions now. It is readily given in terms of purchasing power and access to the resources and knowledge. The rampant consumerism has led to choicelessness amidst identity crisis. While the consumption patterns of the West are significantly changing for most of the people, majority of the third world countries still languish in poverty. In the absence of purchasing power, there is a growing tendency among the masses to resort to the non-normative mode of existence or to go criminal for the quick material gains (Trentman 2016). The electronic network can do sell illusions for these poor masses, which hardly afford two square meals. The phenomenon of de-realization in these developing countries has made life more expensive in the villages, for the youth searching employment beyond rural areas had created scarcity of rural labour. The uneven distribution of wealth is on the increase. The countries, which have become the new states and have experienced the transition directly, are beset with the challenges of nation making. The tasks are many-fold, besides stabilizing the economy, institutions are to be built afresh and a lot of restructuring of its productive capacities is to be carried out. The nature of the social crisis in the developed countries is of different nature. The initial brisk economic growth and reliance on the vast non-western market has now limitations. The growing market of the non-western countries and their own cheap labour for the productive purpose have placed some challenge to the developed countries in the bargain, the consequences are evident, the enormous conditional ties are put by the various loaning agencies of the world bodies for the development of the these developing countries. A new type of colonization is in the process of replacing the older one. It has raised fresh debates, moral as well as economic. Bauman (1999: 100) puts it aptly, 'The cultural hybridization of global may be creative emancipating experience. But cultural disempowerment of the locals seldom is an understandable yet unfortunate inclination of the first to confuse the two and so to present their own variety of false consciousness' and a proof of the mental impairment of the second…the relevance of high technology which has created different type of exclusions on the one hand, and increase in the unemployment with the disappearance of the traditional base of the employment on the other again remains to be understood. The main failure of the late capitalism is to reduce 'necessary labor time' and focus on the widespread image culture so that a consumer world with specific interaction is constituted. A new class of screen viewers with unbridled individualism living private lives depending more on non-place connection is on rise. The loss of social with the fragmentation of the primary groups has led to the resurgence of identity crisis, especially with the end of geography. The dream of becoming a civil society in the post Cold war scenario is turning out to be a mirage. The notion of civilizations might be too early a prediction, but the rise of religious fundamentalism is notably on increase in the societies where the nation making has been a problematic. The religious connectivity is perilous than the ideological connectivity of yester years. Bauman (1994: 3) puts it aptly, 'An integral part of the globalizing processes is progressive spatial segregation, separation and exclusion. Neo-liberal and fundamentalist tendencies, which reflect and articulate the experiences of the people on the receiving end of the globalization, are as much legitimate offspring of globalization as the widely acclaimed hybridization of culture at the global top'. It is a mixed consciousness that is at work. The awareness of the side effects of industrial development on ecology and environment has made the people aware about oil spills, soil degradation, erosions in land and spoiled woods, the construction of big dams, power projects and deforestation have eventually caused the loss of biodiversity (Kumar 2003; Park and Labys 1998). The facts that nature is not the environment and all natural resources are not renewable are overlooked by the multinationals in the developing countries, where it finds new pastures. Multinationals seem to be dictating the terms. The nation states no longer hold control over their dynamics. There is no doubt that defending the environment and rescuing nature has raised a serious debate in both the developed and the developing countries. The new challenge is to bring back the social and collective solidarity with interfaith and inter-group accommodations in a plural framework in the absence of fading concept of geographical community. What is pressing emergent concerns are to find the meaning of the purpose of human existence in this divided digital world. Family is still the most basic unit of human group though it is struggling as an institution. The divisions of roles to sharing of roles have created more problems in trust generation and home making. The marriages are not lasting. The very institution based on the role differentiation is thought to be a patriarchal monoculture. The role sharing does not go with the essences of the family and pure relations are yet to come up to the institutional frame. Denaturalization, co-modification and de-traditionalization have to be defined properly in the context of the multicultural world. It is a period of history that seems to be directionless. Globalization has opened the opportunities. But for the richer few, who are making quick profits, leaving the vast poor languishing. It is a paradox: while it is beneficial to a very few, it leaves out two-thirds of the world's population in the margins.

The Journey from the Global to the Local in Search
of Sustainability

Hence, the global world is a digital world with invisible apparatus of surveillances (Lindgren 2017). It has become very difficult to live in the same community based geographies as used to be. What has emerged fresh are the common human concerns that can bind humans together are the safeguards of mother Earth. It is sustainability of ecosystem and conservation of energies. The coming generations would rely more on the natural resources and the preservation of the ecosystem. Therefore, a safe future is to be thought about. There has to be a comprehensive holistic approach, which lies in the integration of nature as well as the tradition that can withstand the challenges of multicultural accommodation. The social conditions are its starting point. The sustainable development goes beyond economic realms. It revolves round peace, economy, environment and parity. Democracy and an integrated developmental approach for the survival of the humanity and assured future for the coming generation have to be in consonance with the nature and the indigenous evolved cultures. A human centred theory is needed to move on the paradigm of multiple-modernity. That focuses on the dynamic role of information technology in bringing out the social institutions and the cultural values in the developmental process cannot come from one nation or one culture. It has to be complimented from the historical societies that have survived over thousands of years in view of different made onslaughts and natural calamities. This would take care of less developed societies and marginalized communities. One common debate concerning nature and the ecosystem is 'our common future'. There is common consensus that 'development that meets the goals of the present without compromising the ability of future generations to meet their own needs', this could be sustainable development. The 'club of Rome' debate while accepting the limits of the growth raised questions about the scientific policies in practice. Since the first major conference on environment in Stockholm in 1972, some concerns were raised about the tropical deforestation. Acid rain, global warming and the disposal of the ever-increasing waste material could find some space in the emergent debates. Besides there should be a concern about the decline of amphibians, the bleaching of coral reefs and sea species that are fast vanishing. The mode of development has caused the discontinuities and the synergisms. The diagnosis is there to quote Giddens, 'the paradox is that nature has been embraced only at the point of its disappearance, we live today in a remolded nature devoid of nature and this has to be our starting point for a consideration of green political theory'. This requires a political will with a human face on the part of the developed countries to feel for the concerns of the developing countries also. It might need a holistic transformation that would mean a new economic and social order. Since, it may not be possible to go back to nature and to the tradition for a tradition is a power, the idea is to seek to 'limit violence in a world of manufactured uncertainly'. It has led to the resurgence on the mode of understanding of the philosophy of science not in isolation of human philosophies. The era of specialization should need continuities not disruptions, hence we have to go back to the mythic, mimic and inductive-deductive traditions so that under the protection of nature culture flourishes. Nature is different than environment and any tradition cannot be taken for granted. There is a need for human sciences that should be based on ethical grounds, not on the politics of 'unequal exchange'. High technology is not much of use for the poor masses of the world in the less developed countries. The communities lack the minimum requirements of civic life like clean drinking water two square meals and safe dwelling. These requirements can be met through relevant technology, possibly by the use of the intermediate technologies. There is a dire need for such intermediate technologies and their availability to the less privileged people of the world (Shiva 2005, 2016). The high technologies have deprived the common people of the developing countries to live with reasonable degree of material comfort. Its costs are enormous. The developing countries cannot afford these technologies and then strive be welfare states. The other choice is to draw lessons from history and allow indigenous culture to be the core in the realm rather than isolated margins for the museum sake. That would be to study the people who have survived over the centuries, with contended lives and are close to nature. And these communities are proving to be better saviours for the future generations. Therefore, more researches are needed to understand the social organizations of these communities. These communities have survived despite facing the onslaught modernization process but what is missing in the process of conceptualization is the acknowledgement of their history and of complex political economy (Pathy 1999). In a way, it has proven prudent that these communities have not written their own histories but they have stories to tell. These narratives have been handed down the generations. They are so powerful in their assertion that they become their living history. It is here that their culture and the tradition are embedded. The discovery of these traditions that stand on the universal accommodation would lead to a better world. The most accessible unit of the study in cultural studies is narrative, in particular the life history. The content of these narrative will stand to the test of all those values in present-day life which make it possible to live. No doubt, these communities are having a contented life, but their requirements are also to be taken care of and that could be within the realms of those traditions which can accumulate and adopt the intermediate technologies. But, before such measures could help the local to become universal and the universal local, as a harmonious new construction, it has to overcome the historical. Thus, the present crisis is more moral and ethical, which goes beyond the realms of economics. That is possible only if that 'unfinished tasks of the enlightenment' is completed. The basic assumption of the Enlightenment was that 'ignorance is the basic source of all human misery and its elimination is possible through scientific knowledge'. The thrust has to be to produce complimentary notions that shall fill back the left over spaces by Scienticism.

The Unfinished Project of Modernity

The epochal shifts make paradigm change also. If the developments in the sixteenth century led to the finalization of modern project in the Eighteenth century, the happenings in early twentieth century has been a complete decentring of the mind set what previous centuries had produced. The modern project in sense was break with the past in defining and locating the European centrality. It enabled European world view to materialize with the consensus on binaries and judgement on cultures and nature. Specialization and pieces were identified to differentiate it from the whole. The triumph was nexus of ideas and power. The fundamentals of modern project had tacit notions of European superiority in setting the peace for the postulates of rationalization, industrializations and linearity of history with the universal theories of conflict solving and equilibrium maintaining. The tacit notions were based on the European centrality and its referent monoculture. Nature and spaces beyond the boundaries of Europe were to be conquered. The common resource property of forests, mountains and water to be subjugated for the good of human beings, an apparent realization and applications of resources for the prosperity and development, which had European subjects in mind. However, subjugating earth, it's measuring and owning by the European powers led the dynamics of migration to greener pastures from emergent countries and mass migration of slaves from the subjugated countries not only to the European countries but also to other European subjugated lands for the appropriation of work force. The principles of the modern project for the parity and rationalization of power and positions on institutional merit became twisted by the matrix of political economy and domination. Since the Order Making right from the human dawn in this world had one specific notion of comprehension of existence was of community. This community was linked with geography. This has been since living in the caves to the boundary of a nation state. The circumference of the boundary had increased with the instrumentality of the capacity of its management. This principle of centring had remained the moving principle of order making.

Sustainability as a Policy and Planning Issue

There is now a raging debate in the international community concerning the emerging issue of conflict and synergy between global capitalism and sustainable development. The phenomena of global capitalism and sustainable development are frequently depicted to be the polar opposites. In fact, both share a common emphasis to lead us towards a sustainable future in a globalized world. But how to make globalization work for sustainable development still remains a question mark and everyone keeps on pondering over this.

> Rather than leading to economic benefits for all people, global capitalist development has brought environmental catastrophe, social unrest that is unprecedented, increase in poverty

& hunger, increased inequality between and within nations. The experiment may now be called a failure (Sienna Declaration, http://www.toonside.org.sg/title/sienacn.htm).

On the question of environment, the logic of efficiency in economics and technology is found to be an antithesis to environmental values and ecological health, as environmental protection cut into profits (Sen and Grown 1987; Shiva 1989, 1991, 1999).

There are three major perspectives on the present order of global development advocating the economic benefits accrued to the developing countries: the neoliberal cornucopian (Thurow 2000; Gilpin 2000; Ohmae 1990, 1995; Friedman 1962), Sceptics (Callinicos 1994; Giddens 2012), (Held et al. 1999; Huntington 1996) and the Global Justice Group (e.g., Peet and Watts 1996; Agrawal et al. 1999; Sachs 2005). The neoliberal cornucopian school concentrates on globalization's economic benefits and technological imperatives and point to the extraordinary growth in world wealth and the rapid improvement of technology, as evidence of globalization's beneficial impact. They even go further to believe that globalization has accelerated transition towards democratization. The alliance of liberal democracy with market economy, especially in the area of development theory and policy, has given birth to the benefit that 'less government and more market power would inspire an era of global prosperity'. Not only democracy but also environmental restoration is stimulated by a global market. 'Environmental problems are projected to fade with the accelerated use of new clearer technologies as a result of free trade' (Byrne et al. 2002: 9). Reform-minded sceptics point that a majority of global economic activity involves North America, Western Europe and the Asian Pacific region; only marginal roles are available to Africa, Asia, and Latin America. Secondly, globalization is seen as producing differential shaping of national and regional economies, favouring particular sectors and activities that benefit the West. The expanding world markets are the outcome of continued western imperialism and monopoly capitalism facilitated by the nation states. Strengthening of economic relationships through globalization across the globe also has the consequence of cultural disintegration. Instead of a global civilization, cultural and ethnic blocks are seen to be emerging globalization is reducing the opportunities for democracy and leading to increasing economic inequality and fragmentation of governance. Global Justice Group holds that the exploitative and volatile features of industrial era are found to persist in globalization, albeit in new forms. Supposedly post-industrial institutions like science and technology, the media and the Internet are seen functioning along lines that deepen existing inequalities and create new ones. Specific research work on implications for indigenous cultures, environmental quality and civil society underscores the continuation of patterns of unequal development, environmental deterioration and hegemonic cultural influence.

The promises of high growth, distributive prosperity, high end green technology have failed to deliver justice, and equality. When profits come to drive the economy, it is but logical that, inequality will widen, nature will be unscrupulously appropriated and human misery will multiply. This is borne by recent global pandemic which infected around 10 million populations across the world till date and has killed at least 5 lakhs people till date. This underlined the failure of the transnational and retrieved the relevance national-local solution to global problems. However, the global economy can become an agent for prosperity through institutional reform, while observing limits of ecological carrying capacities. Replacing 'free trade' with 'fair trade', 'liberal democracy' with 'democratic participation' and 'optimality' with 'sustainability', these reformers seek to turn the failures of globalization into projects for better, more sensitive, management of the global path way. Re-centring global justice including environmental justice would return a measure of power to communities and elevate 'life world' values above those of the 'gospel of efficiency'.

Conclusion

Sustainability and its normative demand for combining modern with the tradition, therefore, is a problematic that the inherent consumerist logic of late capitalism need to resolve at the earliest if the human world is to hold its own ground in future. What are the ways forward? Not many of course, and not without any disagreement on any one of them. The time has come not only for recognizing weaknesses in the current process of globalization but to ensure that the benefits of globalization reach all parts of the world and all social groups. What we need today is not a global society but 'a sustainable global society founded on respect for nature, universal rights, economic justice and a culture of peace' (The Earth charter Initiative, http://www.earthcharter. org/draft/charter.htm), which will lay the ground for the development and progress of future generations.

Governments and other institutions around the world, as a matter of fact, have adhered to a successful concept like sustainable development. Since the 1992 'Earth Summit', sustainable development planning has become an inalienable part of government programmes for natural action as are embraced under the guidance of the Summit's Agenda 21 task. However, the rhetoric of sustainable development is not enough. What is needed is ecological modernization—a reform of economic institutions, and technologies in response to ecological needs that is based on the idea of 'green' society, realized by the application of appropriate legal, policy, and management tools (WECD 1987). Global growth must be tempered by international, national and local regulation to minimize its tendencies to create social inequalities and ecological harm. Democratic guidance is necessary to keep markets from reproducing unsustainable patterns of consumption and production. Ecologically sustainable development seeks to realize the desired harmonization of economy and environment—the coexistence of market-driven growth, and democratically applied ecologically sensitive brakes. Reorienting the politics closer to 'ecological democracy' movement and building pressure on the state and the international community to demand action beyond the boundaries of corporatist and liberal reform is the need of the hour. The political reorientation must recognize the 'common character of the ecosystem' (Shiva 1991) and guide us through the meadows of market economy towards the goal of sustainable development.

Notes

The paper has inputs from a key note address presented at the symposium on sustainable development at the 14th International Conference of system Research, informatics and Cybernetics held from July 29 to August 3 2002 in Baden-Baden, Germany.

References

Adorno, T. W., & Horkheimer, M. (Trans.). (1997). *Dialectics of Modernity*. Verso.
Agrawal, A., et al. (1999). *Green politics: Global environmental negotiations*. New Delhi: Centre for Science and Environment.
Anderson, B. (1983). *Imagined communities: Reflections on the origin and spread of nationalism*. London: Verso.
Bauman, Z. (1999). *Globalization the human consequences*. Polity Press.
Bhargava, R., Kumar, B. A., & Sudarshan, R. (2007). *Multiculturalism, liberalism and democracy*. New Delhi: Oxford University Press.
Blewitt, J. B. (2008). *Understanding sustainable development*. UK: Earthscan.
Byrne, J., Glover, L., & Martinez, C. (2002). *Environmental justice: Discourses in international political economy, energy and environmental policy*. New Brunswick, USA: Transaction Publisher.
Callinicos, A. (1994). *Marxism and the new imperialism*. London: Book Marks.
Chaudhury, L. et al. (2015). *A new economic history of colonial India*. Routledge.
Friedman, M. (1962). *Capitalism and freedom*. Chicago: University of Chicago Press.
Fulekar, M. H., Bhawana, Pathak, & Kale, R. K. (Eds.). (2014). *Environment and sustainable development*. New Delhi: Springer.
Giddens, A. (2012). *Runaway world: How globalisation is reshaping our lives*. London: Profile Books.
Gilpin, Robert. (2000). *The challenge of global capitalism: The world economy in the 21st century*. Princeton: Princeton University Press.
Goldin, L., & Winters, A. L. (1995). *The economics of sustainable development*. California University Press.
Gupta, D. (2014). *Mistaken modernity: India between the worlds*. Delhi: Harper Collins.
Held, D., et al. (1999). *Global transformations: Politics, economics and culture*. Stanford, California: Stanford University Press.
Huntington, S. P. (1996). *The clash of civlisations and the remaking of the world order*. New York: Simon and Schuster Paperbacks.
Inglis, F. (1993). *Cultural studies*. Blackwell.
Jameson, F. (1991). *Postmodernism or the cultural logic of late capitalism*. Duke University Press.
Kumar, A. (2003). *Dimensions of environmental threats*. Delhi: Daya Publication.
Leidman, S. E. (1997). *The postmodern critique of the project of enlightenment*. Amsterdam: Rodopi.
Lindgren, S. (2017). *Digital media and society*. Sage.
Ohmae, K. (1990). *The borderless world*. New York: Harper Business.
Ohmae, K. (1995). *The end of the nation-state: The rise of regional economies*. New York: Free Press.

Park, H. S., & Labys, W. C. (1998). *Industrial development and environmental degradation.* UK: Edward Elgar Publishing.

Pathy, J. (1999). Tribe, region and nation in the context of the Indian state. *Sociological Bulletin, 48.*

Peet, R., & Watts, M. (1996). *Liberation ecologies: Environment, development and social movements.* London: Routledge.

Pieterse, J. N. (2007). *Ethnicities and global multiculture: Pants for an octopus.* UK: Rowman and Little field.

Sachs, J. D. (2005). *The end of poverty: Economic possibilities for our time.* London: Penguin.

Sachs, J. D. (2008). *Commonwealth economics for a crowded planet.* Penguin.

Sachs, J. D. (2015). *The age of sustainable development.* Columbia University Press.

Said, E. W. (1979). *Orientalism.* Vintage.

Sen, G., & Grown, C. (1987). *Development crises and alternative visions.* New York: Monthly Review Press.

Shiva, V. (1989). *Staying alive: Women, ecology and development.* London: Zed Books.

Shiva, V. (1991). *Ecology and the politics of survival: Conflicts over natural resources in India.* New Delhi: Sage.

Shiva, V. (1999). *Biopiracy: The plunder of the nature and knowledge.* London: South End Press.

Shiva, V. (2005). *India divided: Diversity and democracy under attack.* Seven Stories Press.

Shiva, V. (2016). *Stolen harvest: The hijacking of global food supply.* University Press of Kentucky.

Smith, S. B. (2016). *Modernity and its discontent: Making and unmaking of bourgeoisie.* Yale University Press.

Thurow, L. C. (2000). Globalization: the product of a knowledge-based economy. *The ANNALS of the American Academy of Political and Social Science, 570*(1), 19–31. https://doi.org/10.1177/000271620057000102. Accessed March 4, 2020.

Trentman, F. (2016). *Empire of things:…from fifteenth century to the twenty first.* UK: Penguine.

Wagner, P. (2012). *Modernity: Understanding the present.* Polity.

Watson, C. W. (2000). *Multiculturalism.* Open University Press.

Chapter 5
Explaining Rural Development in Contemporary India: A Paradoxical Situation

K. L. Sharma

Abstract Rural development is not simply an offshoot of policies and programmes initiated by the Indian State and public-sector agencies and NGOs. It is far more inclusive and comprehensive as the rural people themselves aspire for their betterment and put efforts to that end. Undoubtedly, the constitutional provisions, policies and programmes for rural upliftment have weakened some of the institutional bottlenecks, and granted a voice to the deprived and excluded sections of rural India. 'Urban bias' is one such potent factor, obstructing egalitarian rural development. Onslaught on traditional obstructive institutional social arrangements has paved a way for 'new actors', individuals and families to assert for their shares in the process of development. That led rural development to proceed in an inclusive form, both materially and socio-culturally, ensuring incorporation of all sections of society, particularly the poor and the deprived people. Thus, it is both a policy and a process, encompassing activities in public and private domains. Therefore, rural development is a change for a desired type of society and hence, has a human face as well. The above insight is drawn by the author from his study of Six Villages in Rajasthan in the 1960s and again after half a century in 2015-16, with a view to grasp the nature and direction of social change and development. A comparative analysis of the same villages, covering a period of five decades, indicates that 'inclusive development' is the real concern of the people, implying access to assets, markets and opportunities on an equal basis, reduction in disparity of income, maximum benefits for the poor, increased focus on agriculture, employment, health care and education, and equitable sharing of public goods and services.

Keywords Indian village · Institutional bottlenecks · Inclusive development · Distributive · Shares · Rural-Urban divide · Urban bias · New rural middle class

K. L. Sharma (✉)
Jaipur National University, Jaipur, India
e-mail: klsharma@hotmail.com

Jawaharlal Nehru University, New Delhi, India

Rajasthan University, Jaipur, India

Introduction

Indian society is a complex of collectivities, groups and sub-groups, sharing common-alities, specificities and differences. Hence, it is not simply a conglomeration of social and cultural entities and identities. It can be characterized as a unique panorama, having a semblance of both continuity and change. Some people adorn continuity, while some demand vehemently change, and at times ask for a quite rapid social trans-formation. Despite such a paradoxical process of change, India persists as a plural society, though at times, its plural character is jeopardized by the forces opposed to a synthesis of cultures, languages, regions, religions and empowerment of the weaker sections and women.

Often India is viewed as a country, inhabiting a large population in its villages, which are generally deprived of amenities available to the urban people. A divide between village and town is a key point of debate in discourse on development and elections in India. The Indian State has sponsored development programmes with a view to extend benefits to the weaker sections. Officially, the State claims to make an onslaught on inequality, poverty, unemployment, illiteracy and other pathologies of the rural people, with a view to empower them to cherish a dignified life. The methodology for effecting such a vision becomes a serious concern. The question is: Can it be realized by way of projection and enactment of egalitarian social order in a deeply entrenched hierarchical Indian society? The other problematic is: Can the State ensure equality among people by way of competition, in which people engage themselves according to their capabilities, to have an equitable access to resources of the society? But then the next question is: How those people become capable, who have remained incapable for ages due to 'institutional' bottlenecks and 'unfreedoms'? These questions are particularly pertinent to village India, as it is quintessentially different from urban India.

Defining Rural India

Today, the notions about Indian village, such as a 'holistic' entity, an 'isolate', a 'self-sufficient' formation, etc., have become quite redundant and obsolete. It is also not an 'ambiguous entity' as observed by Das' (2012: 187–203). The Indian village has always been a semblance of continuity and change, through the nature and dynamics of its formation have never been uniform. The patterns and practices of social relations embodied inequality and equality, dependence and freedom and conflict and cooperation. However, it is also an undeniable fact that Indian society, being a hierarchical system, had an overriding value load of inequality, unfreedom and clash of interests between the higher and lower castes and social groups.

During the Mughal and the British periods, and after Independence, Village India has/had witnessed both structural and cultural transformations with regard to land tenure systems, administration, styles of life, education and cultural ethos. Habib

(1974: 264–316) and Cohn (1968, 1969: 53–121) provide a graphic description of the cultivators, and the share of the State in the land produce, during the British period. Two hierarchies, one based on the caste system, and the other based on agrarian relations, were a well-established reality. In Rajasthan, feudal order and caste hierarchy existed simultaneously, not corresponding at different layers and levels (Sharma 1998: 26–43). The idea and practice of 'contra-priest' as given by Gould (1967: 26–55) implies the practice of 'equality' within the overall gamut of 'inequality' based on the caste system.

Gupta (2005: 751–758) substantiates the above portrayal of the Indian village by saying, that it was never 'unchanging' and 'idyllic'. But to say that the Indian village is withering away (Gupta: ibid), is an untenable perception. Agriculture and culture as the two cornerstones of the Indian village have changed to a considerable extent. However, the village remains a concrete spatial entity, having still agriculture and culture as its main features, though both are quite transformed. Indian village has not 'vanished' as it has happened in the West. Country-town nexus is far more different today, compared to the situation on the eve of Independence in 1947. New forms of inequality, new patterns of migration and mobility, and transformed modes of inter- and intra-caste relations speak of change and dynamics in rural India. Since village as a spatial entity persists, its cultural and social ethos too remain intact to a considerable extent. Enhanced role of individual and family has weakened the significance of caste-based activities and social ties. New 'actors' have taken over the role of the traditionally entrenched upper caste(s) elderly people.

Onslaught on Traditional Institutional Social Arrangements

Onslaught on the entrenched institutions, such as caste system, orthodox values and norms, by way of the constitutional provisions, land reforms, and education, has brought out basic structural and cultural changes in rural India.

Broadly speaking, we can explain such a transformation as 'development'. Besides such a state-initiated development agenda, development also occurs due to efforts made by individuals and families, who aspire for a cherished living. Development, however, does not imply an egalitarian social order in toto, eliminating inequality, subjugation, exploitation and discrimination. Thus, Sharma (2019: 267–281) asks: What is development? Whose development? Who are beneficiaries of development? Who are capable for realizing their cherished goals? How some people acquire more skills and capabilities than others, and who remain deprived?

The impact of urbanization on rural India calls for a redefinition of the terms, such as 'rural', 'rural-urban divide' and 'urbanity'. Inroads of non-farm sources of income and decreased dependence on agriculture have weakened caste-based pursuits and intercaste relations, in particular the jajmani system (patron-client ties) has become non-functional. Jeffrey (2015: 106) rightly observes that development brings about equitable benefits from a sustained economic growth, through a comprehensive, institutionalized state-directed system of social welfare. The requirement is that development needs to be for a large number of rural poor people. Reduction in socio-economic inequalities by way of rural development would strengthen democratic fabric of India. Such a view is also voiced by Harriss (1982: 15–34). Access to public goods and services would lead to reduction in inequalities and poverty. Thus, rural development is a positive intervention in India.

Rural Development as a Process

Let us make it clear that rural development is not simply a state-sponsored programme, relating to agriculture, irrigation, electrification, transport and communication. It also implies initiatives and innovative activities by particular individuals and families with a view to enhance their economic standing and social status. Hence, it is an unending process of change. Some people move out of their village for lucrative employment and entrepreneurship. Sharma (2014: XXV) states: 'Rural India has changed a lot as rural-urban divide has considerably transformed into a new nexus'. Today, there are new village networks, a new middle class, differentiated structures and new patterns of social mobility. Rural development is not restricted to the public sector programmes alone. In fact, by way of continuously changing networks, new non-farm sources of income, 'caste-free' activities, realization and assertion of dignity and honour, particularly by the lower sections, have enhanced role and significance of individual and family as social entities, and pronounced space for the middle peasantry (Ibid.).

As a process, rural development has witnessed land reforms, differentiation of peasantry, 'capitalistic' mode of production in agriculture, green revolution, panchayati raj, decentralization of power, education, migration, mobility, non-farm income, etc. Besides these significant shifts, awareness regarding environment, mother care, concern for girl child, weakening of inter- and intra-caste relations and appropriation of 'caste' (casteism) in politics are other indicators of rural development. According to the World Bank (2004), rural development is a strategy designed to improve the economic and social life of a specific group of people—the rural poor. Thus, international and national agencies, NGOs, individuals, concerned citizens and reformers also contribute to the process and policy of rural development.

Rural development implies interventions in the life of the rural poor. It is not limited to agriculture or economic activities alone. Actual interventions based on a well-stipulated policy would ensure a transformed egalitarian rural social order. Role of markets and connectivity with external agencies can reduce 'urban bias'. The poor people remain poor due to 'urban bias' in India's development policy and programmes (Lipton 1977). But such a view is negated by Byres (1977: 258–274) as he talks of 'rural bias', because a majority of policymakers come from rural India, whose advocacy is for capitalist agriculture, by seeking loans, subsidies and favours for the top-ranking rich farmers, kulaks and capitalist landlords.

Rural development is not limited to agriculture or economic activities alone. Technological and environmental factors and social and cultural relationships have affected the principal rural producers, such as Yadavs, Jats, Kurmis, Patels, Marathas, Jat Sikhs, Reddys, Kamas. Tradition is under severe stress. Agriculture is no more an attractive means of livelihood. However, a small number of agri-business cultivators is visible. Industry is favoured. More than agriculture and industry, a job in government/public sector is preferred as it ensures stability, security and social status. Since most educated rural young people do not get jobs in government/public sector, their second preference is for a job in private/corporate sector. However, within the village, non-farm income is valued a lot more compared to income from agriculture.

Singh (2009: 178–195) talks of rural development in terms of 'social praxis', based on his understanding of a village in eastern Uttar Pradesh. He refers to two levels of social praxis: (1) the state policies of development and (2) a new resurgence in entrepreneurial ventures. Such a situation has paved a way for an accelerated process of social change in the village.

Different perceptions and observations have been made about rural development. For example, in Uttar Pradesh, government plans (Chauhan 2009: 147–165), interference by dominant caste(s) (Lieton 2003) and discourse on **izzat** (dignity) (Kolenda 2009: 1833–1838) are the main criteria of rural development. Freedom in Kerala and concern for the poor and downtrodden in West Bengal are the main characteristics of change and development. There is a long range of development concerns, such as the role of caste, class and gender, natural calamities, land reforms, human development, weaker sections, child labour, education and aspirations for mobility and higher social status.

Since, rural India has witnessed vast structural and cultural transformation in its economy, infrastructure, education, health care, life-style, social and political awakening and participation in political activities (elections), the discourse on development requires a paradigm shift. The ideas of 'modernization ideals' (Myrdal 1968: Vol. I) or 'development as freedom' (Sen 2000) do not bring out the entire range of rural development. Development is seen today not just in terms of 'economic growth'. Growth without development, social justice and distributive shares with reference to the poor and deprived sections is being negated (Gupta 2009). India can't grow without Bharat (Acharya 2007). Guha (2008: 605–719), though talking about India as a whole, talks of the understanding of Rights, Riots, Rulers and Riches (the Four R's) to know the sharp divides between the beneficiaries and the deprived ones, based on caste, class, rich, poor, gender, and rural and urban people.

Inclusive Rural Development

The idea of 'inclusive development' implies a check on lopsided, ad hoc and uneven nature of development, particularly caused by constraints of caste hierarchy and institutional bottlenecks. Apparently, the notion of 'inclusion' is opposite of the idea of 'exclusion', but both are simultaneously two sides of the caste system. Inclusion of some as members of their own caste amounts to exclusion of the members of other castes, from caste-specific activities. Dumont's (1970) idea of pure is just a principle of inclusion-exclusion syndrome. In the context of development, the notion of 'inclusive growth/development' indicates access to benefits of development to the 'excluded'/deprived and neglected social groups. The main aim is to include the poor, the weaker sections, minorities and women in the process of development. Sharma (2019: 277–280) asks: (1) How people have been excluded? (2) What are the exclusionary devices? How the inclusive measures ensure benefits to the deprived ones? What is the nature of consumer–expenditure distribution? People are segregated in terms of caste, ethnicity, gender rural-urban divide, etc. The provisions of inclusion of the deprived sections of society ensure equitable distributive shares in the process of development. Some notable indices of inclusive development are:

1. Access to assets and thriving markets and enhanced equity in the opportunities for the deprived people;
2. Reduction in disparity of income among different groups of people;
3. Maximum inclusion of the poor in the development process;
4. Increased focus on agriculture, healthcare, and education;
5. Empowerment of the excluded;
6. Equitable sharing of benefits.

Thus, 'inclusion' focuses on adequate participation of the poor, marginal and women. This could be done by elimination of constrations, such as caste, gender, patriarchy and political dominance of a select group. Sen (2005) rightly talks of three factors, namely (1) Reach, (2) Range and (3) Reason. These are three R's of Reform as stated by Sen. Access, extent and selection of priorities are implied in these factors relating to participation in developmental process. Despite R's of Reform, economic explanation of inclusion/exclusion, and caste-based discrimination are the root causes of exclusion from sharing of resources. Discrimination can be seen in regard to jobs, education, health care and social relations. Inequalities in these domains are thus further, increased due to caste-based discrimination (Thorat and Umakant 2004).

Politics of inclusion (Hasan 2009) is found in public domain for seeking support of the SCs, STs, OBCs and minorities. But in the private domain and informal factor, groups suffer due to social exclusion. No social protection is made available to the excluded people. Their participation in public domain is negligible as they are deliberately kept excluded from the institutions and activities that would empower them. Gender discrimination is quite obvious in rural India. Inclusive development implies exercise of rights on an equal basis, with transparency and accountability.

Rural development creates a voice for assertions among the weak, marginalized and excluded sections. A demand for 'social justice', a 'fair justice' (Rawls 1971) to create a 'just society' is the call of rural development. Justice is a natural instinct and not a matter of political bargaining by political parties and outfits. India Rural Development Report 2012–13 (2013) refers to development of rural India in terms of rural social change, livelihood and inclusion, infrastructure, sustainability, local governance and Mahatma Gandhi National Rural Employment Guarantee Act (MGNREGA). According to this Report, 'a new rural India' is on the anvil. However, rural India comprises of diversities, both natural and social, and therefore, a uniform pattern of development seems to be somewhat imaginative and unrealistic. Maximum benefits from rural development programmes are corned by the undeserving people and the better off sections, from among the SCs, STs, OBCs and minorities.

A Revisit to the Six Villages in Rajasthan After Half a Century[1]

Rural Rajasthan provides a vivid picture of social change and development based on my study of six villages in the mid-1960s (Sharma 1974) and after half a century in 2015–16 (Sharma 2019). Being a formation of 22 princely states, a case of 'Indian India', it was a citadel of feudal malpractices of appropriation of human and material resources (Sharma 1974, 2019). Rajasthan was also quite distinct before Independence as feudalism coexisted with colonialism (Sharma 1983). Since Independence, rural Rajasthan has witnessed vast changes, however, it is observed that the legitimized rulership in the princely states or the 'traditions of Rajput domination' have survived even after Independence (Rudolph and Llyod 1984: 7–25; Narain and Mathur 1990: 1–58). Sharma (1998: 168–184) considers such a continuity more of a myth than a reality in the post-independence period.

The twenty-two Princely States of Rajputana were deprived of their rulership after Independence. They had no more authority to levy 175 *lagbags* (cesses), innumerable taxes and *begar* (forced labour) (Sharma 1998; Singh 1964) that were in practice before the abolition of Jagirdari and Zamindari systems. Structured inequality in terms of caste and feudal hierarchies characterized social relationships. People were denied freedom, education, migration and mobility. Avenues of sociopolitical awakening were blocked by such 'unfreedom'. Even if there was reinstance to feudal atrocities and cruelties, it was often crushed by the jagirdars and princes.

Besides the abolition of feudalism, institutional initiatives to transform rural Rajasthan, by way of adult franchise, elections, panchayati raj institutions, development programmes, education, means of transport and communication, etc., were immensely effective in the first two decades. Sharma (2019: 304) observes: 'During the early post-independence period, we have noted multiple contradictions and discontinuities, obverse structural process of transformation, congruities and incongruities, and structurally induced and self-generating process of change'. This was an initial march as the basis of reconstruction of Rajasthan.

The two visits to the six villages, the first in 1965–66 (Sharma 1974), and the second in 2015–16 (Sharma 2019)[2], clearly indicate sea-change over the five decades. In 1965–66, despite cleavages between caste, class and power, land reforms, adult franchise and PRIs, there was an overwhelming correspondence between various determinants of social status. The decades of 1950s and 1960s witnessed downward mobility of the privileged and entrenched sections of people, and upward mobility of the downtrodden at the same time, implying loss and gain of social status, respectively. The six villages in 2015–16 showed a tremendous change and development, in regard to caste, class, power, networks, education, employment, means of transport and communication[3]. The social and human face of the villages was immensely transformed, even beyond recognition, compared to their social and economic morphology in 1965–66. Far more egalitarian relations could be seen in 2015–16.

However, dominance of the new rural middle class clearly indicates uneven distributive shares in the process of development. The new social formation is an evidence of a renewed formation in rural Rajasthan. The double synchronic study of the six villages (Sharma 2019: 305) shows the following patterns of change and development:

1. The constitutional provisions and the initial change;
2. 'Equalization' process in the decades of the 1950s and 1960s, and the effective role of education, migration and mobility in reshaping social relations in the succeeding decades;
3. Political power as a decisive factor in determination of social status;
4. Hazy agrarian hierarchy, quantum jump in non-farm income, preference for government jobs, trade and commerce, and ambition to have access to power and authority.

These dimensions of development are indicative of a new transformed form of inequality, without completely eliminating the traditional bases of inequality, such as caste, gender, patriarchy.

However, today, structural transformation can be characterized (Sharma 2019: 305) in terms of the following patterns of change:

1. Downward social mobility of the previously entrenched sections of society;
2. Upward mobility of the deprived castes, peasants, artisans and workers;
3. Non-farm income as a source of status and social mobility;
4. Transformed country-town nexus as a means of mobility and change.

Based on the study of the six villages in 1965–66 and 2015–16, Sharma (2019: 305–306) observes that the initial change emanated from the policies and programmes of the Indian State in the 1950s and 1960s. In 2015–16, we observed that various developments, schemes and programmes and efforts of the people themselves had brought about notable structural and cultural changes. Today, stratification and difference among the village people can be evidenced in terms of rich, well off and poor families, varying rural-urban networks, uneven access to non-farm sources of income, high and low salaried government functionaries, and the weak and the influential families and individuals.

Change and development were not always necessarily determined by the caste system. Extra-caste factors, in particular, impacted horizontal differences within castes and sub-castes. Economic, political and cultural distinctions also could be found within the families and individual members of different castes as well as within the same castes.

The Six Villages in the 1960s[4]

We studied six villages from Sikar, Jaipur and Bharatpur districts in 1965–66. Land-use, cropping pattern, irrigation, country-town nexus and social and economic formation were not uniform in the six villages. Two villages were selected from each district, one in a remote area, and other in close proximity of the district headquarters. The main purpose of such a comparative study was to understand the changing face of the villages in relation to urbanization, migration, mobility and modern jobs. Besides the role of caste in shaping social relationships, our main task was also to know the role of family and individual. With the weakening of the role of caste in village economy, the middle castes began to assert in political and economic affairs. A new type of class-differentiation could be seen challenging the twice-born social order. Caste alone as a determinant of socio-economic status was questioned, particularly by the upcoming families and individuals. 'Difference' in social and economic realms across the caste hierarchy paved a way for some families and individuals as resourceful and influential actors to have a say in the village economy and polity. The poor, weak and voiceless became marginalized. Thus, the idea of 'composite status' (Tumin 1952) was found by us quite useful (Sharma 1974) in knowing the nature of a synthesis of tradition and modernity or that of continuity and change.

In the 1960s, the beneficiaries of the land reforms and panchayati raj institutions were jubilant, as they could opt for education, migration and urban living. In the six villages, about one-third of male-workers were engaged in new occupations, some within the village, and mostly outside the village, even in far off towns and cities. Villagers were thus quite differentiated within the villages, and inter-village differentiation was also quite marked.

In some studies (Epstein 1962; Kolenda 2009), a lack of correspondence between social change and economic development or a congruence between economic development and cultural change has been reported, negating village as a monolithic entity. Scarlett Epstein observed in a wet village (irrigated), economic development without corresponding social change, and in a dry village (unirrigated), she observed social change without corresponding economic development. A different pattern of change and development was witnessed by Pauline Kolenda in a West Uttar Pradesh village. When the people had improved their economic condition by way of development programmes, over a period of time, they asserted to ensure dignity and honour, particularly for their womenfolk. Village networks (Shah 1991) and emergence of a new middle class in rural Gujarat (Shah 1998) explain the changing face of rural India.

No unilinear development has occurred. It has been observed that the six villages (Sharma 2019: 312) have both cooperation and conflict, selfhood and collective identity, unity and factionalism, and hierarchy and individualism/segmentation. These are found in varying permutations and combinations. Despite such a situation, the six villages have retained their 'ethos', though in varying measures.

The Six Villages in the 1960s and in the 2020s

1. Sikar District

In 1965, Roopgarh, a remote village, had a middle (upper primary) school, a branch post-office, an ayurvedic dispensary, a village panchayat, a credit cooperative society, and the headquarters of Patwari and Village Level Worker (VLW). Today, Roopgarh has a Senior Secondary (10 + 2) School, Primary Health Centre (PHC), Animal Husbandry Health Centre, two Anganbari Centers, two private medical practitioners, two private English medium schools, one private Industrial Training Institute, in addition to the institutions that were there in the 1960s.

Roopgarh and its adjoining smaller hamlet together comprised of 215 households, with a population of 1495 persons. There were 411 male workers, and out of them, 163 were engaged in agriculture, 71 were manual workers, 60 were shoe-makers, and only 43 were engaged in white-collar jobs, as school teachers, clerks, officers and peons. No woman was engaged in a gainful job. Some people were masons, priests, artisans, etc. Only 74 persons worked outside the village, and of them 35 were Brahmins, an upper caste. Out of 24 Brahmin families, 19 depended on income from outside the village.

Today, Roopgarh is well-connected by a tarred road with the tehsil, the district and the state headquarters and with several nearby villages and towns. In the 1960s, no one even had a bicycle. Use of camels and bullock carts was a common means of transport. People hardly knew about telephone. At times, telegrams were received or sent to share generally information regarding a death in a family.

Two Jain families were engaged in shopkeeping and moneylending. A Brahmin also owned a shop, though not with success, partly due to rivalry by the Jain shopkeepers. The Jains, being moneylenders, practised Bohra-Dhuriya (lender-borrower) system, in the mould of Jajmani system. The Jains moved out of the village in 1970s and 1980s to far off places, like Kolkata, Ranchi, Raipur, Sambhar Lake, etc., for greener pastures. Earlier, the village's bazaar was both a mercantile and social space for the people. It is today a barren square, having depleted shops and a neglected place. Today, family and individual alone matter in social and cultural affairs.

Today, in Roopgarh alone, there are as many as 51 shops, located on the roadside, owned by different castes, including middle and lower cases. There are a couple of tea stalls, hair-cutting saloons, sweetmeat shops, fruits and vegetable vendors, tailoring shops, and shops dealing with hardware, stationery goods, clothes, shoes and chappals, computer services, electric goods, mobile services, etc.

Nearly 800 people have moved out of the village during the last four decades. Sikar, Jaipur, Delhi, Kolkata, Guwahati, Kanpur, Raipur, Ranchi, Mumbai, Hyderabad, Ahmedabad, etc., are the well-known designations. Even two persons have settled in America and Finland. Many work as artisans, masons and drivers in Gulf countries.

In the 1960s, there were only four graduates and one post-graduate in the village. In 2016, there were as many as 10 Ph.D. degree holders and 18 engineers. Some engineers have their education from BITs, Pilani, MNIT, IIT, etc. Of the 66 graduates, 21 are women. Of these highly educated persons, some are in All-India and state services, and many are serving in universities, colleges and schools. However, as a result of reservation policy, both men and women from among the SCs, STs and OBCs have been benefited. One SC and one OBC women were elected as Sarpanch. 35 persons from among the SCs are in government jobs, whereas in the 1960s only one person was a school teacher.

A lot more data are available. For our purpose here, it is suffice to say that change and development have restructured social relations and reduced social inequalities and cultural barriers between the higher and the lower castes. Adult franchise, elections, panchayati raj, reservation policy, education, migration and mobility are the main decisive factors in rural development.

Sabalpura, a 'sub-urban' village in Sikar, was at a distance of 5 km from the district headquarters. Compared to the remote village, the suburban village (Sabalpura) had weak jajmani system and less dependence on hereditary occupations. It was a multi-caste village. Institutions, like a post office, a primary school and a credit cooperative society, were there. Brahmins, Rajputs and Jats were the dominant castes. Out of 436 working male members, 165 were engaged in government jobs and 174 persons were cultivators. The remaining ones worked in Sikar town as manual labourers and shop assistants.

Today, in 2015–16, Sabalpura has practically submerged with Sikar town, though technically it remains a village. Its population has increased seven times, compared to the 1960s. The village has several tea stalls and nearly 150 grocery shops. It has become, in fact, a segment of Sikar town. A lot of residential colonies, offices and institutions are situated in Sabalpura due to expansion of Sikar town.

The primary school has been upgraded up to senior secondary (10 + 2) level. One Kendriya Vidyalaya (Central Government School) is also there. A private college for girls and some NGOs are other important institutions. Community centre, health clinics, private schools and employment in Gulf Countries are other notable developments in the present-day Sabalpura. Rajputs prefer to work in Police, Army and CRPF. Muslims are inclined to work in the Gulf countries. The members belonging to castes, such as Brahmins, Jats and Balais (SC), work mainly in Sikar, and some of them have government jobs. Unlike Roopgarh, people in Sabalpura are not inclined towards higher education. People have a lot of assets in the form of vehicles, land, gold and houses (for rental income). A small-town culture and life-style have pervaded Sabalpura. A feeling of being 'urban' dweller is often expressed by its inhabitants.

2. Jaipur District

Bhutera, a remote village, like Roopgarh, had several institutions with a population of 1717 people. There were 210 houses, divided into several castes, such as Brahmins, Rajputs, Jats, Ahirs, Kumhars, Malis. The village was isolated from other villages and towns due to lack of road and public transport. Only 10 persons worked outside the village. Agriculture was the main source of livelihood.

Bhutera has witnessed tremendous change over the years since the 1960s. Its population has increased five times. The primary school, like the schools in Roopgarh and Sabalpura, is now a Senior Secondary School, with Science Stream (Biology). The river-bed, which kept the village cut off in the 1960s, has become today a resource for irrigation and drinking water, even for the nearby villages.

Besides enriched agriculture, Bhutera has witnessed enormous economic and social change. A lot of people are employed as school and college lectures, army and police functionaries, bank manager, school teachers and in other public-sector departments. The SCs have also been benefited from the reservation policy.

No one is unemployed today in this previously remote and isolated village. In numerable shops, tea stalls, photo studio, beauty parlour, tent house, 15 milk dairies, hair-cutting saloons, medical shops, poultry, etc., are employed in Bhutera. The villagers have been taking utter benefits of the government schemes, such as MNEREGA, E-Mitra, Sahakari Kraya-Vikraya Samiti, Kisan Card, Atal Seva Kendra. Along with such a new village scene, due to enhanced irrigated cultivation, traditional institutions, such as Jajmani, village council, have not disappeared, as it is in the case of Roopgarh and Sabalpura, the previously remote and suburban villages, respectively in the Sikar District.

Harmara, a peri-urban village, in the surroundings of Jaipur, is situated on national highway 11. In 1965–66, it was connected by a single tarred road, today, it is connected by a well maintained four lane road. A village with meagre means, rudimentary institutions, a few grocery shops, Harma was also like other villages, a multicaste formation. However, nearly, half of the working male members worked in Jaipur.

The coming up of an industrial complex, known as Vishwakarma and some localities, such as Murlipura and Vidyadhar Nagar, have bridged up the distance of 18 km. between Jaipur and Harmara. Today, Harmara does not bear its identity of a village as it had in the 1960s. Harmara is also connected by the Ajmer-Delhi Byepass. Harmara and its adjoining hamlets have been merged with Jaipur Municipal Corporation, and now two wards have been carved out as parts of Jaipur Nagar Nigam.

Nearly, one lakh people inhabit the two wards of Harmara. Impersonal relations, urbanization and 'individualism' have swept away the idea and practice of **bhaichara**. Today, the face of the village is that of a crowded township. The two sides of the highway are surrounded by nearly 700 shops. Of these, 50 are sweetmeat shops, 80 haricutting saloons, 50 grocery shops, 50 medical stores, etc. Watch repairing, photography, computer/cyber services, cooking gas agencies, banks, ATMs and many other amenities are available in Harmara. Today, Harmara is more of a sub-city of Jaipur. Being a part of Jaipur, the people of Harmara enjoy civic benefits like the residents of Jaipur.

3. Bharatpur District

In Bharatpur district, Bawari, a remote village in Bayana tehsil, agriculture was the main source of livelihood. Some people were educated, like Roopgarh, a remote village in Sikar district. Jats and Gurjars owned and controlled most of the agricultural land. Today, Barwari is a completely transformed village. Its population has doubled. Earlier, no institutions were in existence, and today, a middle school, ANM Health Centre, Anganbadi, shops, etc., are there. The village is connected by a pucca road with Bharatpur, Bayana and Dholpur towns. Enchantment for higher education and lucrative government jobs is evident from the fact that today there are medical doctors, veterinary doctors, lectures, Rajasthan Administrative Service (RAS) person, a woman engineer, Block Development Officer (BDO), a nurse, school teachers and clerks belonging to Bawari. Most of them are Jats and Gurjars, who are the two dominant castes in Bayana tehsil. Some persons are also employed in Indian army and police and other departments. Agriculture is no more a lucrative pursuit as it was in the 1960s. Several people have been benefited by the government schemes, such as Indira Awas Yojana and MNREGA. Nearly, 20 persons are working in private schools and colleges at Bharatpur. Thus, today Bawari is, in fact, no more an interior village. It has connectivity by pucca road with other villages and towns. Its face is really a new one, compared to its situation in 1965–66, as a backward village.

Murwara, a peri-urban village, had poor institutional structure, like Bawari, in the 1960s, except that it had an easy access to Bharatpur town, being in its proximity of 4 km. Half of the people depended on agriculture, and the remaining ones worked in Bharatpur as manual, industrial and white-collar workers, Jats and Brahmins were influential groups. Today, the village is practically merged with Bharatpur city. It has not changed much, like other five villages. Problem of drinking water is quite acute due to salinity. Drug addiction among youth has become a menace. Many people grow Ganja (opium) plants. However, Murwara is quite different today. There are 55 persons in government jobs (in army, police, education department).

A SC Woman is **Sarpanch**. A ST Woman is a Lower Division Clerk (LDC). The Principal of the school is a woman. An ANM Nurse, a Saini by caste, is married to a Brahmin. Jatavs (SC) are no more engaged in their hereditary occupations. The game of **Kho-Kho** is quite popular in Murwara. Spatially, Murwara is merged with Bharatpur, but it remains a village in terms of village Panchayat and other institutional networks.

Policy Implications

A policy on rural development must address questions relating to its meaning, background of beneficiaries and nature and direction of change and development. The ultimate requirement of development is to see that the poor people become legitimate beneficiaries of programmes initiated by the state and non-state agencies. A clear parameter for judging inequality and equitable benefits from schemes of development needs to be in place.

The paradox is that reforms in rural India have received a setback from both 'rural bias' and 'urban bias' policies and programmes. In the first case, the rural rich corner maximum benefits; and secondly, 'urban bias' causes minimal attention to the rural people, particularly the poor ones. A well-stipulated policy would ensure a transformed egalitarian rural social order. In this paper, we have argued for transformation of the poor rural people in both structural and cultural contexts. A 'social praxis', comprising of a renewed set of development programmes and avenues of self-employment, may be two cornerstones of rural development. Growth with justice and distributive shares for the poor and deprived sections must be ensured. Inclusive rural development should provide more space to the poor. Reach of the poor to the development programmes would give them voice by way of their participation in the processes of development. Assertion for *izzat* (dignity) and participation in development processes may substitute the nation of 'reforms' by a claim for 'rights' in rural India. Ultimately, rural development must aim at empowerment of the poor and deprived/excluded people.

Concluding Remarks

The six narratives indicate a clear tendency towards inclusive development as people have been exercising their civil, civic and political rights. Migration, mobility, salaried jobs, assertive exercise of adult franchise, participation in panchayati raj institutions are symptomatic of strengthening of democracy by way of development. No inclusive development is possible without distributive shares on an equal basis regarding employment, health care, educations, etc. Rural development has paved a way for social justice and empowerment of the deprived/excluded sections of society. Though uneven pace of development between different states, districts, and social categories is also a stark reality to a considerable extent. The Indian state, being a soft entity, has not been harsh enough to plug the holes in its legal and administrative enactments. Hence, casteism, gendering and marginality raise their ugly faces, at times. Crisis of governability persists, though protests are also quite frequent and effective as well. Bottlenecks in empowerment of the excluded/deprived are not the same as were in the 1960s. However, a lot more is required to bridge the gap between the rural and the urban worlds, the rich and the poor, the resourceful and the helpless people. A paradigm shift is in sight. But it has not fully matured and galvanized.

Notes

1. A study of Six Villages in Rajasthan was carried out in 1965–66 as a part of my doctoral study at the University of Rajasthan (1968). The doctoral thesis was published in 1974, and several papers were published during 1968–1974 in national and international journals. As a National Fellow (ICSSR) during 2014–16, it was planned to revisit the Six Villages, with a view to understand the gamut of change and development over half a century. Author's book—*Caste, Social Inequality and Mobility in Rural India*, 2019, New Delhi, Sage, contains a chapter on a comparative analysis (pp. 302–332). I have heavily drawn from my earlier work (1974) and the present one (2019) for this paper.

2. The fieldwork was done in 1965–66 and in 2015–16 in the six villages. The names of the villages are real ones. The details of the villages are available in author's books (1974; 2019).
3. Revisits to the Six Villages were an eye-opener to have a gaze of tremendous change and development. In this paper, information has been appropriated, eclectically.
4. The study was planned based on definite criteria, such as proximity to district headquarters, impact of urban centres, land-tenure systems, irrigation, cropping pattern. The idea was to go beyond single village studies, which were quite popular in the 1950s and 1960s. A departure was felt necessary realizing diversities based on the above variables in rural Rajasthan.

References

Acharya, S. (2007). *Can India grow without bharat?*. Academic Foundation: New Delhi.

Byres, T. J. (1977). Agrarian transition and the Agrarian question. *Journal of Peasant Studies, 4*(3), 258–274.

Chauhan, B. R. (2009). *Rural life: Grass roots perspectives.* New Delhi: Concept Publishing Company.

Cohn, B. S. (1968). Notes on the history of the study of Indian society and culture. In S. Milton & B. S. Cohn (Eds.), *Structure and change in Indian society.* Aldine Publishing Company: Chicago, IL.

Cohn, B. S. (1969). Structural change in Indian rural society: 1596–1885. In R. E. Frykenberg (Ed.), *Land control and social structure in Indian history* (pp. 53–121). Wisconsin: Regents of the University of Wisconsin.

Das, G. (2012). *The elephant paradigm, India wrestles with change* (pp. 187–203). New Delhi: Penguin Books.

Dumont, L. (1970). *Homo hierarchies.* London: Paladin Granda Publishing Ltd.

Epstein, T. S. (1962). *Economic development and social change in South India.* Manchester: Manchester University Press.

Gould, H. A. (1967). Priest and contra—Priest: A structural analysis of Jajmani relations in the Hindu Plains and the Nilgiri Hills. Contributions to Indian Sociology (New Services), No. 1, pp. 26–55.

Guha, R. (2008). *India: After Gandhi* (pp. 605–719). Picador: New Delhi.

Gupta, D. (2005). Wither the Indian village: Culture and agriculture in rural India. *Economic and Political Weekly, 40*(8), 751–758.

Gupta, D. (2009). *The caged phoenix: Can India fly?*. New Delhi: Penguin/Viking/Oxford University Press.

Habib, Irfan. (1974). Social distribution of landed property in the Pre-British India: A historical survey. In R. S. Sharma (Ed.), *Indian society: Historical probings* (pp. 264–316). New Delhi: People's Publishing House.

Harriss, J. (1982). Capitalism and peasant farming, Agrarian structure and ideology in Northern Tamil Nadu, Bombay/Delhi. Oxford University Press.

Hasan, Z. (2009). *Politics of inclusion: Castes, minorities and affirmative action.* New Delhi: Oxford University Press.

India Rural Development Report, 2013 (2012–13).

Jeffrey, D. S. (2015). *The age of sustainable development.* New York: Columbia University Press.

Kolenda, P. (2009). Micro ideology and micro utopia in Khalapur: Changes in the discourse on caste over thirty years. *Economic and Political Weekly, 24*(2), 1833–1838.

Lieton, G. K. (2003). *Power, politics and rural development: Essays on India.* Maohar: New Delhi.

Lipton, M. (1977). *Why poor people stay poor: A study of urban bias in world development.* London: Temple Smith.

Myrdal, G. (1968). *Asian drama* (Vol. I). New Harmonda-worth: Penguin Books.

Narain, I., & Mathur, P. C. (1990). The thousand year Raj: Regional isolation and Rajput Hinduism in Rajasthan before and after 1947. In F. R. Frankel & M. S. A. Rao (Eds.), *Dominance and state power in modern India: Decline of a social order* (Vol. II, pp. 1–58). Oxford University Press: Delhi.

Rawls, J. (1971). *A theory of justice.* Cambridge: MA Harvard University Press.

Rudolph, L., & Llyod, S.H. (1984). Essays on Rajasthan. Concept Publishing Company: New Delhi.

Sen, A. (2000). *Development as freedom.* New Delhi: Oxford University Press.

Sen, A. (2005). The three R's of reform. *Economic and political weekly, 40*(19), 1971–1974.

Shah, A.M. (1991). The rural-urban-networks in India. In K.L. Sharma & D. Gupta (Eds.), *Country-town Nexus* (pp. 11–42). Rawat Publications: Jaipur.

Shah, G. (1998). Caste sentiments, class formation and dominance in Gujarat. In K. L. Sharma (Ed.), *Caste and class in India (reprint).* Jaipur: Rawat Publications.

Sharma, K. L. (1974). *The changing rural stratification system (a comparative study of six villages in Rajasthan).* Orient Longman: New Delhi.

Sharma, K. L. (1983). Agrarian stratification: Old issues new explanations and new issues and old explanations. *Economic and Political Weekly* (Vol. 18, Nos. 42 & 43).

Sharma, K. L. (1998). *Caste, feudalism and peasantry: The social formation of Shekhawati* (pp. 26–43). Manohar: Delhi.

Sharma, K. L. (2014). Introduction: Reconceptualising the Indian village. In K. L. Sharma (Ed.), *Sociological probings in rural society* (Vol. 2, pp. XXI–XL). Sage: New Delhi.

Sharma, K. L. (2019). *Caste, social inequality and mobility in Rural India* (Chapter 15, pp. 302–332). Sage: New Delhi.

Singh, D. (1964). *Land reforms in Rajasthan (a study of evasion, implementation and socio-economic effects of land reforms).* New Delhi: Planning Commission of India.

Singh, Y. (2009). Social Praxis, conceptual categories and social change: Observations from a Village study. *Sociological Bulletin, 58*(2), 178–195.

Thorat, S., & Umakant. (2004). Introduction. In S. Thorat & Umakant (Eds.), *Caste, race and discrimination, discourses in international context* (pp. XIII–XXXV). Rawat Publications: Jaipur and New Delhi.

Tumin, M. M. (1952). *Caste in a peasant society.* Princeton: Princeton University Press.

World Bank (2004) Cited from Anand Kumar, 2014, "People, power and paradigm shift". In S. Yogendra (Ed.), *Indian sociology* (Vol. I, pp. 377–380). New Delhi: Oxford University Press.

Chapter 6
Man-Forest Interaction in a Metropolis: Perspectives from Hermeneutics

Tapan R. Mohanty

I love the forest. It is bad to live in cities: there are too many of the lustful.
—Nietzsche (1940)

Abstract The looming threat of climate change, terrorism and rising trend of conspicuous consumption have necessitated a process of global change that has not only brought death, destruction and damage but has also questioned the very ethics of civilization and human conduct. Our greed has long overridden our needs and hurtled us towards an economy and lifestyle that has little concern for ecology, environment or ethics. Indeed, the need to underline a social policy that would instil environmental consciousness would automatically make it imperative to take account of social diversities and human angularities. Hence, the need is to go beyond the instrumentalities of reason and competence of empiricism for that they do not enable us in depicting and comprehending the reality besides failing to take into account the variability of context. The WTO regime has seen plenty of conflict among the developed and developing nations over the issue of standardization of global environmental norms, for many developing nations it is not merely a choice between ethics or economics but the very survival in a competitive market, needless to add that it has also its corollary in local context as well. The dominance of developed states and developed communities in exploiting natural resources is a case in point. In this background, the paper is an attempt to use the hermeneutic framework to interpret man-forest interaction in an unlikely context, i.e. metropolis and develop a framework for understanding current environmental concern.

Keywords Environmental consciousness · Sustainable consumption · Hermeneutics · Tribal communities · Sustainable development

T. R. Mohanty (✉)
Centre for Socio-Legal Studies, National Law Institute University, Bhopal, India
e-mail: tapanmohanty@gmail.com

Introduction

The looming threat of global terrorism and rising trend of conspicuous consumption have necessitated a process of global change not only in terms of death, destruction and damage that usually accosted wars but more than that it has questioned the very ethics of civilization and human conduct. The escalation of war and its cascading cost will not only affect the warmongers from both side but also all of us directly or indirectly. We will become victims of war though most of us have little to do with it precisely because in times of bridged borders and interconnected identities it has been very much impossible to remain either isolated or insular. The cost of war will be dearer to every citizen of the globe literally as well as figuratively, at least its environmental cost will be huge, may be irreversible. War necessarily a product of human greed has always spelled doom for the nature and we must try our level best to protect the fragile planet from the mindless activities of *homo sapiens*. But then human agencies have their own limitations and one can only do his best for himself and the society and it is in this context the present paper must be understood.

Global environmental concern may be described a 'postmodern condition' to borrow a conceptual cavalry from Loyotard's lexicon. Though one can trace the efflorescence presence of 'nature' in all human activities that includes both physical and psychic but a strong environmental concern can be envisaged in the 'urbanism' of Chicago School in early twentieth century in the context of city from the perspective of sociology. It is interesting to note that similar sentiments were echoed in India if one cares to glance through Ramkrishna Mukherjee's writing at the same period. But serious environmental research and studies began well after 1972, i.e. in the year of 'International Conference on Human and Environment' at Stockholm and took deep strides after the publication of 'Our Common Future'. However, the study of environment for long has been viewed both as an elitist and eccentric project and only of late filtered down to the population living in the margin. In fact, the whole debate-encapsulating environment versus development is the product of distrust and myopia arising out of people living into vertically splinted zones. The hallowed debate of modern times highlighting the discrepancy between conservation and consumerism sounds hollow unless we take the cultural dimension involved in it rather seriously. Similarly, the economics versus environment diatribe too needs a systematic analysis as all these issues are interrelated, interdependent and interpenetrated, to speak in the language of functionalist paradigm. However, the context of the present paper is syntactically different but semantically rests in similar plane. In this paper, an attempt is made to explain and explicate the convergence and divergence of human action, its form and content to underscore the interrelationship between man and nature in an unlikely yet important setting. The paper is divided into three parts. Part one is the introduction where an effort is directed to familiarize the reader with the subject. Second section deals with the theoretical issues at hand and in the final one I have tried to put some empirical observations with an intention to justify my theoretical standpoint.

Theoretical Issues

It is believed that global ecological crisis is a product of unbridled consumerism immanent in capitalist economy. The forces of consumerism bolstered by globalization and liberalization have commenced a witch-hunt to plunder natural resources. Ruthless exploitation of nature to thrive economic imperialism probably has constraint nature to retaliate. Hence, the crisis, it is in fact, is the vengeance of nature against the nefarious nexus of capitalism and consumerism, and their nasty design to exploit and exterminate.

Continuous assault of man over nature has resulted in serious difficulty and damaged irreversibly the natural eco-system of many parts of the globe. It has denuded Himalayas, depleted ozone layers, damaged millions of hectares of forest cover, deserted thousands of acres of land and has exterminated many a species from the earth. The serenity of Siberia, the scenic beauty of Savanna and the sublimity of Himalayas are replaced by smokes emanating from the chimneys of Europe, dungeons of slums from Africa and burgeoning population of Asia.

At this critical juncture of human history, the intensity and enormity of environmental degradation ring the alarm bell for the civilization. Negligence of this may well end in writing the obituary of the fragile planet. Rio conference and subsequent concern at the grass-roots level have somehow been able to foster the belief of a rising consciousness among the governments as well as laymen regarding environmental degradation. Simultaneously intellectual exercise continues to explore the subtleties of relationship that exists between man and nature. Emphasis is being made at all level to promote the spirit of eco-friendliness and protect the nature from further degeneration. Various public and private organizations are also in the fray to spread the message of eco-friendliness across the limits of caste, class, nations and nationalities.

The topic concerning the relationship between man and nature is a subject matter of many a disciplines notable among them being philosophy, sociology, anthropology, ecology, forestry, etc. But these disciplines view this aspect rather differently and with a particular perspective in mind and that have created a problem of being extremely specific and narrow in scope. The onset of interdisciplinary approach has brought new dimensions and directions to this highly sensitive and ever-expanding area.

One has to take into cognizance of the topic, which delves into the dynamics of relationship between forest and metropolis, and here it would be pertinent to define these concepts. Forest, as is often understood does not necessarily means a large canopy of trees and definitely not always refers to dense, deep and dark habitat of wild, dangerous animals. Similarly metropolis are not antonyms of forest as it often referred to, rather it needs to be emphasized that people and park are often found to be complementary. The exigencies of ethereality and instrumentality of civilization presuppose an interrelationship that needs to be explored. But the quintessential question is how to delve and design such an expedition. Whether global environment concern is restricted to mainliners or marginals? Do we have a role to play? Does one stick to understanding of environment in the context of deep forests and wildlife or one has to go beyond that? Further how important it is to assess the need of the people and understand the nuances of human actors and agency while delineating an issue that affects everybody's life? These are some questions that a present paper intends to address and which cannot be possible without an appraisal of methods.

Methods have a very special significance in social sciences generally and especially in sociology. In fact, sociology's assertion to the hall of fame in the cerebral complex of social sciences lies solely on its methodological sophistication and analytical superiority. From Durkheim to Derrida, the history of sociology is replete with methodological nuances and perspicacity of explanation. Sociologists have essentially tried to understand and explain human behaviour, their cogency and competency in various social setups not only in terms of the purity of social structure but also from a variety of angles. As human agencies are autonomous and guided more often by centrality of conditions and morbidity of mores, it is difficult to enmesh their entirety into stereotyped structures of pigeonholes. Hence, sociology is the storehouse of methodology of social sciences. Though I have focused on hermeneutics but one need to go by the textual meaning of hermeneutics and stop only at interpretive method, and in fact, the method is open ended and I have tried to go beyond hermeneutics and absorbed the essence of meaning inherent in most human actions.

Hermeneutics is a rather recent development in the history of methods in social sciences and is increasingly related to the methodological individualism propagated by Weber and the tradition of phenomenology. Translated, hermeneutics refers to the science of interpretation and depends on the cognitive and intellectual capacities of the concerned individual. The shift from structure to meaning in anthropological discourse has poignantly captured the historiography of hermeneutics best exemplified in the writings of E. E. Evans-Pritchard and Claude Levi-Strauss. Evans-Pritchard's monumental study on Nuers and his interpretation of oracle, magic and witchcraft among Azande are credited with breaking the barrier of outsider-insider dichotomy. Later Levi-Strauss's treatment of myths and moieties as well as his exploration of meanings in seemingly simple things through semiotics opened new vistas in the study of human interrelationships, their conditions and contexts. But hermeneutics is not mere interpretation, neither it aims at discovering implicit meanings as has been the hallmark of anthropological research spearheaded by Mary Douglas and Max Gluckman. The history of hermeneutics often traces its beginning in the writings of Husserl, Heidegger and Gadamer. But it has blossomed in the intellectual pursuits of Ricoeur and Habermas. Today it boasts not only of a strong historical mooring but also of having a phenomenal following. Confined to the library of linguistics and semantics of semiotics, hermeneutics has the propensity to develop methodological snags but unleashed into the vast repertoire of ethnography it blooms.

In my earlier descriptions of hermeneutics, I had drawn attention to the issues of anthropology and anthropological research rather deliberately. The reason is the immense relevance of anthropological research for an understanding of social realities, their conditions and consequences. The theoretical underpinning and practical relevance of anthropological research at least in this context is of immediate help to us as it is needless to add that anthropologists have usually concerned themselves with the primitive societies and its members usually are located in dense forests, mountainous terrains and deserts, in conditions otherwise difficult to survive.

As and when we talk about social policy and instilling environmental consciousness in all and sundry it is imperative that we should consider social diversities and angularities of human conduct. Hence, the need to go beyond the instrumentalities of reason and competence of empiricism for that they do not enable us in depicting and comprehending the reality besides failing to examine the variability of context. According to Weber, judgements of action need to be derived from an assessment of the internal logic of a situation. The distinctive characteristic of a problem in social policy is that it can only be approached and solved through normative standards of value because the problem lies in the domain of general cultural values (Weber 1949: 75).

Similarly, when we talk of global conventions and international treaties the issues of inequities among nations and inherent power relationship often come into picture. The north–south divide over environmental issues is a clear testimony to such fracas. The WTO regime has seen plenty of conflict among the developed and developing nations over the issue of standardization of global environmental norms, for many developing nations it is not merely a choice between ethics or economics but the very survival in a competitive market needless to add that it has also its corollary in local context as well. The dominance of developed states and developed communities in exploiting natural resources is a case in point. Reacting to this situation one can say that it amounts to third word exploitation of a fourth world colony. I am sure tribal community and marginal communities across the world feel the same way. It has led to such consternation that people often question the very ethics of environment and feel threatened by the environmental activities and activists. The ideological state apparatus threatens them into submission and throttled their access to land, water, fuel and fodder creating serious sociological problems. Looking at the picture it would not be wrong to comment that what Marx had visualized in a different context has a lot of contemporary relevance. The ideas of ruling class in every era of human existence are ruling ideas, i.e. the class which controls the instrument and forces of production too controls its mind and is indeed, its ruling *intellectual force* (1970: 64, original italics).

There is a theoretical and epistemological dimension that forms the background through which we understand our actions, their underlying assumptions and consequences. The task of delineating and discovering social realities are more often than not based on moral and existential considerations about the social world. Consequently, the methods chosen will depend administration of necessary tools and techniques of research competent enough to produce desirable result. Therefore, there is both epistemological and ontological aspects inherent in the process of social enquiry and production of knowledge. Needless to add the process of carrying out the research and completing has both its origin and growth in the context of society where both interacts with each other, their context and content.

This brand of research requires critical reasoning and analytical framework. Critical reasoning is not merely a negative connotation of rejecting the existing social order, its structure and dynamics, its orientation and operation for instance, caste or a particular ideology. Rather a critical perspective implies a mode of engagement with reality based on moral and political considerations, their logic and strands. As far as critical theory is concerned it leads to an understanding of the relationship between knowledge and human interests (Habermas 1972).

More than any other thing it treats human beings as subjects of research and focuses on their moods, morals and margins of thought and action. But this is not all about individual experiences but also the effort towards finding clues of commonalities in varied and divergent situations and segments of population. The ability to acknowledge, analyze and absorb the degree and kind of human experience emanates from a search of commonality, from shared experiences that forms the core of empathy and underlines the ability of the researcher. This is in a sense can be described as 'objective realization' of subjective phenomena. The approach to an objective social reality passes through the ability to comprehend the subjective dimensions of human interaction, the shared meaning of their actions, their intention and purpose viz. values, myths, symbols, etc. This understanding surmises the broader commonalities and similarities of shared human experience and the basic requirement of an investigator to begin his journey of social inquiry from the standpoint universal common cerebral Rubicon.

Martin Heidegger (1889–1976) restated the very foundation of phenomenological inquiries by shifting the focus from epistemology to ontology and thereby limiting the ideas of Kant. For Heidegger, human beings are not merely passive observers of an external world that is mediated and sorted by our consciousness (Husserl), rather they are active participants with an independent existence who are members of that simply because they exist as 'being-in-time'. This new approach with its focus on 'being and time' orchestrated as 'Dasein' meaning 'Pre-understanding', a reiteration of a priori consciousness challenges the basic tenets Kant and Dilthey takes social research to the realm of acute subjective interpretation. In the words of Ricour it depicts, 'the place where the question of being arises, the place of manifestation-the centrality of Dasein is simply that of a being which understands being' (Ricoeur 1982: 54).

Another exponent of subjective interpretation of objective world in the intellectual tradition of hermeneutics is German philosopher Hans Georg Gadamer who believed understanding bridges the gap between history and future possibilities which finds a distinct resonance in the C. W. Mill's concept 'Sociological Imagination'. Gadamer (1975) influenced by the ideas of Heidegger used the hermeneutic language of 'text' for this purpose, arguing that understanding is not mere reading and grasping of text but contextualizing it its cultural cauldron. The text is simply expression of a historical reality and beacons an interpretation from the investigator to give it's a meaning and context, locating it in the cultural milieu.

What the author has tried here is to use the hermeneutic framework to interpret man-forest interaction in an unlikely context, i.e. metropolis. Even from the perspective of urban sociology the theme seems interesting. The first thing is to question the assumption and attitude towards exotic locale to find an expression to refer to man-nature relationship. This is an arena which has been perennially neglected both by the natural and social scientists.

In this project emphasis is given to understand local communities and interpret their action. It is essential to know that why do they act in a particular way and can we impose out life-worlds and worldviews on them? And if so, then can we save us from the deleterious effects of cultural ethnocentrism, of homogenization and hegemonization? Conceding that all things go well then can really sustain the agenda even if we for a moment believe that we have implemented it? Law acts its best only when people recognize its significance and also when it has emerged from below rather imposed from above (Weber), so is about ideology, belief system and non-native mechanisms of human interaction. Hence, the need for interpreting the action and its inherent meaning in the insiders' *lebenswelt*, their travail and trauma in a larger theoretical and methodological niche.

However, extraordinary it may sound I must insist that the universe of my study is Delhi—the much maligned and misconstrued metropolis of modern India. Despite its dirt, disease and nefarious disturbances the city has a special attachment for me as it reminds me of my intellectual development, its diversity and vivaciousness has taught me a lot about human nature and its manifestations. Keeping in mind these variations and variability I have tried to outline this paper. It must be noted that Delhi has a strong history and cultural landscape apart from development of amenities explained in terms of narratives in modernization, development and infrastructure in comparison with any other city in India. Delhi has moved well beyond the sleeping urban village to become a modern bustling metropolis.

Gadamer accepted that the meaning of text is not dependent upon their original or intended sense, but on such factors as the language, norms and traditions in and through which subjectivity finds itself constituted. Interpretation according to Gadamer is an engagement which takes the form of a reciprocal relationship between reader and text. Interpretation is unlimited, open-ended process.

Discussion and Analysis

The throes of theory: Some empirical observations about Asola-Bhatti Wildlife Sanctuary situated in the southernmost part of Delhi touching Haryana border near Suraj Kund. It has an area of 6873 acres of arid and semi arid land. The border of the sanctuary forms contiguity with villages like Fatehpur and Chandan Hula in east, Anangpur in west, Deoli, Sangam Vihar and Tughlakabad on north, and its southern part stretches to Gurgaon-Faridabad Highway. Geographically speaking, it falls on the Aravali Mountain range, the most ancient mountain range of India. Asola and Bhatti constitute the major part of Delhi ridge, which is an out-stretched terrain of Aravali range. From bed of river Yamuna at Shahdara to the deserts of Rajasthan via the state of Haryana the ridge has a vast but sporadic cover. However, due to various ecological as well as socio-economic and cultural factors the ridge has never remained at its natural state. As far as the Delhi ridge is concerned, we can divide it into four parts, the northern ridge which is represented by the green patches of Kamala Nehru ridge, near Delhi University and Bara Hindu Rao hospital. Central ridge constitutes a canopy of trees in around Rajendra Nagar and Buddha jayanti Park. The Green belt we see at Dhaulakuan is referred as south-central ridge while the Asola-Bhatti area is known as southern ridge. Asola wildlife sanctuary assumes greater significance in any reference to the greenery of Delhi because of certain factors. First area wise it is in advantageous position, as it is a total area of 28.71 km². The area is vast one in relation to other parts of the ridge. Secondly, unlike other parts of the ridge this area falls under the supervisory authority of Delhi government's department of Forest, Environment and Wildlife and this status gives the area a definite edge over others. The third most important aspect of this area is its contiguity where rest parts of the ridge are fragmented and uneven.

As Delhi's pollution level is growing alarmingly and gradually turning it into a virtual gas chamber, the protection of ridge becomes imperative. Not only ridge will act as a natural 'carbon sink' zone but will also protect the metropolis from wanton winter and scorching summer like a natural air conditioner. The presence of a unique natural eco-system in the periphery of a metropolis definitely has larger significance and benefits than we presume at its sight. Ecologically speaking it may not have a subsistence value but it certainly has got sustenance value.

However, in an attempt to understand the socio-economic profile of the area I would like to highlight the life style and life chances of the population living near and depending on the sanctuary. The first thing that strengthens the idea to have a socio-economic survey of the area is the breaches in wall, which amounts to the number of 126, a substantial number considering the length of the wall which is only 4.6 km as the proposed two more kilometres are yet to be built due to legal hassle. Natural decaying of the wall is not withstanding as the sheer number of the breaches confirms beyond doubt that there has been illegal human interference with the wall resulting in these fissures. This also certifies that there have been certain administrative loopholes in appraising the situation. It may be that lack of alternative with the people constraints them to use the limited resources of the area or it is the sprit of augmentation of profit that plays the trick. This becomes the social scientists' arena to unravel the reality. Besides the villages of Fatehpur, Chandan Hula, Satbari, Maidangarhi, Tughlakabad, Bhatti Kala and Anangpur, there are four unauthorized colonies which exists right inside the heart of the sanctuary that threatens the very existence of sanctuary itself. Sanjay Nagar, Balbir Nagar and Indira Nagar are the three colonies established as early as 1975 in the Bhatti area. Another unauthorized colony known as J. J. Camp near Sangam Vihar is also made its way into the vicinity of the sanctuary. In fact, the dependence of these people over sanctuary for various activities siege the possibility of any improvement. Not only these three settlements but the villagers of Chandan Hula, Fatehpur and Anangpur also exert tremendous pressure on this extremely fragile eco-system. Mainly for the purposes of collecting fuel, fodder and wildlife, people use the sanctuary besides coming for defecation, grazing, hunting. They also use the sanctuary as a through fare for different connecting places.

Our survey of the region reveals that 5099 families reside in the three colonies with a population of 14,728 peoples. In the fourth unauthorized colony, there are 246 families with a population of 2386. As the data indicate, taken from a sample of

300 families, cent percent of these four settlements use the ridge as a public lavatory. Besides 73% of the population of colonies of Sanjay Nagar, Balbir Nagar, Indira Nagar and of Chandan Hula collect fuel wood and cow dung from the sanctuary. The cause behind the settlement of these colonies goes to the time of active mining and quarrying in this area. In fact, when mining operation was on full swing people far and near places gathered at the site in order to get employment and with a ready employment in their hand, they gradually set up their hutment. Cows and other cattle put enormous pressure over the grazing land of the sanctuary. The villagers of Anangpur with their camels, cows and buffaloes and villagers of Fatehpur with their sizable number of cattle are almost threatening the rest of the sanctuary with extinction. A simple glance at the sanctuary reflects the true picture that the decline of the sanctuary is directly linked to the phenomena of over grazing and trampling.

Defecation is one of the major reasons which forces people to use the sanctuary. Lack of proper sanitation facility and financial exigency had left the option of using this vast tract as common lavatory. The study also explains that more than 99.33% of the total population are aware of the fact that it is declared as a wildlife sanctuary but they continue to use it as they do not have an alternative.

In rainy season, people tend to use it more as everybody comes to collect fodder and grass for their livestock. Surprisingly, many people have admitted during research that although they can afford to provide food and fodder to their animal without depending over the sanctuary, but they use it as they have nothing to lose rather more to gain as they have almost free access to the area. During one of the visits to the site we find 24 women with head logs of wood with a mean weight of 32 kg on a Sunday. The inquiry revealed that due to the sparse presence of forest staffs in working days they have changed their timing. The forest staffs also admitted the illegal encroachment of land, stealing of wood and occasional poaching. They also mentioned that on many occasions few villagers even threaten them with dire consequences if they take any action against them, they failed to get the help of the police as they do not have any formal power vested with them. In one of the field observation, the researcher himself caught three persons red handed while hunting a hare and in other instance discovered the remnants of a peafowl, and the circumstantial evidences clearly indicated human involvement in the gory incident.

Going back to the problems of grazing it can be said that the Gujjars of Anangpur and Fatehpur are really throttling the sanctuary with over grazing. While many of the residents of Fatehpur do it for sake of profit and boosting their income level than any other genuine cause, the villagers of Anangpur do it for the fulfilment of partial necessity as far as rainy season is concerned. The topographic condition of the village catch them stranded in rain and due to that they leave their cattle in the sanctuary to feed as well rest. Now even the villagers of Bhatti are using some of the pits as their common dustbin. When this natural yet extremely fragile eco-system is under severe threat, necessary precautions must be taken to preserve this unique land with natural vegetation for the best of the mankind.

Looking at the socio-economic status of human settlements in and around the wildlife sanctuary and the prevailing condition of the forest, it becomes prime concern to search for an alternative. If the socio-economic standard of the people of the area is appalling then the status of the forest is nothing but precarious, therefore, the plight of the people and the predicament of the forest are interlinked to each other. As far as our understanding of the place and its surroundings is concerned, I believe that the improvement of the forest and the population can be simultaneously achieved through an integrated forest management approach.

When the concept of forest management and integrative approach comes into discussion, it automatically delineates with peoples' participation and involvement in conservation and sharing of responsibility and privileges. However, much depends on the execution of the planning and whether it is properly defined. If we highlight on what we exactly desire would be essentially a conservationist approach to deal with the contradiction between ecology and economy. This is a classic case where the poverty of the people and their economic necessity is at conflict with ecological requirements. Definitely we are not intending for a poverty alleviation programme but are occupied with a project that ensures the life support system. Keeping that in perspective, logical steps needs to be taken to attain popular participation and support over which the success of the planning heavily depends. We firmly believe that in the paradigm of food-water-fuel crisis, human beings are the fundamental link and their involvement can not be over looked. Management plan would not only deal with the unidirectional approach of the protection of endangered species from extinction but also protecting the economic benefits of the traditional users of the natural resources, if not quite the same way but through providing alternatives. In fact, the evolution of forest management in the history of conservation is regarded as watershed as it restores the ecological balance of the area without jeopardizing the economic benefits of the people. In all, management of forest is a balancing chord that perfectly harmonizes the correlation between man and forest. In this context, we suggest to provide fodder and grass to the local people during winter besides solving the acute water crisis of the area at the earliest either with the involvement of governmental and non-governmental agencies or through forest department initiatives. Subsequently, in the long run the presence of dense forest will ensure the preservation of rainwater and an end to the problem. Similarly, education and awareness related to forest conservation should be imparted to the local people in order to generate environmental consciousness among them. To solve the employment problems of the locals, some concrete steps should be taken. As the majority of the population is illiterate and semi literate and marginal labourers, we propose for the setting up of an eco-friendly labour intensive industry. The research also proposes to bring an interpretation centre in the sanctuary to provide vital information regarding the sanctuary and its ambience. The interpretation centre, in course of time, may also open a refreshment camp, which will require local help and thereby generate employment. Of course, the need of the local people should be kept at the core while planning any development and beautification of the area in future.

So, with the efforts of an all-round and sustainable development by encapsulating the needs of both man and environment, holistic change would take shape in the area. Therefore, a solid eco-development package for the people and a stringent action plan for the safeguard of the environment will resolve the perpetual conflict between the forces of exploitation and the voices of survival. Hence, what I intend to present as the main conclusion of the paper is that people must be kept the cornerstone of any policy if we are serious about its success. No policy can work unless and until it is people centric. This is in fact, the idea that is fast emerging from principles and practices of good governance, decentralization and democratization. There is need to translate the logic and language of development to the people for whom it is intended. It requires both commitment and accountability from the policy executing agencies to comprehend the human dimension of sustainable development. If we are keen to protect the bio-diversity and our natural resources, then we ought to develop a strategy for that, a strategy that keeps people in mind at the micro-, meso- and macro-level. And if this is going to be the gospel of globalization centric model of development, then hermeneutics is the only methodological tool that is available at our disposal. We can ignore only at our peril.

Policy Implications

The relationship between theory and praxis finds its true reflection in policy implications of studies. This gives guidance to government action and administrative direction for effective implementation of an actionable idea. In this context, the current article has the potential to distinctively alter the way policymakers and lawmakers perceive the relationship between law, environmental governance and urban planning. It envisages that forests are not significant not only for ecological reasons or for food, fodder and fuel consumption requirement of immediate inhabitants but also acts as a 'carbon sink zone' as emphasized in Kyoto Protocol and have tremendous ecotourism potential especially in metropolis. The advantage of some metropolis and urban centres of having forest covers within or beyond city limits that if effectively managed they will reduce their pollution levels, water woos and act as an eco-interpretation centre as well as natural parks. In the extremely commercial, industrial and corporatized world the interaction between man and forest in metropolitan cities may sound odd but it is precisely the solution to critical problems of climate change and environmental hazards. As weekend zones of entertainment, oxygen inhaling centre and zones of detoxification forests have a solution to many of the problems of cities and their populations. Centring the experience in cities like Delhi and Mumbai, the author has drawn timely attention to the possibilities of a new urban planning and renewal method for better and healthier living. A warm up call to our policymakers.

References

Gadamer, H. G. (1975). *Truth and method*. Originally published in 1960. London: Sheed and Ward.

Habermas, J. (1972). *Knowledge and human interests*. London: Heinemann.

Marx, K., & Engels, F. (1970). *The German ideology*. Lawrence and Wishart.

Nietzsche, F. (1940). *Thus Spake Zarathustra tr. Thomas Common*. New York: Carlton House.

Ricoeur, P. (1982). *Hermeneutics and the human sciences* (J. B. Thompson, (Ed.) and Trans.). Cambridge: Cambridge University Press.

Weber, M. (1949). *Methodology of social sciences* (E. A. Shils & H. A. Finch, (Ed.) and Trans.). Glencoe, Ill.: Free Press.

Part II
Issues of Environment

Chapter 7
Environmental Consequences of Dams: A Study of Select Hydroelectric Projects in India

Namita Gupta

Abstract The fundamental assumption behind the emergence of development studies after the Second World War was overall well-being for the mass of people. Pushed by this development strategy, the construction of large dams was justified on economic, social and political grounds. Few studies have been carried out to determine the ecological, economic and cultural importance of rivers before they are dammed. Similar has been the case of long-term environmental and social consequences of areas after the construction of a dam. However, the ongoing struggle against the large dams across the globe challenges this dominant model of development that holds out the promise of material wealth through modernization but perpetuates an unequal distribution of resources and wrecks social and environmental havoc. Once classified by Jawaharlal Nehru, the first Prime Minister of independent India, as the 'temples of modern India', the large multi-purpose dams and river valley projects have today become the focus of widespread agitation in India. The Tehri and Pong dams in the north; the Kosi, Gandhak, Bodhghat projects in east; the Narmada valley project in central India; Bedthi, Bhopalpatnam and Ichampatti in the west; the Tungabhadra, Malaprabha and Ghatprabha projects in south, all have faced resistance from the local community. The issues raised are location and design of the project as well as the social and ecological consequences of these projects on local community. The present paper, in this background, discusses the long-term environmental consequences of some select hydroelectric projects in India. It also touches upon the issue of inadequate compensation and improper rehabilitation of those displaced, as it is another core issue of discontent against these major projects.

Keywords Dams · Development · Displacement · Environmental consequences · Hydroelectric projects

N. Gupta (✉)
Centre for Human Rights and Duties, Panjab University, Chandigarh, India
e-mail: namita4rights@gmail.com

Dam, Development and Environment

The academic discourse on 'Development' began in the latter half of the twentieth century. The most basic assumption underlying the study of development in the south was that the term focused narrowly on bringing down the absolute poverty and consequently led to general well-being in the society. These fundamental assumptions led to the emergence of development as a subject area as well as an issue worth international deliberations. The term development became more complex with large numbers of newly independent countries came into being in 1950s (Haynes 2008: 186). By the beginning of the third millennium, there was an emergence of two polarized expositions on how to improve the development levels of post-colonial less developed countries, i.e. radicals and reformists (Rai 2005). In most debates, however, the question of how to accomplish this goal remained unsettled leading to often implicit tension between these two viewpoints.

The processes of economic globalization made easy availability of cost-effective larger options of goods and services to people across the world, irrespective of the season (Stiglitz 2002). However, during this process of development, the environmental costs were not only overlooked; rather, they were 'externalized' and not included while doing economic calculation. These costs nevertheless accrued and emerged as environmental hazards such as climate change, biodiversity loss and toxic pollution, with major impacts on human health and society. This economically driven displacement of environmental costs principally affected the human rights of two constituencies: today's poor and future generations (Barry and Woods 2013: 390).

The construction of large dams has been justified on economic, social and political grounds as a part of development strategy. The benefits highlighted by the government agencies were major and could not be overlooked by the general public, especially in developing and less developed countries. These included hydropower generation, irrigation, water supply, flood management, navigation and recreation to name a few.

Given such a large benefit, it was very difficult to justify the stand against construction of large dams. However, it was since 1960s, practitioners already had started questioning, in fact, criticizing the undesirable and avoidable environmental and social impacts of dams. The large dams were declared as part of a 'flawed paradigm that caused an increasing disconnection between the necessary environmental health of river basins and the current needs of people and governments for the provision of water, energy and food' (Scudder 2005: 16). It further led to inequitable development and environmental degradation.

The dilemma for planners over the short term has been that dams may be the most desirable option to fulfil the requirements of rising global population, especially in the case of crisis or near-crisis situations. In such circumstances, the construction of large dam was seen as the only way to deal with current economic needs ignoring the irreversible longer-term social or ecological consequences of such projects. Large dams seemed to be the only solution available to deal with the rising demands of the expanding human population, which already has surpassed the carrying capacity of earth's ecosystem.

Few studies have been conducted to determine the ecological, economic and cultural importance of rivers before they were dammed. Similar has been the case of long-term environmental and social consequences of areas after the construction of a dam. The ongoing struggle against the large dams across the globe has challenged this dominant model of development. Though the current development model assures material well-being, it perpetuated inequality by widening the gap between the rich and the poor and leading towards social and ecological catastrophe (Scudder 2005). Many large hydroelectricity projects in India have faced lots of resistance from displaced people as well as from the civil society organizations. The Tehri, Pong, Kosi, Gandhak, Bodhghat, Narmada, Bedthi, Bhopalpatnam and Ichampatti, Tungabhadra, Malaprabha and Ghatprabha, all these river basin projects have faced resistance (Biswal 2015: 465). The multiple issues raised were: location and design of the project, social and ecological consequences of these projects as well as the issue of inadequate compensation and improper rehabilitation of those displaced. The environmental consequences of two such projects have been studied in this paper. The first one is Bhakra Dam Project as it was one of the largest investment schemes in newly independent India and there was least resistance from displaced people. The second one is Sardar Sarovar Project, a project which generated lots of controversies both at the national and at the global level.

Bhakra Dam Project: An Untold Story

Bhakra Nangal project is something tremendous, something stupendous, something which shakes you up when you see it. Bhakra, the new temple of resurgent India, is the symbol of India's progress

These statements were made by India's first Prime Minister Pandit Jawaharlal Nehru while inaugurating the Bhakra Dam in 1963. Built on the River Sutlej in Himachal Pradesh, just before it enters into Punjab, the project was considered the largest single investment scheme in the newly independent India. The dam has a height of 226 m. A reservoir named Gobind Sagar is spread over an area of 168.35 km^2. The dam has two powerhouses with an installed capacity of 1325 MW. The earliest reference to the Bhakra project can be traced back in 1908 when Sir Louis Dane, the then Lieutenant Governor of Punjab, made a reference regarding the same in a note. The proposal was revived in the year 1915, and a comprehensive project report was prepared in 1919. The initial literature mentioned it purely as an irrigation project to overcome the devastating effects of famine in that area. Due to the conflict between the princely states and later due to partition of India in 1947, it took many decades before the project could see the light of the day. After the partition of Punjab, almost 80% of the irrigation system went to Pakistan along with a large portion of fertile agricultural land. This provided another justifiable reason to go ahead with the project. The work of the dam was completed in 1963.

The evaluation of environmental impact of Bhakra Dam has been a difficult task as long time has passed since the project came into being and it was difficult to retrieve baseline data to compare the ecological characteristics of the area before and after the construction of the dam. This is one of the reasons why there has been scarcity of research on the environmental as well as socio-economic impacts of the dam. One literature available is the report released by Bhakra Beas Management Board (BBMB) titled *'Socio-Economic and Environmental Impacts of Bhakra Beas Project—An Assessment'* in 2001 (Duggal et al. 2001). The report listed many benefits of the project; however, it is interesting to note that the report did not mention even a single negative impact of the dam. In a populist government structure, this seems to be the natural and expected outcome of any report released by a government agency. There is another interesting work which analysed the socio-economic and environmental consequences of the dam. This is Shripad Dharamadhikary's *'Unravelling Bhakra: Assessing the Temple of Resurgent India'* in 2001. The study contradicted the report of BBMB and enumerated multiple devastating impacts the project had on local community as well as on ecological system of that area. Though the projected outcomes of any hydroelectric projects such as irrigation, flood control, employment generation, land reclamation, fishing and tourism could not be ignored, it would be important to note some of the major negative environmental impacts of the project.

As true with other hydroelectric projects across the globe, the biggest environmental impact of Bhakra Project has been the submergence of around 168.35 km^2 of land area to create a reservoir 'Gobind Sagar'. Furthermore, 1000 acres were acquired to establish a township in Nangal. The area had been very fertile and was well irrigated by River Sutlej as well as due to the flow of many natural streams flowing from the mountains in that area. Wheat, corn and cotton were cultivated, and local community had big orchards and cattle. Out of the total submerged area, 5750 ha consisted of forest land which was rich in biological diversity (Singh et al. 2000). Another consequence of the construction of dam has been the loss of native fishes. Various studies substantiated the fact that '32 species of fish have almost threatened, 20 species became vulnerable, and 12 and 2 species have been put into the category of endangered and critically endangered respectively due to the project' (NBSAP in Dharamadhikary 2001: 194). One of the most famous native species of fish *'Mahseer'* came into the threatened list. It has to be further noted that commercialization and licensing of fishing rights have taken away the livelihood of local community who had been carrying out fishing in that area in an unorganized and informal manner since long.

Furthermore, the reservoir has led to the significant changes in the micro-climatic conditions in that area which has adversely affected the health and well-being of the local community. The dependency of people on lake water and public water supply system, instead of freshwater earlier, has increased the intensity of many diseases such as gastroenteritis, enteric fever, viral hepatitis and malaria (due to mosquito breeding). The area covered by the reservoir as well as nearby areas has thick fog almost throughout the day in winter season which led to respiratory diseases (Dharamadhikary 2001).

When the studies of ecological impacts of large dams started getting carried out, the initial emphasis was on immediate impact on the ecology of the local area. However, the recent studies on dams have studied thoroughly the consequences of dam construction both upstream and downstream. During the construction of Bhakra Dam, the diversion of river water was started in mid-1880s with opening of Sirhind Canal out of River Sutlej. The impact of dam on downstream areas was much obvious and could not have been ignored by the policymakers at the time of the inception of the project as well. This could be substantiated from the fact that the First Five Year Plan mentioned the need to protect down streams in its chapter 26 on 'Irrigation and Power' as early as in 1950. The same, though, has not been put in practice while conceptualizing the hydroelectric projects in India till date. The most visible downstream impact of the Bhakra Dam has been the disappearance of grey canals. Out of the total 17 million acre feet (MAF) flowing into Gobind Sagar, only 1.4 MAF remains in the river to flow down to Harike, which is just 8% of the total inflow into the reservoir (Anand 1956). This led to the loss of natural fertility of agricultural land due to annual floods. This has further led to the increased dependence on tube well irrigation and excessive use of fertilizers to retain the productivity of the land. This along with populist government policy of free electricity and water to farmers has led to the depletion of groundwater resources as well as the high incidence of cancer and other diseases due to excessive use of chemical fertilizers and pesticides. This raises another big question mark on the current development paradigm which has long-term serious and irreversible environmental and health consequences.

The last and most significant issue regarding the large dam projects has been the sedimentation and siltation in the reservoir area. The issue is important from economic perspective as well, as it impacts the performance of the dam. The government reports substantiate the fact that 'Silt deposited in Bhakra Reservoir in live storage has been 9.70% of live storage capacity' (Duggal et al. 2002). The 10% reduction in storage capacity has serious repercussions not only on the life of the dam but also poses questions such as 'how high will the river then flow, how much will it spread, what will happen in downstream area and what will happen when the life of the dam will be over' (Dharamadhikary 2001: 200). Most of these questions not only remained unanswered but were also ignored at policymaking level. According to 1999 report from the International Water Management Institute, 'In the Bhakra Irrigation System, the practice of allocating and distributing canal water supplies…leads to the current high productivity of water. The long-term sustainability of agricultural productivity seems threatened. In some areas, saline water tables are rising, and soils are becoming sodic, while in areas that have fresh groundwater, water tables are falling. There is an urgent need for the farm, regionally, and system wide' (Sakthivadivel et al. 1999: 20–21). There has also been a dam-induced flood management problem. The sudden release of large flows from dam may have devastating impacts on downstream communities and especially on those, who, believing themselves protected by a dam, have moved onto previously flooded alluvial soils. In the case of Bhakra Project, McCully wrote 'In 1978, nearly 65000 people in Punjab were made homeless by floods exacerbated by forced discharges from Bhakra Dam... Eleven years later a similar flood occurred' (McCully 2001: 147). The year 2019 again witnessed similar flood; due to heavy rains, dam crossed its critical mark of 1680 ft. The flood gates of the dam were opened putting the life of people staying in 300 villages in the downstream areas at risk.

Besides the above-stated environmental consequences of the Bhakra Project, another untold aspect of the project, which is worth mentioning here, is the improper resettlement of those displaced. It is interesting and unique that there was negligible resistance to the project even from the oustees. The biggest reason for the same was that the dam was constructed at the time when the excitement for the newly found independence was quite high accompanied by patriotic sentiments. 'Bhakra project was implemented under unique circumstances. These were the circumstances due to which it got not just the cooperation of the oustees but also their blind faith to the extent that no other project in India got' (Dharamadhikary 2001: 224). However, whether the future generations will also be under the same rhetoric about the dam or not is a question to ponder upon.

Narmada Valley Development Project: A Vision or an Ecological Disaster

Since the last few decades, the debates on the viability of large hydroelectric projects in India have been dominated by the most controversial Sardar Sarovar Project. The Narmada is the largest west-flowing river in India traversing 1312 km. Originating from Madhya Pradesh, it flows through the states of Madhya Pradesh, Maharashtra and Gujarat. The project was first mooted in the year 1946 by the then provincial government; however, the project kept on lingering due to interstate dispute among the affected states. The Narmada Water Disputes Tribunal (NWDT) was set up in the year 1969, which in its final report, submitted in 1979, made recommendations for the rehabilitation and resettlement of oustees of the project. As per the plan, the project consisted of 30 major, 135 medium and 3000 minor dams. The project has the potential of irrigating 1.8 million hectares area, ensuring potable water to around 4 crore people and an installed capacity of 1450 MW of power (Raj 1990). One of the major dams to be constructed was Sardar Sarovar Dam in Gujarat.

The strong civil society-led movement has opposed the project to tooth and nail. These opponents have pointed out that the resistance to the construction of huge dams on the River Narmada signifies the struggle for a society based on the principle of justice and equity in India (Friends of River Narmada 2008). They have further questioned the presumptions made by the state and asserted that its planning has been unfair, and the construction has caused large-scale abuse of human rights along with serious and irreversible ecological consequences. On the other hand, there have been a large number of people, especially those staying in Gujarat, who supported the project. This support has been so strong that any view expressed against the project evoked sub-national feelings in the state. This is evident in the fact that Gujarat has made the second highest number of TADA (Terrorist and Disruptive Activities Prevention Act, 1985) arrests, more of them in order to suppress 'anti-development project' agitations (Judge 2000). The government despite the controversies around the project, since the beginning, has continued to go ahead with the project by promising cheaper electricity, irrigation and potable water for the people of desert-prone areas. The proclaimed promises of the state and counterarguments of the civil society organizations as well as pro-dam people have led to a series of judicial pronouncements on the issue of rehabilitation of the displaced as well as on the ecological consequences of the project.

The World Bank signed an agreement with the Indian Government in 1985 to partially finance the Sardar Sarovar Project (SSP). However, as the construction of dam started, the resistance against the project also flared up. The mounting criticism of the World Bank for ignoring the social and ecological impact of the dam before funding it led to the establishment of first ever Independent Review by the Bank to assess the project. The report submitted was highly critical about the project. The report condemned the gaps in the social and ecological impact analysis of the project, besides other issues. It suggested that the World Bank should withdraw itself from the project (Morse and Berger 1995). The Bank officially did so in 1993. The withdrawal of the World Bank, however, could not demotivate the government and the work on the site continued till 1995 when the Indian Supreme Court halted the work of the project till the further investigation. The construction work was restored by the order of Supreme Court in the year 1999. The subsequent judgements were like a roller coaster for the civil society activists, raising hope and despair from time to time.

The Sardar Sarovar Project was conceptualized without any comprehensive environmental impact assessment (EIA). The project was initially declined by the Government of India in 1983 due to serious gaps in the facts presented about ecological impacts of the project. It was in 1987 that the Ministry of Environment and Forests provided approval to the project with a precondition that the concerned state governments will carry out a comprehensive ecological impact assessment of the project by 1989. 'The studies were tied to the construction schedule, so the compromise became known as the *pari passu* principle: environmental impacts were to be determined and mitigation measures put in place in concert with construction' (Kapur et al. 1997: 694). However, in reality, majority of these conditions were not complied by the project authorities.

The Sardar Sarovar Dam submerged around 13,744 ha. of forest land, 11,318 ha. cultivable land and 14,072 ha. land was designated for other uses. The worst environmental impact of the project will be visible in Gujarat in the coming decades, as majority of the area to be irrigated is either moderately or severely prone to waterlogging and salinity. Hence, approximately one million hectares of fertile land would be lost. The initial plan stated that submerged forest area would be compensated by afforestation of equivalent amount of non-forest land. This does not seem to be a satisfactory solution of submergence as reforestation cannot replace a natural forest. 'There is no feasible way to completely recover the loss of these forests, or of saving much of the biological diversity that they contain. This was heightened by the fact that compensatory afforestation in the case of SSP is being done in Kutch, an ecological zone completely different from the Narmada Valley. There will therefore be an inevitable loss' (van Gelder et al. 2002: 81). It further lacks in biodiversity and does not provide forest corridors enabling wildlife to cross over. The change in existing ecosystem will lead to the extinction of rich flora and fauna of the area. The experts designated by the Narmada Control Authority to gauge the plantation stated that it would not be possible to recreate the tropical deciduous forests submerged due to Sardar Sarovar Project, in the arid district of Kutch, and any such effort will be 'mitigatory' and not 'compensatory' (Ojha 1989; Kushalapa 1992). The efforts made to ameliorate the adverse impacts have emphasized only on restoring the number of commercial species of fishes while totally overlooking the need to restore the native species of fauna and flora.

Another important issue is the annual rate of siltation per 100 km of catchment area of the dams, which is quite high. A study by the Bengaluru-based Indian Institute of Science revealed that at least 100,000 ha of irrigated land will be affected by severe waterlogging as well as salinization of the soil (Gokhale 1991). The data from several large reservoirs in India show that the actual rate of siltation is on the average 200–400% higher than the siltation rate assumed while planning the project (Irrigation Committee Report 1972; Public Accounts Committee Report 1993).

Considering the accelerated deforestation due to dam construction, a comprehensive Catchment Area Treatment has to be carried out to reduce silt load and to restore environmental equilibrium of that area. Besides this, Command Area Development must have focused on prevention of waterlogging; optimization of water utilization; and preservation of the quality of water resources. This could have been done by establishing a command area by providing adequate drainage system (Department of Environment and Forests 1986), which has not been done in the case of SSP. The impacts of the project on ecology and local economy of people staying downstream have been overlooked; environment gaps were ignored; and an adequate environmental impact analysis was found to be missing.

"If rejuvenation of rivers is to receive a central place in water resources development and the Narmada is to remain alive, these issues need to be brought into discussion and resolved as soon as possible" (Economic and Political Weekly 2014:10).

Policy Implications

The notion of due process of law is inherently interlinked with the principle of natural justice. It is based on the assumption that a state will work in accordance with the law (Udombana 2006). As far as environmental impacts of large dams are concerned, due process contemplates that due attention must be paid as far as ensuring environmental balance is concerned. WCD in its report 'Dams and Displacement' (2000) called for a reduction in negative impacts of dams by increasing the efficiency of the use of existing assets and by avoiding and minimizing ecosystem impacts. The prerequisite of an environmental impact assessment before a development project is conceptualized is based on the pretext that ecological imbalance can adversely impact the lives of the community depends upon it (Adeola and Viljoen 2018). Large-scale hydroelectric projects have faced strong resistance due to the injustice attached to these dams. The scenario can be improved with the inception of new decision-making processes and by building confidence in these processes. 'In seeking to build this confidence we do not, in many cases, begin with a clean slate, but with a difficult legacy that needs to be recognized. This legacy can only be overcome if there is a rapid investment of confidence in the legitimacy of the processes that are put in place' (World Commission on Dams 2000: 210). Hence, there is an urgent need to have serious intervention by policymakers, planners as well as administrators to understand the long-term ecological consequences of these mega projects along with exploring alternatives to such projects. We all have to acknowledge the fact that a sustainable development model must be based on the principle of equality and natural justice.

Conclusion

A point which needs to be noted down in the whole discourse of controversies, agitations and judicial intervention is that the prime attention of every agency has always been more on the issue of compensation and resettlement as well as on restoring the rights of the tribal population. Environmental criticisms have been usually not integrated into the primary resettlement reform campaigns. The environment was not a prominent issue of concern within the World Bank until early 1980s, and the office of Environmental Affairs was comparatively less effective in the Bank's power structure. The loan agreement of Sardar Sarovar Project only mentioned that there should be an environmental work plan in place by December 1985. What constituted an 'environmental work plan' was not discussed in detail. Only few statements mentioned the need to provide an adequate training to the staff, and studies on some environmental issues must be carried out (Wade 2011). It was true for the Narmada Water Dispute Tribunal as well, which did not explicitly address the environmental issues. However, in its subsequent hearings in next ten years, environmental issues along with rights of indigenous people have found a place in the procedural and organizational structures within India as well as in the World Bank (Khagram 2004).

There is no denying in the fact that large dams remain necessary development option for providing water and energy resources to the increasing population across the world. Even then, the decision to build a large dam must be based upon open, transparent and participative social and environmental impact assessments of such projects. It is also pertinent to ensure that institutional inadequacies in planning as well as in implementation must be plucked out to realize the full potential of such projects.

Many alternatives to large dams have come up across the globe in the last few decades. Various NGOs in India are making an effort to rejuvenate the centuries-old practices of water harvesting along with new technologies to utilize rainwater in an ecologically sound manner. Most of these efforts are focusing on local approaches which are environmentally sound as well as ensure fair benefit sharing. Countries like Japan have created multiple sub-surface dams. Many countries are experimenting with run-of-river hydroelectricity projects as an alternative approach to utilize water resources. Such projects may run without storing water and with small piece of land, with minimal impact on ecosystem of the area.

There is an urgent need to have serious deliberations on finding alternatives for sustainable water resource development. Such decisions cannot be taken at one level. Large dams involve powerful project authorities with generally strong support of the state. Hence, challenging the current ideology and suggesting the alternative way of development are not always welcomed. There has to be an active participation of all stakeholders such as government agencies, private sector, donors and civil society organizations. Secondly, feasibility of these projects is generally carried out with an already established 'high- modernist ideology' which asserts the supremacy of science and technology for the ordering of nature and society (Scott 1998: 4). The outcome of the same can be destructive for late-industrializing countries where this untested vision may be superimposed more easily upon a 'not so strong' civil society. Though NGOs have played by far the most important role in advocating the plights of project affected people, it is high time that they realize their potential by conducting research as well as by mobilizing local administration to plan and implement the projects in a socially and ecologically responsible manner. The non-profit organizations can also play an effective role in conducting independent long-term monitoring of the projects to better assess outcomes and lessons learnt can be applied elsewhere.

In the end, it is important to answer certain questions before going ahead with the proposal of large dams, as raised during the 2001 Symposium on the benefits and concerns of dams in Dresden (Scudder 2005: 294). 'What sort of should we be aiming at? How can we ensure that this development involves the equitable distribution of benefits and burdens, opportunities and risks? How can we ensure that human rights are observed? Similarly, how can we ensure that account is taken of the ecosystem's limited ability to cope? And lastly, up to what extent the issue of 'development induced displacement' is addressed and mitigated' (Verma 2004a, b, 2015: 245–275, 2016: 23–48).

References

Adeola, R., & Viljoen, F. (2018). Climate change, development projects and internal displacement in Africa. *Journal of African Law, 62, 3.*

Anand, R. L. (1956). *Punjab agriculture facts and figures.* Government of Punjab: Economic and Statistical Organization.

Barry and Woods. (2013). The environment. In Michael Goodhart (Ed.), *Human rights: Politics and practice.* UK: Oxford University Press.

Biswal, T. (2015). *Human rights, gender and environment.* New Delhi: Viva.

Department of Environment and Forest. (1986). *Narmada Sagar and Sardar Sarovar multi-purpose projects*. Geneva: International Environmental Law Research Centre.

Dharamadhikary, S. (2001). *Unravelling Bhakra: Assessing the temple of resurgent India*. Madhya Pradesh: Manthan Adhyayan Kendra.

Duggal, S. K., Bhalla, J. K., & Dogra, R. S. (2001). *Socio-economic and environmental impacts of Bhakra Beas project—An assessment*. Chandigarh: Bhakra Beas Management Board.

Duggal, Er. S. K., Bhalla, J. K., & Bhatia, N.K. (2002). Flood and sedimentation status of Bhakra Reservoir—A case study. In S. Dharamadhikary (Ed.), *Unravelling Bhakra: Assessing the temple of resurgent India*. Madhya Pradesh Manthan Adhyayan Kendra.

Economic and Political Weekly. (2014). Dialogue needed on Sardar Sarovar. *Economic and Political Weekly, 49*(37).

Friends of river Narmada. (2008). Introduction. http://www.narmada.org/introduction.html. Accessed on August 19, 2018.

Gokhale, V. (1991). India's Narmada Valley hydro project. *Eco decision, Environment and Policy Magazine*. No. 1.

Haynes, J. (2008). *Development studies*. New Delhi: Rawat.

Irrigation Committee Report. (1972). *Irrigation committee report*. New Delhi: Government of India.

Judge, P. S. (2000). Resettlement and rehabilitation of the displaced. In S. N. Chary & V. Vinod (Eds.), *Environmental management: An Indian perspective*. New Delhi: Mc Million.

Kapur, D., Lewis, J. P., & Webb, R. C. (1997). *The world bank: Its first half century (vol. I & II)*. Washington D.C: Brookings Institution Press.

Khagram, S. (2004). *Dams and development*. New Delhi: Oxford University Press.

Kushalapa, K. A. (1992). Special monitoring of compensatory afforestation in Kachh District—Gujarat: A report. Annexed to the Agenda for the 15th Meeting of the Environment Sub-Group of the Narmada Control Authority.

McCully. (2001). Silenced Rivers: The ecology and politics of large dams. London. In T. Scudder (Ed.), *The future of large dams: Dealing with social, environmental, institutional and political costs*. 2005. Earthscan: London.

Morse, B., & Berger, T. (1995). Finding and recommendations of the independent review. In William F. Fisher (Ed.), *Toward sustainable development: Struggling over India's Narmada River*. M.E. Sharpe: Armonk, N.Y.

NBSAP. National Biodiversity Strategy and Action Plan in Dharamadhikary Shripad. (2001). *Unravelling Bhakra: Assessing the temple of resurgent India*. Madhya Pradesh: Manthan Adhyayan Kendra. Downloaded from http://www.sdnp.delhi.nic.in/nbsap/states/punjab/draftbsap.html. Accessed June 14, 2002.

Ojha, R. K. (1989). Letter to the chief conservator of forests (Central)', Bhopal: Ministry of environment and forests, western region office, Bhopal, regarding inspection of compensatory afforestation sites in Gujarat.

Public Accounts Committee Report. (1993). *Planning process and monitoring mechanism with reference to irrigation projects*. 141st report (1992–93), Seventh Lok Sabha. New Delhi: Ministry of Planning, (Planning Commission), Lok Sabha Secretariat.

Rai, S. (2005). Gender and development. In J. Haynes (Ed.), *Advances in development studies* (pp. 226–246). Basingstoke: Palgrave Macmillion.

Raj, P. A. (1990). *Facts: Sardar Sarovar project*. Gandhinagar: Sardar Sarovar Narmada Nigam Ltd.

Sakthivadivel, R., Thiruvengadachari, S., Amerasinghe, U., Bastiaanssen, W. G. M., & Molden, D. (1999). Performance evaluation of the Bhakra irrigation system, India, using remote sensing and GIS techniques. Research Report 28. Colombo. International Water Management Institute in Thayer Scudder, *The Future of Large Dams: Dealing with Social, Environmental, Institutional and Political Costs*. 2005. Earthscan: London.

Scott, J. C. (1998). *Seeing like a state: How certain schemes to improve the human condition have failed*. New Haven: Yale University Press.

Scudder, T. (2005). *The future of large dams: Dealing with social, environmental, institutional and political costs.* London: Earthscan.

Singh, S., Mehta, R., Uppal, V., Kabra, A., Taneja, B., Rao, P. (2000). Environmental and social impacts of large dams—The Indian experience. *Prepared for the world commission on dams.* New Delhi: Indian Institute of Public Administration (21).

Stiglitz, J. (2002). *Globalization and its discontents.* London: Allen Lane.

Udombana, N. J. (2006). The African commission on human and peoples' rights and the development of fair trial in Africa'. *African Human Rights Law Journal, 6*(2).

van Gelder, J. W., van der Valk, F., Dros, J. M., & Worm, J. (2002). *The impacts and financing of large dams.* A research paper prepared for WWF International. Living Waters Programme: Amsterdam.

Verma, M. K. (2004a). *Development, displacement and resettlement.* Jaipur, New Delhi: Rawat Publications.

Verma, M. K. (2004b). Development induced displacement: A socio-economic study of thermal power projects. *Man in India Journal.*

Verma, M. K. (2015). *Globalization and environment: Discourse, policies and practices.* Jaipur, New Delhi: Rawat Publications.

Verma, M. K. (2016). Development induced displacement, SEZs and the state of farmers in India: Some insights from the recent experiences. *Journal of the Human Rights Commission, 15*, 23–48.

Wade, R. H. (2011). Muddy waters: Inside the World Bank as it struggled with the Narmada projects. *Economic and Political Weekly, 46*(40).

WCD (World Commission on Dams). (2000). *Dams and development: A new framework for decision-making.* The report of the World Commission on Dams. London: Earthscan.

Chapter 8
Impact of Efforts on Ganga Restoration and Conservation

Ravindra Kumar

Abstract It is a well-known and recognized fact that water impacts everyone. During recent years, India's water usage has increased and diversified, creating both increased water shortages and water quality degradation in rivers and aquifers, thereby threatening broader environmental sustainability. River healthcare gaps blight the quality of life of the average Indian who have been relegated to the periphery. Whereas governments are expected to restore and maintain the wholesomeness of the rivers ensuring environmental flows and preventing the pollution ingress into the water bodies, it is also noteworthy that the protection of water and environmental infrastructure is a social responsibility of every citizen—both individually and collectively. In this background, the paper examines the Ganga conservation and rejuvenation strategies and its impact on environment and drinking water. While government is committed for conserving and rejuvenating National River Ganga, and also addressing interrelated issues like sustainable agriculture, basin protection against floodplains disasters, river hazard management, urban river management, wastewater management and revival of water bodies for providing environmentally safe sanitation and through a well-designated scheme of afforestation in riparian zones to purify base flows and run-off draining into the river, they are also working on to enhance the ecosystem services of our rivers and water bodies that remain healthy for downstream users. In this regard, institutional network of integrated water resource management plan, policy and regulatory governance provides synergy and helps other key stakeholders, experts, investors and well-wishers. On micro-level, key points of river restoration include aspects of flows (*aviralta, nirmalta*) and functions of river as geologic entity and ecological entity. One may track river science, engineering and operations including afforestation and biodiversity to suggest ways of improving the overall efficacy of aquatic ecology, ecological restoration (lateral, longitudinal and vertical connectivity) and geological safeguarding (sediment transport, assessing quantity, quality and nutrient value). Critical success of recovering wastewater and restoration of drains are components of urban river management.

R. Kumar (✉)
WWF-India, New Delhi, India
e-mail: ravindra53@yahoo.co.in

DST Centre for Policy Research, BBA Central University, Lucknow, India

© The Author(s), under exclusive license to Springer Nature Singapore Pte Ltd. 2021   115
M. K. Verma (ed.), *Environment, Development and Sustainability in India: Perspectives, Issues and Alternatives*, https://doi.org/10.1007/978-981-33-6248-2_8

Decentralized infrastructure can greatly enhance the speed of water treatment leading to one city-one operator through reuse of treated sewage/trade effluents. In order to have a successful water economics, it essentially requires creating enabling environment for sustained infrastructure management through water valuation, pricing and effective implementation of the urban river management strategies, and well-functioning water markets. Ganga basin plan prepared including environmental flows allocations suggest three-pronged strategy for its implementation.

Keywords Environmental sustainability · Ecological integrity · Restoration and conservation · Afforestation · Sustainable agricultural · Urban River

Introduction

The National River Ganga has been engineered and shaped to serve humanity by controlling flows, digging riverbeds, encroaching floodplains and disturbing even stream-aquifer equilibrium in the basin. Hydropower is the main focus of development in the upstream mountainous regions, whereas in plains, the main river stem is regulated with dams and barrages to divert water in associated irrigation canal systems. These alterations are largely for food security, flood protection, domestic and industrial supplies and power generation. All these head control structural developments, and abstractions affect the river's flow regime, which, in turn, impacts downstream water availability, water quality and riverine ecosystems. While the services a river renders are well recognized, it is equally critical that such interventions onto the river systems either substantially reduce its flows or bring in huge diurnal, daily or seasonal variability in flows, which is contrary to the natural regime of flows, as a result the river struggles to maintain its might and glory. In order to maintain the capacity of a population to safeguard water-related disasters (floods and droughts), and for preserving ecosystems in a climate change regime, it is important to find your own way to flows.

The Ganga is a trans-boundary and international river having total length of whopping 2525 km and a basin area of 1,087,300 km^2. It runs through China (3%), Nepal (14%), Bangladesh (4%) and India (79%). There are concerns about water sharing conflicts between countries, visible at Farakka Barrage and interstate issues regarding River Interlinking Projects of Government of India under study.

The Ganga Basin area in India is 861,404 km^2, spread in 11 states which include five main stem land states—Uttarakhand, Uttar Pradesh, Bihar, Jharkhand and West Bengal. The Ganga Basin covers 26% of India's geographical area and 57% of fertile agricultural landmass and acts as a major source of lifeline around it. It caters of 43% of India's population livelihood and contributes 28.1% of India's water resources. Around 143 different freshwater fish species live and depend on its water. The river is considered as mother due to the multiple functions it plays in the lives of the people, pertaining to ecological, socio-cultural and livelihoods, which synonymies it as the most holy and worthy of being worshipped.

The longest middle stretch of 1136 km of River Ganga and 28.8% of its basin falls in the state of Uttar Pradesh which unfortunately is extremely polluted in terms of biochemical oxygen demand (BOD) and Faecal Coli form. Most ironically, this is true in case of several other rivers as well. Not only River Ganga but its tributaries are struggling to survive due to various kinds of pollution, and therefore, it is a major cause of concern. Due to the notion of holiness and purity attached with the River Ganga for Hindus (at Gangotri, the river has maximum level of dissolved oxygen than any other river system in the world), its good quality water is essentially desired along with its channel depths to fulfil religious and cultural activities of holi dip (*astha ki dubki*) and *aachman* (*a drop of holi water in the mouth*).

Governments regularly host the world fame Kumbh Mela (fair) at Prayagraj where about 30 million pilgrims take bath in a single day in Triveni Sangam, which is a largest religious congregation in the world. Propelled with tradition or the urge to seek salvation, pilgrims, saints and cultural tourists take dip in the holy confluence of the Rivers Ganga and Yamuna during Mahakumbh (organized after every 12 years during January–March for 45 days at Prayagraj) to wash away sins of lifetime that help attain salvation from rebirth and sufferings. Other visitors are drawn by desire to experience the power of congregation, for ritual bath, offer prayers and move out during auspicious period on *ghats* (river bank) extended over 2 km on both banks; one has to jostle for space. Kalpvasi (religious tourists who reside for a month in a make-shift tent city on the river's bank) reside there for a month enjoying bliss of religious discourses (*satsang*) and two-square meal (*prasad)* served by *Akharas* (saint's Ashram established in a big tent by various sects) to all devotees irrespective of caste, creed and religion. More water is allowed by state to flow in river during Maghi-Kumbh (a Hindu calendar month falling in January—a pot filled with elixir) mela (fair), even at the cost of irrigation water.

Even though human beings are interacting with rivers for fulfilling their basic needs since millennia, river's role in shaping earth's surface is recognized only during twentieth century when various researches started on it. Ecological studies on Ganga conducted at Varanasi in 1950s, River Yamuna at Delhi (Palla to Okhla) in late 1950s and many other such researches held post-1974 reported poor health of the rivers. Therefore, modelling the hydrology of the Ganga Basin is critical for estimation, planning and management of present and future water resources sustainability. In this background, on the basis of the study conducted in the state of Uttar Pradesh (SWARA 2020), the paper attempts to critically analyse and comprehend the desirability of maintaining environmental flows in the River Ganga along with its process, if it has to be done. Environmental flows mimic the natural pattern of river's flow variation. To recognize the physical limit beyond which a natural water body suffers irreversible damage to its ecosystem functions, environmental flows are, therefore, one of the central elements in water resources planning and management for sustainable development to meet future water needs. To achieve the objective of restoration and conservation of the river Ganga, assessment of environmental flows and its maintenance throughout the year is one of the most important aspects towards river health.

River Health

The term health is used for the first time by Leopold and Wolman (1957) for addressing land issues. Ecosystem health since 1980s and river health debate from 1995 onwards gained attention of freshwater biology. The word stream health is used as an analogy, a metaphor that provides inspiration and insight into the understanding of river health. An analogy between human health and ecosystem health oversimplifies the complex issue (Boulton et al. 1998) of human existence and survival on earth. The term river health is useful since it is readily interpreted by the general public and evokes societal concern about human impacts on rivers. Health implies a flourishing condition, well-being, vitality or prosperity. A healthy organism is resilient, able to recover from many stresses (Karr and Chu 1998). No single indicator reveals river health unequivocally (Boulton et al. 1998).

A healthy stream is an ecosystem that is sustainable and resilient, maintaining its ecological structure and function over time while continuing to meet societal needs and expectations. Often, integrity is used synonymously to symbolize river health assessment. Integrity implies an unimpaired condition or the quality or state of being complete or undivided; it implies correspondence with some original condition. Biological integrity refers to the capacity to support and maintain a balanced, integrated, adaptive biological system having the full range of elements (genes, species and assemblages) and processes expected in the natural habitat of a region. Ecological integrity is the sum of physical, chemical and biological integrity (Karr and Chu 1998). For ecological integrity, the base flows in the rivers are critical, especially during pre-monsoon season.

A study suggests that 'since 1970s the groundwater over abstractions for irrigation and other uses within Ganga Basin has led to decline of base flows (through aquifers) by 59%' (Nature 2018). Many smaller tributaries of River Ganga have dried up due to groundwater over abstractions. The net flows reduction in Ganga River water could jeopardize domestic water supply, crop water requirements, river transport, ecology, etc., of densely populated northern Indian plains. In other words, the consumption of River Ganga water and its impact in terms of waste discharge both lead to deteriorating health of the river. Therefore, it needs immediate attention. While the water pollution is an issue that needs to be dealt at source, the reduced flows in the river should be addressed by maintaining environmental flows (E-Flows) in the River Ganga.

The Government of India, by becoming aware and sensitized about the significance of uninterrupted water flows, is promoting various programmes and activities for Ganga rejuvenation based on the principle of 'Aviral Dhara' (continuous free flow) and 'Nirmal Dhara' (pollution free stream) (GRBMP- 2015). Moreover, apart from that, several other studies are also commissioned either by the government or by other agencies and institutions through which assessment of environmental flows for River Ganga has been done (Central Water Commission 2015; WWF-India 2012, 2013, 2019, cGanga and NMCG 2018, etc.) in order to provide some number (minimum or maximum) requirement of flows. In this background, the paper attempts to answer the second-generation question that 'if the E-Flows are to be maintained in the river Ganga, then how this can be done, especially in the light of existing committed water allocations for the rejuvenation of Ganga river in Uttar Pradesh'. The paper ends with suggesting various ways and means through which E-Flows' implementation in the River Ganga can be achieved at an investment cost of $10,000 million towards integrated water resources management in the study area of the Ganga Basin.

Study Area for Strategic Initiatives Under Ganga River Basin Management Plan in Uttar Pradesh

Main stem Ganga Basin area falling in Uttar Pradesh is selected for the water management, wherein 80% of its waters are diverted for irrigation in Ganga-Yamuna doab. Ganga River's largest stretch from Bijnor to Ballia passes through Uttar Pradesh, covering an area of 67,723 km^2 (i.e. 7.88% of the total Ganga Basin and 28.1% of the state). Location map of Ganga Basin (main stem) and its sub-basin in Uttar Pradesh is shown in Fig. 8.1. The basin covers (fully or partially) 27 districts of Uttar Pradesh.

Brief profile of Ganga Basin as compared with other basins of Uttar Pradesh is given in Table 8.1. Apparently, Ganga Basin and Yamuna Basin each cover over 28% area of the state. Ganga basin is characterized by highest population of 64.63 million (Census 2011) having population density of 954 persons/km^2. Land under cultivation is 67.63% and population involved in agriculture activity remains 57.85%. Net sown area irrigated is 94.20%. The area irrigated by source is canal 24.19%, groundwater 71.87% and by other sources 3.95%. Cropping intensity stands 161.76% (2015). The literacy rate of the area marks 69.00%, whereas the marginal landholding is 81.80%.

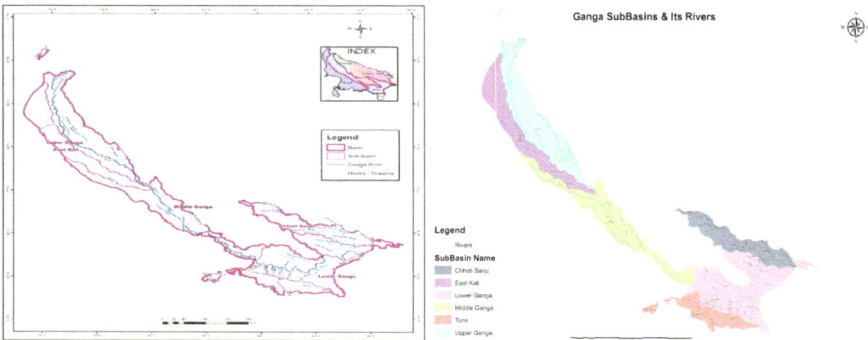

Fig. 8.1 Location map of Ganga Basin (main stem in Uttar Pradesh) and its sub-basins. *Source* Ganga Basin Plan 2020 prepared by M/S TAHAL Consulting Engineers Limited, Israel (90 m SRTM data) with which author is associated

Table 8.1 Brief profile of Ganga Basin as compared with other Basins of the state of Uttar Pradesh

Basin	Area (km²)	Land under cultivation (%)	Cropping intensity (%)	Net sown area irrigated (%)	Gross sown area irrigated (%)	Net area irrigated by (%)			Landholdings (%)				
						Canals	GW	Other	Marginal (<1 ha)	Small (1–2 ha)	Semi-medium (2–4 ha)	Medium (4–10 ha)	Large (>10 ha)
Gomti	31,434 (13.0%)	64.35	163.12	90.59	88.66	21.80	78.17	0.03	84.58	10.59	3.96	0.85	0.03
Gandak	974 (0.4%)	71.60	137.13	76.60	75.10	23.76	73.51	2.73	90.11	6.97	2.39	0.51	0.02
Rapti	14,658 (6.1%)	71.10	155.71	75.71	63.47	7.09	88.94	3.98	84.58	10.98	3.69	0.72	0.03
Ramganga	**20,417 (8.5%)**	78.79	161.61	97.08	93.85	4.00	95.53	0.47	75.63	15.56	7.00	1.74	0.07
Ghaghra	31,503 (13.1%)	70.03	156.55	85.92	78.53	5.75	93.61	0.64	84.63	10.40	4.01	0.93	0.04
Sone	5093 (2.1%)	19.16	115.79	17.74	17.99	17.63	32.31	50.07	67.58	17.26	9.61	4.69	0.87
Yamuna	69,327 (28.7%)	71.24	151.33	82.39	69.63	22.55	73.75	3.70	67.85	18.26	9.76	3.85	0.28
Ganga	67,773 (28.1%)	67.63	161.76	94.20	85.15	24.19	71.87	3.95	81.80	11.92	4.91	1.29	0.08
Source	Agriculture department data, 2014–15								Agriculture census, 2010–11				

(continued)

Table 8.1 (continued)

| Basin | Area (km²) | Population | | | | Density, per sq.km (India: 382) | Literacy rate (%) | SC (%) | ST (%) | BPL house hold | Towns | | | | Population >1 lakh |
		Involved in agriculture activities (%)	Total, million	Rural (%)	Urban (%)						Total	Statutory	Census		
Gomti	31,434 (13.0%)	62.50	27.72	81.50	18.50	882	68.70	26.70	0.09	42.77	86	68	18	6
Gandak	974 (0.4%)	75.60	0.99	97.70	2.30	1013	50.10	16.00	2.13	52.03	1	1	0	0
Rapti	14,658 (6.1%)	72.31	13.92	88.70	11.30	948	62.70	17.70	0.74	43.68	41	29	12	2
Ramganga	20,417 (8.5%)	58.70	18.29	73.40	26.60	896	60.10	15.70	0.04	29.10	93	72	21	5
Ghaghra	31,503 (13.1%)	72.98	25.86	91.94	8.06	821	59.88	19.10	1.07	42.60	70	58	12	5
Sone	5093 (2.1%)	67.00	1.25	78.30	21.70	245	62.00	18.19	27.50	44.38	16	8	8	0
Yamuna	69,327 (28.7%)	56.58	47.17	68.67	31.33	680	71.74	21.50	0.21	27.75	280	197	83	22
Ganga	67,773 (28.1%)	57.85	64.63	75.59	24.41	954	69.00	20.40	0.40	32.16	327	216	111	24
Source	Census of India, 2011													

Ganga Basin is bestowed with large perennial rivers, and underlay is large groundwater resource too. Rainfall, subsurface flows and snowmelt from glaciers are the main sources of water in River Ganga. Ganges inflow at confluence of Bhagirathi and Alaknanda is 23.90 BCM. At Narora, it becomes 31.40 BCM. In route, the flow augmentation by major tributaries like Ramganga 15.62 BCM and Yamuna 93.02 BCM contributes flows at Allahabad to 152 BCM. In between, withdrawal of water for irrigation is committed in the order of 1606 cubic metre per second.

The first irrigation canal system in Uttar Pradesh dates back to Mughal dynasty in between 1719 and 1748 when Eastern Yamuna Canal was constructed. It was remodelled by British during pre-independence year 1830 for capacity 85 m^3/s. The calamitous events of the year 1837–38 and the extent of human misery then caused by the utter failure of the crops in the central provinces, leading to famine in its most aggravated shape. Upper Ganga Canal System (UGC) commissioned in 1854 Bhimgoda Head works, Haridwar, across the great Ganga River, initiated by P. T. Cautley (1839–45). Post-Tehri Dam, the capacity of PUGC has been increased to 13,500 cusec (382.43 m^3/s). In the backdrop of 1866 severe famine, Narora weir commissioned in 1878 that was modernized as Narora Barrage in 1961–68. The combined capacity of LGC/PLGC shall be 17,400 cusec (492.92 m^3/s) post-Tehri Dam.

Ganga Canal Capacities on Increase

The various irrigation canal systems have undergone modernization over time and their capacities are on increase. Post-Tehri Dam, the capacity of Kharif channels such as Eastern Ganga Canal (EGC) and Madhya Ganga Canal (MGC) has been increased along with PUGC, LGC and PLGC.

UGC = 6500 (1854), 6750 (1938), 10500 (1951), PUGC = 13500 (1982)

LGC = 6000 (1878), 8500 (1974), PLGC = 4200(1982), 6900 (2015), 8900 UPWSRP-II.

EGC = 4850 (1980-92), 5850 (2009) post-Tehri Dam

MGC-I = 8280 (1998-2001), MGC-II = 4200 (still under construction).

Running days for UGC and LGC in Kharif is 183 and Rabi 182. Running days for EGC and MGC in Kharif is 154 only. For crop pattern, 46.5% in Kharif (wet crops) and 36% in Rabi (dry crops) the above canal systems are adequate. However, overuse of irrigation water has been less productive compared with other states of India.

Major and medium pump canals of seven numbers having combined capacity of 7410 cusec (209.92 m^3/s) are established on the banks of River Ganga in tail end of gravity canals to meet the irrigation demand in the district of Raebareli, Unnao, Mirzapur, Chandauli, Ravidas Nagar and Ghazipur. There are four minor lift pump canals situated in districts Kanpur, Fatehpur and Pratapgarh having combined capacity of 55.5 cusec (1.57 m³/s).

River Behaviour: Loosing and Gaining Stretches of Ganga River in Uttar Pradesh

To understand river behaviour about losing (where river water disappears in recharging groundwater) and gaining stretches (where groundwater augments river's flow), long-term annual water flow volume of Ganga River for 55 years (based on Central Water Commission data from 1861 to 2015) indicating average and 75% dependability at various locations in Uttar Pradesh is summarized in Table 8.2 and shown in Fig. 8.2.

Ganga Basin flow volume at 75% dependability comes to 10,797 MCM against average annual flow volume of 13,108 MCM.

Perusal of Fig. 8.2 suggests that there are three catchment zones at Fatehgarh, Shahzadpur and Varanasi in which Ganga River's annual flow volume diminishes (loosing stream) due to excessive pumping of groundwater resulting into reduced base flow and lowering of the groundwater table, lower than river top or bottom surface water. The details are as follows:

1. Kachhla bridge (Badaun)-Fatehgarh Zone: dip in annual flow volume from upstream Kachhla bridge 1419 MCM–1260 MCM (i.e. difference is 159 Million Cubic Metre) is noted.
2. Bhitaura-Shahzadpur (Kaushambi) Zone: there is dip in flow volume from 2017 MCM upstream Bhitaura to 1536 MCM at Shahzadpur (i.e. difference is 481 MCM). The missing water again re-appears at Prayagraj as catchment contribution often known as invisible Sarasvati River's confluence into Ganga, which is popularly known as Triveni Sangam.
3. Mirzapur-Varanasi Zone: dip in flow volume from upstream Mirzapur of 6379 MCM–6136 MCM is noticed (i.e. difference is 243 MCM).

So, long distance release of water (from Tehri Dam, Bhimgoda and Narora Barrage downstream) is less effective towards E-flows' maintenance in the Ganga River as being presently practised during Kumbh mela. This is at the cost of loss of water for irrigation. Better way will be decentralized local water resource management and merely 400 cusec (11.33 m³/s) release of Sharda-Sahayak canal water from Bhadri escape, 40 km upstream Sangam nose at Prayagraj, would be more effective.

Table 8.2 Catchment area, 75% dependable flow and average annual flow volume at various locations in Ganga Basin (Uttar Pradesh)

Gauge-discharge site	Catchment area (km^2)	75% dependable flow volume (MCM)	Average flow volume (MCM)	Details of water diversion/abstraction structure in the zone
Garhmukteshwar (Moradabad)	29,709	1068	13,389	From Narora Barrage existing lower Ganga Canal system draws 226.6 cubic metre per second (m^3/s) water for irrigation
Kachhla bridge (Badaun)	34,446	1097	1419	
Fatehgarh	40,096	893	1260	
Ankit Ghat (Kanpur Rural)	82,209	1198	1855	
Kanpur	87,650	1269	1974	5.7 m^3/s water is diverted from Lav Kush Barrage at Kanpur for drinking
Bhitaura (Fatehpur)	90,444	1237	2017	Dalmau pump canal A & B combined draws 32.3 m^3/s water at Raebareli and Unnao
Shahzadpur (Kaushambi)	93,604	1127	1536	
Prayagraj	463,971	4047	5931	Yamuna joins Ganga
Mirzapur	485,277	4877	6379	Narayanpur lift canal draws 40.8 m^3/s water for irrigation
Varanasi	489,087	4427	6136	Gyanpur pump canal water draws 38.2 m^3/s water for irrigation
Ganga Basin	**67,923.73**	**10,797**	**13,108**	

Source Estimated by author

Ganga Basin Water Plan: Managing Demand and Supply of Water

Agriculture consumed 96% of water among all uses in the base study year of 2015. So, based on the analysis of past trends and the need for crop intensification and diversification, considering the available resources and options, the following Agriculture Growth Scenarios (Agr.Sc) have been considered.

- Agr.Sc-1: Projected Crop Areas as per the prevailing Trend (BAU) limited to Cultivable Area.

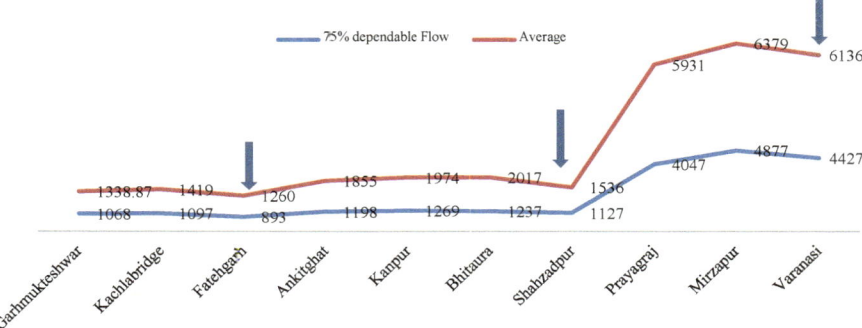

Fig. 8.2 Average annual and 75% dependable flow volume (million cubic metre) in Ganga River. *Source* Constructed by author based on Central Water Commission data (1861–2015)

- Agr.Sc-2: Projected Crop Areas as per the prevailing Trend limited to Cultivable Area along with crop diversification, implementation of conjunctive use management, equitable distribution of water and micro-irrigation in 10% of cropped area.

Future water balance scenarios calculated in Ganga Basin by considering agricultural scenario-1 and scenario-2 with additionally required infrastructure for groundwater use in conjunction with surface water and restriction on groundwater extraction within annual replenishment, maintaining minimum environmental flows with crop intensification/diversification—replacing paddy with SRI in 20%, 30% and 40% of canal command area in years 2025, 2035 and 2045, respectively, and introducing micro-irrigation in 10% of cropped area. The agricultural data used in water balance model are given in Table 8.3.

For anticipated crop yield in year 2025, the maximum yield achieved in last 10 years has been considered for all the crops. Further, for years 2035 and 2045, 20% increase per decade, i.e. 2% increase per year (over 2025 values) in yield of wheat and rice, has been considered, while, for others crops, 10% increase per decade, i.e. 1% increase per year (over 2025 values), has been considered. For rice (SRI), the yield has been considered as 1.5 times that of traditional rice cultivation. Still, these anticipated crop yields are very much on conservative side as some of the states of India have achieved these crop yields in present scenario and the potential yield for the main crops of rice and wheat of Indo-Gangetic Plains of India is much higher. Increased cropping intensity is based on the assumption that with implementation of conjunctive use and various land resource development and management programmes, more water resource will be available and soil health will improve, resulting into more seasonal fallow into crop area, giving ultimate cropping intensity of 189%.

Table 8.3 Proposed crop areas, crop yield and cropping intensity in Agr.Sc-2

Crop	Area 2025	Crop yield (Qtl./ha)	Area 2035	Crop yield (Qtl./ha)	Area 2045	Crop yield (Qtl./ha)
Rice Kharif	1,432,651	32.76	1,297,453	39.31	1,133,530	47.17
Wheat	3,162,176	45.97	3,317,788	50.57	3,400,341	55.62
Barley	15,185	38.39	7688	42.23	6067	46.45
Jowar	29,283	23.53	24,230	25.88	22,255	28.47
Bajra	467,713	46.35	517,945	50.99	558,041	56.08
Maize Kharif	220,005	33.82	185,387	37.2	163,594	40.92
Gram	48,417	19	33,046	20.9	25,114	22.99
Pea	20,236	22.46	13,574	24.71	9386	27.18
Arhar	64,325	28.02	50,116	30.82	40,286	33.9
Sugarcane	422,021	784.92	411,318	863.41	402,126	949.75
Potato	230,416	394.18	241,460	433.6	251,799	476.96
Other Kharif crops	360,688	12.04	379,270	13.24	397,436	14.57
Other Rabi crops	311,238	20.53	317,640	22.58	334,089	24.84
Rice (SRI)	358,132	49.14	556,055	58.97	755,733	70.76
Jayad	242,045	18.96	306,942	20.86	354,372	22.94
Total	7,384,529		7,659,913		7,854,169	
Cultivable area	5,091,612		5,091,612		5,091,612	
Cropping intensity percentage with respect to 2015						
Kharif	73		75		76	
Rabi	92		95		97	
Jayad	15		16		17	
Total	180		186		189	

Source Ganga Basin Plan 2020, prepared by M/S TAHAL Consulting Engineers Ltd. Israel/India

Outcome of Water Balance Modelling

The outcome of U.P. Ganga Basin Plan (2020) has been discussed for base year 2015 and future water balance scenarios for 2024–25, 2034–35 and 2044–45, with crop intensification and diversification as per prevailing trend (Business as Usual). The present and future sectoral water demand, supplies and shortages have been shown in Table 8.4.

For main-stem Ganga Basin falling in Uttar Pradesh, the water balance model results shown in Table 8.4 depicts that the future water demand in 2044–45 for projected population of 210 million as compared to 68.60 million in base year 2015 (i.e. three times increase) will grow in urban domestic from 978 MCM to

Table 8.4 Summary of annual water balance for Ganga Basin, Uttar Pradesh

Scenario	Business as usual			
Demand in MCM	2014–15	2024–25	2034–35	2044–45
Population	68,634,723	80,193,999	18,863,480	20,990,224
Urban domestic demand	978.1	1322.6	1480.8	1588.3
Urban groundwater supply	714.3	1058.6	1238.9	1332.2
Urban surface water supply	263.8	264.0	241.9	256.1
Wastewater production/treated	547.7/29.7	772.4/52.1	905.5/88.0	1,016.5/98.9
Rural domestic demand	840.5	1314.4	1924.0	2157.4
Livestock demand	422.2	484.7	550.8	616.9
Industry demand	70.6	132.0	193.5	254.9
Power plants demand	318.8	395.3	473.9	473.9
Unmet power plants demand	0.00	18.3	41.3	41.3
Total rural demand	1652.06	2326.4	3142.2	3503.1
Total rural groundwater supply	1352.5	1949.0	2685.8	3044.1
Total rural surface water supply	299.56	359.1	415.1	417.7
Irrigation demand in canal command area (CCA)	15,406.9	17,333.1	18,171.1	18,593.7
Surface water supply in CCA	9160.9	10,121.2	10,087.7	10,190.1
Groundwater supply in CCA	5114.8	5918.7	6577.4	6794.5
Unmet irrigation demand in CCA	1131.2	1293.1	1506.0	1609.1
Irrigation shortage in CCA (%)	7.34	7.46	8.28	8.65
Irrigation demand outside CCA	13,855.6	13,545.1	14,110.9	14,506.8
Groundwater supply outside CCA	5462.7	5218.2	5266.7	5292.0
Surface water miner supply outside CCA	852.7	854.1	849.9	850.7
Unmet irrigation demand outside CCA	7540.2	7472.8	7994.2	8364.1
Irrigation shortage outside CCA (%)	54.4	55.2	56.6	57.6
Total irrigation demand	29,262.5	30,878.20	32,282.00	33,100.5
GW recharge from normal rain	9824.2	9824.2	9824.2	9824.2
GW recharge from other sources	9392.5	9935.3	10,321.8	10,528.1
Total GW availability	19,216.7	19,759.4	20,146.0	20,352.2
Groundwater pumping	12,663.0	14,167.2	15,794.6	16,490.5
Stage of groundwater extraction (%)	65.90	71.70	78.40	81.02
Total demand	31,892.6	34,527.2	36,905.0	38,191.9
Total groundwater supply	12,644.3	14,144.5	15,768.8	16,462.7
Total surface water supply	9724.3	10,744.3	10,744.8	10,863.9
Total surface water miner supply	852.7	854.1	849.9	850.7
Total unmet demand	8671.4	8,784.3	9,541.5	10,014.5
Total shortage in %	27.19	25.44	25.85	26.22

Source Draft Ganga Basin Plan, 2020QS, prepared by TAHAL Consulting Engineers Ltd., Israel for State Water Resources Agency, Uttar Pradesh with which the author is associated

1588 MCM (i.e. 1.62 times increase), rural domestic from 841 MCM to 2157 MCM (i.e. 2.57 times increase), livestock from 422 MCM to 617 MCM (i.e. 1.46 times increase), industrial from 71 MCM to 255 MCM (i.e. 3.61 times increase) and power plants from 319 MCM to 474 MCM (i.e. 1.49 times increase). Thus, total non-agricultural water demand shall increase from 1652 MCM to 3503 MCM (i.e. 2.12 times increase). For cropping pattern under business in usual scenario (i.e. cropping trend analysed for increase/decrease shall remain same in future also), the irrigation demand will increase from 29,263 MCM to 33,101 MCM (i.e. 1.13 times increase). Therefore, total water demand in U.P. Ganga Basin shall increase from 31,893 MCM to 38,192 MCM (i.e. 1.13 times increase), whereas the stage of groundwater extraction shall increase from present 65.91 to 81.02%. Thus, total unmet demand shall increase from 8671 MCM to 10,015 MCM (i.e. 1.15 times increase). The gap in demand and supply could decrease from 27.18 to 26.22%.

Ganga Basin Investment Summary

Water Conservation: Additional water 850.89 MCM will be available with an expenditure of 8072.32 Cr. Rs. on Watershed development, wetland development and rooftop harvesting activities.

Wastewater Generated and Treated

Generated wastewater production of 514.54 MLD (547.7 MCM) for base year 2015 and 861.41 MLD (1016.5 MCM) for the year 2045 with an additional expenditure of 710.76 Cr. Rs. for construction of remaining STPs will provide about 1000 MCM additional water to be considered for irrigation purposes.

This generated volume of water (conservation and treatment) amounting to 1850 MCM yearly will reduce the shortages from 26.22 to 21.38%, with area-specific additional expenditure.

For analysis, water conservation, groundwater recharge, rooftop rainwater harvesting, wetlands development, sodic land reclamation, watershed development, command area development, removing water system distribution deficiencies and drainage system deficiencies along with strengthening water user associations has been considered.

The inference drawn underlines that there is no surplus water in the basin for which any additional plan is required to be formulated. Existing resources are already overstressed and mismanagement is exaggerating problems day by day. It requires proper management with some policy constraint to keep the resources sustainable.

For overstressed urban areas, there is an urgent need to make a compulsory provision of harvesting for private/government buildings and offices along with common utility spaces. Use of recycled water in multi-storeyed colonies and offices may also be considered.

Table 8.5 Summary of development costs in Ganga Basin (Uttar Pradesh), 2017 prices

S. No.	Components	Cost, in INR crores
1	Formation and strengthening of WUAs	814.24
2	Removing canal system deficiencies	3471.86
3	Removing drainage system deficiencies	2603.90
4	Shallow tubewell installations	353.59
5	Watershed development	2136.21
6	Sodic land reclamation	935.00
7	Wetland development	2247.62
8	Wastewater treatment	710.26
9	Rooftop rainwater harvesting	1552.28
10	Micro-irrigation	22,917.40
11	Command area development and water management works	22917.40
12	Preparatory activities @ 5%	3000.00
Total basin development cost		63,659.76, Say, $10,000

Source Estimated by the author and is included in draft Ganga Basin Plan 2020, prepared by TAHAL Consulting Engineers Ltd. Israel for State Water Resources Agency, Uttar Pradesh

Total cost to be invested for the development of different infrastructures and institutions for Ganga Basin in Uttar Pradesh is summarized in Table 8.5.

Pollution in River Ganga

Pollution in River Ganga with respect to only Faecal Coliform is in stretch between Bijnor and Narora downstream, whereas critically polluted stretch in terms of Biological Oxygen Demand (BOD) and Faecal Coliform is between Kannauj, Kanpur, Allahabad and Varanasi to Ballia. Pollution load (BOD) from tributaries by Ramganga and Kali is most critical. Non-point source pollution includes agricultural run-off, open defecation, pious refuse, partially cremated bodies, associated materials, etc. Industrial pollution in enhancing its toxicity by 669 MLD (according to Central Pollution Control Board) through 1109 Grossly Polluting Industries (GPIs) such as tanneries, pulp and paper, sugar mills, textiles and dyeing and distilleries, etc. Sewage is additional source of pollution which is discharging 2953 million litres per day (MLD) in Ganga from 155 drains generated by 97 towns on the main stem in Uttarakhand, Uttar Pradesh, Bihar, Jharkhand and West Bengal.

Namami Gange and Other Restoration Programmes at National Level

Among several Ganga River's restoration and conservation plans, Ganga Action Plans (GAP)-1 of 1985 and GAP-II of 1993, JNNURM started in 2005 and Atal Mission for Rejuvenation and Urban Transformation (AMRUT) began in 2015 had definite impact towards this end. A separate ministry was established for Ganga Rejuvenation in July 2014; thereafter, Ganga River Basin Management Plan (GRBMP) developed in January 2015 by consortium of IITs. A budget outlay of Rs. 20,000 crores approved by government in May 2015 for Namami Gange Project. National Mission for Clean Ganga, an implementing arm, is declared as an authority under Environment Protection Act, 1986, in October 2016. Establishment of State and District Ganga Committees in June 2017 and notification of the minimum environmental flow for River Ganga at different stretches in October 2018 can be called as a visionary step in this regard undertaken to achieve the mission.

Namami Gange Programmes aim at speedily rejuvenating and conserving National River Ganga. The seven thrust areas are: (i) Aviral Dhara; (ii) Nirmal Dhara; (iii) riverfront development; (iv) capacity building; (v) research and monitoring; (vi) protection of aquatic flora and fauna; and (vii) awareness creation.

Twenty-one action points were considered in Namami Gange Programme and most important are: determination and maintenance of environmental flow, upgrading existing STPs, creating additional STPs, industrial pollution abatement, managing agricultural runoff, development of modern Dhobi ghats, creating model cremation ghats, development of Ganga Grams (villages situated on river banks), Ganga task force, Ganga Institute of River Science at Varanasi, establishment of cGanga at IIT Kanpur, afforestation drive of medical plants and native tree species, conserving diversity of Gangetic aquatic life, etc.

The details of the projects sanctioned by Namami Gange Plan for the conservation and restoration of the River Ganga till November 2018 is given in Table 8.6.

These initiatives are moving towards an evidence-based policymaking, accelerating technology transfer, providing a platform for water entrepreneurship, developing market-based mechanism for water trade and making India a global hub for water innovation, global water stewardship, innovative financing models and engaging communities.

Detailed status of existing sewage infrastructure and interventions started in Uttar Pradesh is shown in Table 8.7.

Sewage generation in complete Ganga Basin is estimated 2953 MLD in year 2016 and projected 3603 MLD by the year 2035. Available treatment capacity through 84 STPs in 46 towns till 1 November 2017 is 1584 MLD. The status of existing sewage infrastructure is shown in Fig. 8.3. Perusal of Fig. 8.3 shows that defunct 31 STPs have capacity of 270 MLD. The operational but underutilized 14 STPs have capacity 581 MLD, whereas, recently commissioned and working fine along with 20 years Operation & Maintenance (O&M), 39 STPs have capacity of 733 MLD. Interventions are underway for 45 STPs, wherein 12 STPs (91 MLD) already upgraded and O&M

Table 8.6 Projects sanctioned under Namami Gange (figures in crore)

Projects sanctioned under Namami Gange			Sanctioned cost, November 18	Expenses as on September 18
Pollution abatement projects				
1	Sewage Infrastructure (105 on Ganga, 26 on tributaries)	131	19,742	3708.54
2	Modular STPs-decentralized treatment	1	410	0.0
3	Bio-remediation	11	201.23	0
	Sub-total	143	20,353.23	3708.54
4	Rural sanitation (4465 villages along Ganga)	1	1426.26	1017.99
5	Industrial pollution abatement	12	900.13	0
	Sub-total	13	2326.39	1017.99
	Total	156	22,680.00	4726.53
River front, ghats and crematoria projects				
6	River front development	1	243.27	233.77
7	Ghats and crematoria (Old)	24	921.78	497.10
8	Ghats and crematoria (New)	35	204.39	0
9	Ghats cleaning	3	43.87	8.43
10	River surface cleaning	1	33.53	2.37
	Sub-total	64	1242.45	741.67
Afforestation and biodiversity conservation				
11	Afforestation	16	236.56	127.79
12	Biodiversity conservation	6	33.42	20.31
	Sub-total	22	270.00	148.10
Other projects				
13	Institutional development	6	185.00	37.67
14	Project implementation support/research and study projects	4	126.56	5.49
15	Composite ecological task force	2	167.63	0.06
Sub-total		12	479.26	5.55
Grand total		254	24,672.00	5660.06

Source Namami Gange programme—at a Glance—2018, NMCG

Table 8.7 Status of STPs in U.P. as on 30 November 2018 {capacity in million litres per day (MLD)}

State	Projects Nos.	STP capacity	Completed	Work under progress	Tendering process
Uttar Pradesh	40	1106.18	11	22	7
Total Ganga Basin	**131**	**3869**	**31**	**64**	**36**

Source Namami Gange programme—at a Glance—2018, NMCG

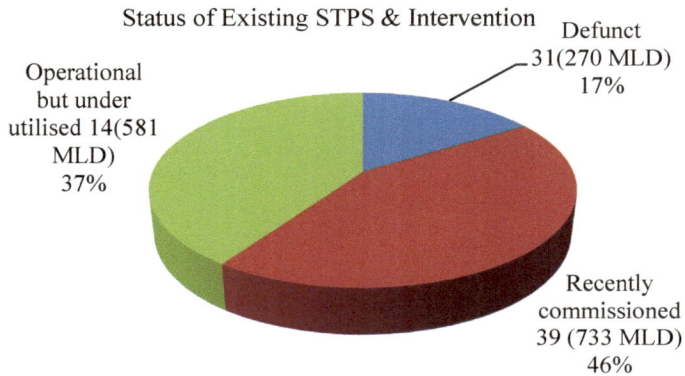

Status of Existing STPS & Intervention

Defunct 31(270 MLD) 17%

Operational but under utilised 14(581 MLD) 37%

Recently commissioned 39 (733 MLD) 46%

Fig. 8.3 Status of existing sewage infrastructure and initiatives in Ganga main stem. *Source* Namami Gange Programme-At a Glance-2018, NMCG

sanctioned, and 8 STPs (530 MLD) are integrated under One City One Operator. For another 23 STPs, tender for upgradation has been placed. STP Projects in four towns (Kanpur, Prayagraj, Mathura, and Farrukhabad) have already been awarded and in rest nine towns are under tendering process.

Industrial Pollution Control

Nine hundred and sixty-one gross polluting industries (GPI) on main stem of River Ganga and major tributaries have been inspected during the period of 12 April 2018 to July 2018 by twelve technical institutes. As on November 2018, polluting industries closed by itself figured out to be 278, whereas closure direction issued vide u/s 5 of Environment (Protection) Act to 10 and, show cause notice issued to 199 polluting industries for non-compliance of the norms.

Industrial Sector Development

By considering that certain industries are bound to emit water pollution due to the mechanisms involved in their production process, the state government formulated separate plan for them (SMCG). The details are as follows:

Tannery: Common effluents treatment plant (CETP) approved at Jajmau (Kanpur) at the cost of INR 554 crore with adoption of cleaner process and reduced water consumption.

Paper and Pulp: Zero black liquor discharge achieved. Estimated reduction in the freshwater consumption and effluent generation is about 45–50% as compared to 2012.

Distillery: Zero liquid discharge achieved in 32 molasses-based distilleries.

Sugar: Effluent generation reduced from 400 to 200 litres per ton of cane crushed.

Textile: Most of the units are in the process of upgradation of existing effluent treatment plant (ETPs)/installation of new ETPs/CETPs (common effluent treatment plants). Along the line, 6.25 MLD CEPT is sanctioned for Mathura.

Water Quality Monitoring Stations (WQMS): 44 real-time WQMS are made under operation to keep water quality in check.

Water Quality Status of River Ganga

Central Pollution Control Board (CPCB) has carried out an extensive survey of micro-invertebrates in the entire length of River Ganga. The study suggests that above two-third of the river length is moderately impacted. Significantly impacted reach lies at certain point locations, say near tannery hub of Kanpur. Near pristine situation is near source of mouth, and slightly impacted reach is in Alaknanda-Mandakini-Pinder system due to hydroelectric project under construction. Downstream Kachhla Ghat to Ghatia Ghat is due to Kali River which is highly polluted.

Impact of lockdown imposed to curb COVID-19 (since 24 March 2020 to 15 April 2020) has resulted in overall improvement in water quality of River Ganga, especially with regard to increased DO, and reduced nitrate concentration (CPCB, April 2020). This may primarily be attributed to the absence of industrial wastewater discharge (300 MLD, i.e. 9% of total wastewater being discharged into the river every day), agricultural runoff and increased freshwater flow. The reduction in BOD and COD concentration was relatively less due to continuous discharge of domestic water into the river. Reduced activities at ghats and entrainment of solid organic waste into the river may also have contributed to better water quality.

Table 8.8 Afforestation targets in Uttar Pradesh

State	Area in ha	INR crore
Uttar Pradesh#	4942	40.51
Total Basin	26,810	199.98

Source Namami Gange programme—at a Glance—2018, NMCG

Afforestation: Forest Research Institute, Dehradun, has prepared afforestation plan of 134,106 ha at an estimated cost of INR 2294 crores. For 2016-19, Uttar Pradesh state details are given in Table 8.8.

#Uttar Pradesh has launched 'Haritima Abhiyan' to provide common platform for all to participate so that the campaign for Cleaning Ganga is promoted as a massive Jan Andolan (mass movement). The scheme has been made a huge budgetary allocation dividing into state and centre. A budgetary allocation of INR 71 crores for states and INR 164 crore for centre (totalling INR 235 cr.) is made for the activities like sponsorship of awareness programmes, plantation by institutions/industries/NGO/citizen groups on common lands and adoption of Ganga villages by corporate/institutions.

On similar line, some more interesting steps have been taken by government in an effort to clean Ganga River. Few of them are as follows:

Ganga Task Force: A battalion of 532 ex-servicemen is engaged in creating public awareness, tree plantation, participating in campaigns and patrolling Ganga River for biodiversity protection and monitoring river pollution.

Ganga Vichar Manch: It is actively involved in public outreach activities, tree plantation and *ghat* cleaning.

Ganga Praharis: Self-motivated individuals who will mobilize others in Ganga conservation efforts. The Ganga Praharis are being trained for ecological monitoring of biodiversity of Ganga River along with tree plantation techniques, awareness generation and community mobilization.

Jalaj Scheme: It advocates innovative steps for Ganga Praharis to combine livelihood improvement with aquatic life and Ganga conservation.

Treatment of sick rivers: Treatment of water depends upon the diagnosis of the biodiversity along with the attaining of desirable 'standard' and the adoption of treatment system. Therefore, realizing it, on priority basis, one river in a district totalling for eight rivers in Uttar Pradesh is selected for treatment purpose. Few of them are, for example, Gomti, Aril, Tamsa, Varuna, Manorama Rivers, etc.

Preventive Measures: To this end, catchment land use, waste treatment, flows, habitat alteration studies are under consideration.

Curative Measures: In this, maintenance of environmental flows, revival of meanders and habitat features, changes in land use practices, restoration of riparian/floodplain habitats, groundwater recharge, controlling exotic species, regulating gravel/sand extraction, etc., have been considered.

Environmental Flows at Critical Locations on River Ganga

E-Flows' assessment was carried out at two locations on Ganga main stem which include (i) downstream Madhya Ganga Barrage and (ii) downstream Narora Barrage. Cross-section (CS) surveys at these two locations were carried out (Fig. 8.4), and E-Flows were recommended keeping in view water requirements of Indian Major Carps (IMP) by WWF-India and cGanga, IIT Kanpur, in May 2019.

The summary of Hydraulic Parameters is given in Table 8.9.

Perusal of Table 8.9 shows that Ganga River top width, average depth and average velocity are increasing from Madhya Ganga Barrage to Narora Barrage. E-Flows estimated and gap with present flows in Ganga River downstream Madhya Ganga Barrage at Balawali and downstream Narora Barrage is given in Tables 8.10 and 8.11, respectively.

The percentage of shortfall in terms of present water availability along with E-Flows' recommendations for downstream Bhimgoda Barrage at downstream of Madhya Ganga Barrage at Balawali is illustrated in Fig. 8.5. Similar gap in recommended and observed flows is shown for downstream Narora Barrage site in Fig. 8.6.

According to Gazette of Government of India by NMCG Order of 9 October 2018 (MOWR, RD and GR 2018), the 20%, 25% and 30% of monthly flows are be maintained at locations downstream of structures at Devprayag to Haridwar in Uttarakhand during non-monsoon and monsoon respectively in upper stretch of River Ganga. In plains, downstream Bhimgoda Barrage 36–57 m³/s and downstream Bijnor, Narora and Kanpur Barrage 24–48 m³/s flows during non-monsoon and monsoon respectively have been issued.

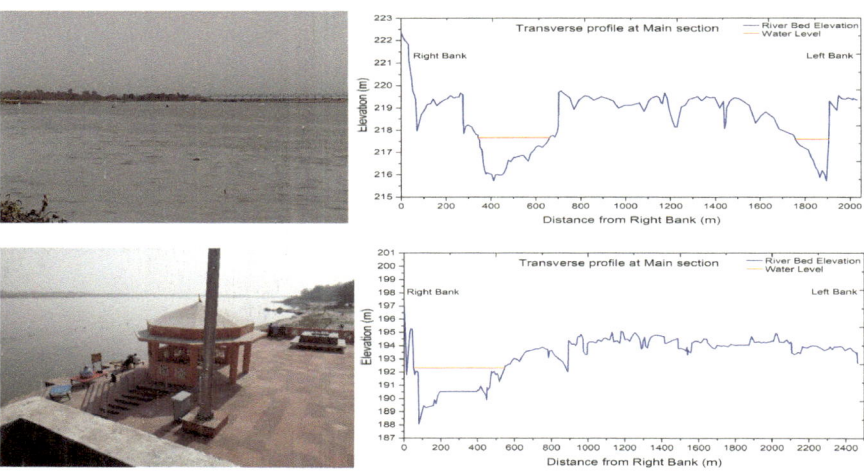

Fig. 8.4 Elevation profile from DGPS: CS-1 at Madhya Ganga Barrage and CS-2 Narora Barrage. *Source* WWF-India Report 2019 for IWMI

Table 8.9 Summary of hydraulic parameters at Madhya Ganga and Narora Barrage

Site details	Cross-sectional Point	Top width (m)	Average depth (m)	Average velocity (m/s)
Madhya Ganga Barrage Cross Section-1	Upstream	257.89	1.38	0.38
	Main	Channel 1: 316.71 Channel 2: 148.02	Channel 1: 0.96 Channel 2: 0.85	
	Downstream	248.32	1.38	
Narora Barrage Cross Section-2	Upstream	407.37	1.55	0.54
	Main	495.26	1.34	
	Downstream	388.56	1.54	

Source Cross-sectional survey of Ganga River at Bijnor and Narora, WWF-India, 2019

Since Lower Ganga irrigation canal off-taking from Narora Barrage receives water perennially from Tehri Dam/Bhimgoda Barrage, and supplemented by Kalagarh Dam on Ramganga through Ramganga feeder Canal meeting Ganga at Brijghat (Moradabad), the actual challenge is ensuring E-Flows downstream Narora Barrage. Secondly, River Ganga is losing stream between Narora to Allahabad particularly during non-monsoon period; hence, the above shown minimum E-flows seem to be inadequate. WWF-India has assessed E-flows at Prayagraj as 225 m^3/s during Kumbh on normal days and 310 m^3/s on main bathing dates for the presence of millions of pilgrims (WWF-India 2013).

Alternatively, if we improve water-use efficiency (WUE) in canal command area, huge amount of water may be freed from agriculture and can fill up the deficiency. An exercise in this regard has been done with Upper Ganga Canal (UGC) system and Lower Ganga Canal (LGC) system, which are depicted in Table 8.12.

Augmentation of flows at downstream of Bhimgoda Barrage for maintenance of E-Flows can also be achieved by having additional releases from the upstream reservoir (Tehri Dam). However, the long-term solution lies in enhancing water-use efficiency, thereby reducing the irrigation demand, leading to reduced withdrawals at the head of the irrigation systems.

Change in Crop Pattern

The potential of crop diversification for E-Flows' maintenance has been considered looking at marginal reduction in area under some of the water-intensive crops while suggesting less water-intensive crops, but with good economic gains.

Table 8.10 E-flows assessment downstream Balawali Barrage on River Ganga and gap with present flows (average and 90% dependable)

Balawali (all values are in cumec, i.e. m^3/s)

| | Avg present flow | 90% dependable flow | For minimum ecological requirement | | | For E-flows | | | |
			Minimum ecological requirement (recommended)	Gap towards Avg present flow	Gap towards 90% dependable flow	E-flows (recommended)	Monthly values in MCM	Gap towards Avg present flow	Gap towards 90% dependable flow
Jan (1–10)	60.46	13.93	28.00	32.46	−14.07	57.9	120.5	2.58	−43.95
Jan (11–20)	60.31	7.70	28.00	32.31	−20.30	32.0		28.31	−24.30
Jan (21–31)	51.19	11.96	28.00	23.19	−16.04	49.7		1.50	−37.73
Feb (1–10)	47.76	7.99	28.00	19.76	−20.01	33.2	110.0	14.58	25.19
Feb (11–20)	70.11	11.33	28.00	42.11	−16.67	47.1		23.05	−35.73
Feb (21–28)	60.34	11.36	28.00	32.34	−16.64	47.2		13.15	−35.83
Mar (1–10)	61.22	6.74	28.00	33.22	−21.26	28.0	111.3	33.22	−21.26
Mar (11–20)	52.22	12.96	28.00	24.22	−15.04	53.8		−1.61	−40.87
Mar (21–31)	60.03	11.33	28.00	32.03	−16.67	47.1		12.97	−35.73
Apr (1–10)	48.80	11.33	28.00	20.80	−16.67	47.1	163.4	1.74	−35.73
Apr (11–20)	70.12	17.05	28.00	42.12	−10.95	70.8		−0.73	−53.79
Apr (21–30)	101.64	17.18	28.00	73.64	−10.82	71.4		30.28	−54.19
May (1–10)	133.91	26.47	28.00	105.91	−1.53	110.0	683.2	23.93	−83.51
May (11–20)	215.37	42.44	28.00	187.37	14.44	176.3		39.07	−133.87

(continued)

Table 8.10 (continued)

Balawali (all values are in cumec, i.e. m^3/s)

	Avg present flow	90% dependable flow	For minimum ecological requirement			For E-flows			
			Minimum ecological requirement (recommended)	Gap towards Avg present flow	Gap towards 90% dependable flow	E-flows (recommended)	Monthly values in MCM	Gap towards Avg present flow	Gap towards 90% dependable flow
May (21–30)	298.33	121.56	28.00	270.33	93.56	505.0		−206.71	−383.48
Jun (1–10)	363.12	126.06	53.00	310.12	73.06	523.7	1488.2	−160.63	−397.69
Jun (11–20)	453.31	112.24	53.00	400.31	59.24	466.3		−13.01	−354.08
Jun (21–30)	725.21	365.37	53.00	672.21	312.37	733.8		−8.54	−368.39
Jul (1–10)	1072.40	489.74	146.93	925.47	342.81	927.2	4033.2	145.16	−437.49
Jul (11–20)	1489.48	736.87	146.93	1342.55	589.94	1451.2		38.27	−714.34
Jul (21–31)	1937.87	1134.74	138.39	1799.48	996.35	2293.2		−355.35	−1158.48
Aug (1–10)	2362.55	1243.81	240.86	2121.70	1002.95	2561.3	6312.8	−198.74	−1317.49
Aug (11–20)	2410.94	1188.78	240.86	2170.09	947.92	2436.9		−25.97	−1248.13
Aug (21–31)	2138.11	1136.15	223.78	1914.33	912.37	2313.9		−175.82	−1177.78
Sep (1–10)	1829.91	874.04	240.86	1589.06	633.18	1725.5	2973.2	104.43	−851.45
Sep (11–20)	1403.63	549.79	146.93	1256.70	402.86	1054.5		349.09	−504.76
Sep (21–30)	930.85	365.51	146.93	783.93	218.58	663.8		267.04	−298.31
Oct (1–10)	545.70	323.38	124.43	421.28	198.95	574.5	1378.0	−28.78	−251.11
Oct (11–20)	457.55	156.31	28.00	429.55	128.31	649.4		−191.87	−493.11
Oct (21–31)	367.07	89.59	28.00	339.07	61.59	372.2		−5.13	−282.62

(continued)

Table 8.10 (continued)

Balawali (all values are in cumec, i.e. m^3/s)

| | Avg present flow | 90% dependable flow | For minimum ecological requirement | | | For E-flows | | | |
			Minimum ecological requirement (recommended)	Gap towards Avg present flow	Gap towards 90% dependable flow	E-flows (recommended)	Monthly values in MCM	Gap towards Avg present flow	Gap towards 90% dependable flow
Nov (1–10)	229.44	41.97	28.00	201.44	13.97	174.4	266.1	55.07	−132.40
Nov (11 20)	113.55	15.23	28.00	85.55	−12.77	63.3		50.29	−48.03
Nov (21–30)	62.22	16.99	28.00	34.22	−11.01	70.6		−8.37	−53.60
Dec (1–10)	62.28	16.99	28.00	34.28	−11.01	70.6	148.5	−8.31	−53.60
Dec (11–20)	76.43	11.40	28.00	48.43	−16.60	47.4		29.08	−35.96
Dec (21–30)	68.85	13.01	28.00	40.85	−14.99	54.0		14.82	−41.03

Table 8.11 E-flows' assessment downstream Narora Barrage on River Ganga and gap with present flows (average and 90% dependable) Narora (all values are in cumec i.e. m³/s)

			For minimum ecological requirement			For E-flows				
	Avg present flow	90% dependable flow	Minimum ecological requirement (recommended)	Gap towards Avg present flow	Gap towards 90% dependable flow	E-flows (recommended)	Monthly values in MCM	Gap towards Avg present flow	Gap towards 90% dependable flow
Jan (1–10)	15.99	3.96	19.00	−3.01	−15.04	19.0	49.2	−3.01	−15.04
Jan (11–20)	15.50	3.96	19.00	−3.50	−15.04	19.0		−3.50	−15.04
Jan (21–31)	9.04	3.96	19.00	−9.96	−15.04	19.0		−9.96	−15.04
Feb (1–10)	4.98	3.96	19.00	−14.02	−15.04	19.0	49.2	−14.02	−15.04
Feb (11–20)	7.15	3.96	19.00	−11.85	−15.04	19.0		−11.85	−15.04
Feb (21–28)	11.18	3.96	19.00	−7.82	−15.04	19.0		−7.82	−15.04
Mar (1–10)	8.79	3.96	19.00	−10.21	−15.04	19.0	73.1	−10.21	−15.04
Mar (11–20)	16.24	5.22	19.00	−2.76	−13.78	25.0		−8.76	−19.78
Mar (21–31)	61.03	8.50	19.00	42.03	−10.50	40.7		20.31	−32.22
Apr (1–10)	15.79	8.50	19.00	−3.21	−10.50	40.7	105.5	−24.92	−32.22
Apr (11–20)	23.23	8.50	19.00	4.23	−10.50	40.7		−17.49	−32.22
Apr (21–30)	35.64	8.50	19.00	16.64	−10.50	40.7		−5.07	−32.22
May (1–10)	51.43	8.50	19.00	32.43	−10.50	40.7	105.5	10.71	−32.22
May (11–20)	51.27	8.50	19.00	32.27	−10.50	40.7		10.55	−32.22

(continued)

Table 8.11 (continued)

Narora (all values are in cumec i.e. m³/s)

	Avg present flow	90% dependable flow	For minimum ecological requirement			For E-flows			
			Minimum ecological requirement (recommended)	Gap towards Avg present flow	Gap towards 90% dependable flow	E-flows (recommended)	Monthly values in MCM	Gap towards Avg present flow	Gap towards 90% dependable flow
May (21–30)	63.91	8.50	19.00	44.91	−10.50	40.7		23.20	−32.22
Jun (1–10)	140.65	8.20	31.00	109.65	−22.80	39.3	294.6	101.36	−31.09
Jun (11–20)	376.84	14.80	31.00	345.84	−16.20	70.9		305.93	−56.12
Jun (21–30)	678.85	151.73	31.00	647.85	120.73	231.0		447.81	−79.31
Jul (1–10)	1292.45	502.43	31.00	1261.45	471.43	765.1	3460.3	527.37	−262.65
Jul (11–20)	1820.62	652.63	150.00	1670.62	502.63	960.8		859.84	−308.15
Jul (21–31)	2623.63	1479.05	247.36	2376.26	1231.69	2282.2		341.43	−803.15
Aug (1–10)	3202.67	1244.07	388.00	2814.67	856.07	1899.6	5408.2	1303.10	−655.51
Aug (11–20)	3297.02	1305.67	388.00	2909.02	917.67	2007.2		1289.83	−701.51
Aug (21–31)	2960.56	1510.91	355.55	2605.01	1155.36	2357.6		602.99	−846.66
Sep (1–10)	2610.53	1294.76	269.00	2341.53	1025.76	1981.2	2605.6	629.29	−686.48
Sep (11–20)	2159.30	473.19	269.00	1890.30	204.19	622.8		1536.47	−149.63
Sep (21–30)	1039.28	306.78	150.00	889.28	156.78	414.0		625.24	−107.27
Oct (1–10)	490.64	68.71	139.20	351.44	−70.49	37.7	158.0	452.93	31.01
Oct (11–20)	252.65	21.83	19.00	233.65	2.83	104.6		148.01	−82.80
Oct (21–31)	213.68	8.50	19.00	194.68	−10.50	40.7		172.96	−32.22

(continued)

Table 8.11 (continued)

Narora (all values are in cumec i.e. m³/s)

	Avg present flow	90% dependable flow	For minimum ecological requirement			For E-flows			
			Minimum ecological requirement (recommended)	Gap towards Avg present flow	Gap towards 90% dependable flow	E-flows (recommended)	Monthly values in MCM	Gap towards Avg present flow	Gap towards 90% dependable flow
Nov (1–10)	119.24	8.50	19.00	100.24	−10.50	40.7	105.5	78.53	−32.22
Nov (11–20)	56.99	8.50	19.00	37.99	−10.50	40.7		16.27	−32.22
Nov (21–30)	14.76	8.50	19.00	−4.24	−10.50	40.7		−25.95	−32.22
Dec (1–10)	12.64	8.50	19.00	−6.36	−10.50	40.7	77.3	−28.07	−32.22
Dec (11–20)	10.79	6.23	19.00	−8.21	−12.77	29.9		−19.07	−23.63
Dec (21–30)	7.07	3.96	19.00	−11.93	−15.04	19.0		−11.93	−15.04

Source IIT Kanpur for WWF-India

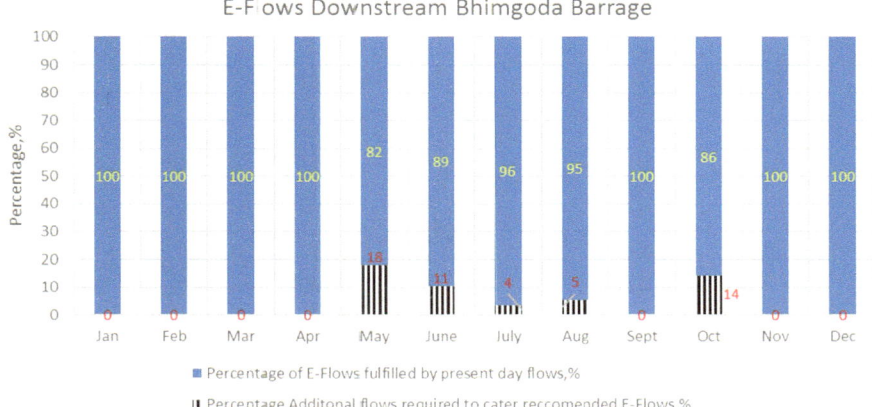

Fig. 8.5 E-Flows d/s Bhimgoda Barrage. *Source* WWF-India Report (2019)

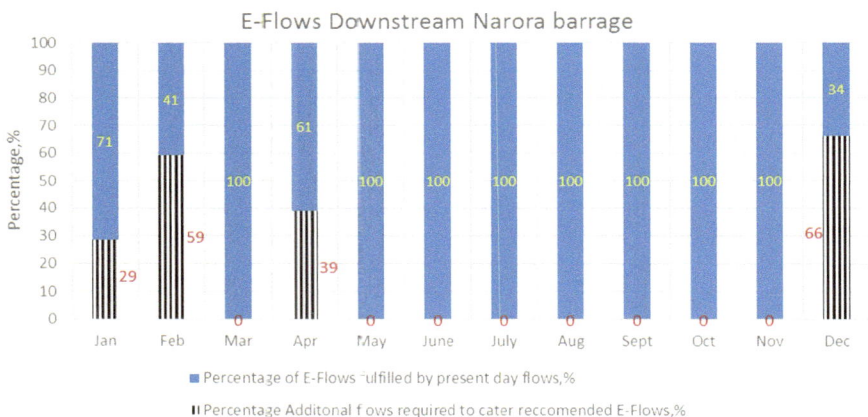

Fig. 8.6 E-flows d/s Narora Barrage

Agriculture Income Gain

The annual value gain for the farmer is estimated in terms of the changes in farm income by calculating the difference in total value of crop production from the business-as-usual scenario to the other water management schemes proposed due to increased crop yields and with increase in cropping intensity to 189%.

Table 8.12 Different WUE scenarios for both UGC and LGC systems for the fulfilment of E-flows in Ganga

	Bhimgoda Barrage			Narora Barrage			
Scenarios	Description	Whether E-flows would be achieved	Percentage of E-flows gap fulfilled (%)	Scenarios	Description	Whether E-flows would be achieved	Percentage of E-flows gap fulfilled (%)
	BAU	No	–		BAU	No	–
A	Water-use efficiency enhanced by 5%	No	35	A	Water-use efficiency enhanced by 1%	No	46
B	Water-use efficiency enhanced by 10%	No	70	B	Water-use efficiency enhanced by 3%	Yes	137
C	Water-use efficiency enhanced by 15%	Yes	104	C	Water-use efficiency enhanced by 5%	Yes	228
D	Water-use efficiency enhanced by 20%	Yes	139	D	Water-use efficiency enhanced by 20%	Yes	910

Source WWF-India Report, 2019

Institutional Efforts

Under Uttar Pradesh Water Structuring Project Phase-I (in Ghaghra-Gomti Basin) and ongoing Phase-II (in LGC command in Ganga Basin), there are various reforms underway to streamline the situation by setting up water users associations (WUAs) through passing and enforcing necessary legislations and executive orders. There is a growing debate within the formal circles that by extending necessary services to farmers, i.e. Soil-Health Card and pressure irrigation (drip and sprinkler), water-use efficiency can be enhanced. And it can effectively be implemented through the involvement of WUAs.

Policy Implications

Since irrigation is the biggest user of Ganga waters, therefore, a much more concerted effort is required to 'free-up' water from irrigation in order to maintain environmental flows in River Ganga. A regulated groundwater use in protected areas will go a long way in supporting the cause of restoration and conservation of the River Ganga through base-flow augmentation, especially during dry season flows. The challenge of implementing environmental flows in Ganga Basin is: (a) enhancing flows in the rivers by demand and supply management in farming which can be achieved by implementing better water management practices in agriculture and irrigation activities; (b) Changed Operation Rule of barrages and dams in alignment with the E-Flows' requirements; (c) recycle and reuse of wastewater generated and (d) groundwater use limited to 90% of annual replenishment. Apart from regulation of groundwater through policies, acts and supply interventions, institutional instrumentation should also be considered for rejuvenation of wetlands, ponds, etc., and water conservation and rainwater harvesting. It would provide myriad of ecosystem services and most importantly recharging the aquifers.

Good impact of lockdown to curb COVID-19 (imposed by India since 24 March 2020) on water quality of River Ganga as reported by CPCB suggests that Ganga is self-sufficient to cleanse herself if human intervention is limited to natural limits. Therefore, it provides institutional mechanism for coordination involving the states, local bodies and elements of civil society regarding rejuvenation and conservation of mother Ganga.

Conclusions and Recommendations

In a tropical country like India, the mismatch in demand and supply of water can be regulated with infrastructural improvement and efficient water use in Ganga Basin. Most importantly, it is noteworthy that demand-side management of water is equally important over supply-side solutions to maintain a balance. The research underlines that for main-stem Ganga Basin falling in the state of Uttar Pradesh, future water demand in 2044–45 for projected population of 210 million is expected to increase by three times as compared to 68.60 million population in base year 2015. The estimations highlight that the water demand in urban domestic areas will grow 1.60 times from existing 978 to 1588 MCM. Similarly, the water demand for rural domestic purposes will rise 2.57 times, from existing 841 MCM to 2157 MCM; for livestock, it will enhance 1.46 times, from 422 MCM to 617 MCM; and for industrial purposes increase recorded is 3.61 times, which is from 71 MCM to 255 MCM. Likewise, the demand of water for power plants will also accelerate 1.49 times, from existing 319 MCM to 474 MCM. Thus, in a nutshell, total non-agricultural water demand shall increase 2.12 times, from 1652 MCM to 3503 MCM. For cropping pattern under business as usual scenario (i.e. cropping trend analysed for increase/decrease shall remain same in future also), the irrigation demand will increase 1.13 times from 29,263 MCM to 33,101 MCM. Therefore, total water demand in U.P. Ganga Basin shall increase 1.13 times from 31,893 MCM to 38,192 MCM, whereas the stage of groundwater extraction shall increase from present 65.91 to 81.02%, which is quite a dismal figure. Thus, total unmet demand of water in the existing situation shall increase 1.15 times from 8671 MCM to 10,015 MCM, which is quite huge in nature and may create water problem. The gap in demand and supply could decrease from 27.18 to 26.22% only.

If we look at the water resource augmentation, reuse of generated wastewater in urban centres appears as the first most viable alternative. The wastewater production is estimated as 861.86 million litres per day (MLD) by the year 2045 in U.P. Ganga Basin. For treatment of wastewater of major towns having population of 20,000 or more, the existing treatment capacity is 443.50 MLD, while 183.95 MLD capacities are under construction. Sewage treatment plants for rest of wastewater treatment of 234.41 MLD in Ganga Basin are to be constructed by UP Jal Nigam department under Clean Ganga Mission or under any other scheme to meet out the water requirement. The approximate present cost of STPs for fulfilment of additional treatment requirement will be INR 710.26 crore. It will provide 314.89 MCM additional water to be considered for irrigation purposes. Further, 840.89 MCM additional water can be made available with an expenditure of 5906.11 crore through watershed development, wetland development and rooftop rainwater harvesting activities. Thus, additional generated volume of water amounting to 1155.78 MCM yearly will reduce the water shortage by 11.54%, with area-specific additional expenditure in Ganga Basin of Uttar Pradesh.

Way Forward

There are an opportunity and challenge for implementing environmental flows in Ganga River Basin. To address the issue, three-pronged approach is considered in Draft Ganga Basin Plan from 2025 to 2045. These are: (a) enhancing flows in the rivers by demand and supply management in farming which can be achieved by implementing better water management practices in agriculture and irrigation activities; (b) Changed Operation Rule of barrages and dams in alignment with the E-Flows' requirements; and (c) recycle and reuse of wastewater generated and groundwater use limited to 90% of annual replenishment. Along the line, apart from regulation of groundwater through policies, acts and supply interventions, institutional instrumentation should also be considered for rejuvenation of wetlands, ponds, etc. It would provide myriad of ecosystem services and most importantly recharging the aquifers.

Since irrigation is the biggest user of Ganga waters, a much more concerted effort is required to 'free-up' water from irrigation in order to maintain E-Flows in River Ganga. At times, the phenomenon of surface water–groundwater interaction is natural one, but a regulated groundwater use in protected areas will go a long way in supporting the cause of restoration and conservation of River Ganga.

References

Boulton, A. J., et al. (1998). The functional significance of the hyporheic zone in streams and rivers. *Annual Review of Ecology and Systematic, 29*, 59–81.

Central Pollution Control Board, MEF & CC, GoI (2020) Impact of Lockdown on Water Quality of river Ganga (24 March to 15 April, 2020), PDF 293_1587978571_mediaph. time-sofindia.indiatimes.com/india/cleaner ganga-cpcb validates-improvement in water quality. Sees bigger change. Accessed on May 20, 2020.

Central Water Commission, Government of India. (2015). Study Report "Environmental Flow Assessment for Himalayan Ganga" by Hydrological Studies Organisation. Hydrology (NE) Directorate, SEVA BHAWAN, New Delhi.

Ganga River Basin Management Plan. (2015). Mission 1: *Aviral Dhara* (Continuous Flows) (pp. 23, 24). The document can be accessed at: http://52.7.188.233/sites/default/files/Mission%201_AD.pdf. Accessed on Feb 20, 2020.

Government of India. Namami Gange Programme-At a Glance. (2018). Publication of NMCG, MOWR, RD and GR now known as Jal Shakti Ministry of GoI, Delhi.

Kar, J. R., & Chu, E. W. (1998). *Restoring life in running waters: Better biological monitoring.* Washington, DC: Island Press.

Leopold, L. B., & Wolman, M. G. (1957). *River channel patterns: Braided, meandering, and straight.* USGS, Paper 282-B, Washington DC.

Ministry of Water Resources, RD & GR, GoI. (2018). Gazette notification, Oct 9, 2018. New Delhi.

National Mission for Clean Ganga and cGanga. (2018). *Flow regimes and environmental flows in Ganga River system.* New Delhi: MOWR, RD & GR, GOI.

Nature, J. (2018). https://www.nature.com/articles/s41598-018-30246-7. Accessed on Feb 20, 2020.

State Water Resources Agency (SWARA), Government of Uttar Pradesh. (2020). Draft Ganga Basin Plan, prepared by TAHAL Consulting Engineers Ltd., Israel/India and submitted in March, 2020.

WWF India. (2012). Assessment of E-flows for Upper Ganga River Basin, 162 page + CD. New Delhi. Printed by Thomson Press, New Delhi.

WWF India. (2013). Environmental flows for Kumbh 2013 at Triveni Sangam, Allahabad. HSBC Water Programme, Printed by Impress, New Delhi.

WWF-India Report-2019: Cross-sectional Survey- Bijnor to Narora, by IIT Kanpur.

Chapter 9
Locating Agricultural Distress in India: Realigning for Sustainability and Nutritional Security

Purba Chattopadhyay

Abstract Contemporary India is a major agricultural producer and exporter. The green revolution in the 70s ensured food self-sufficiency. There has been a fourfold increase in grain yields per hectare, leading to 3.7 times increase in output since then. Presently, we are contemplating the second generation of green revolution. Towards this end, specialists of sustainable agriculture seek to assimilate three main goals into their work: a healthy environment, economic profitability, and socio-economic equity. However, a close scrutiny of facts reveals that entire discussion of sustainable agriculture is relying on the farmers, a large percentage of whom are undernourished, uneducated, barely clothed, mentally broken or in short, whose own existence is unsustainable. Added to this, climate change is tightening its grip and threatening food productivity. The present paper attempts to analyse the existing situation in two parts, wherein the first part, the consumption and production trends of agricultural produce, is examined on the basis of available secondary data. In the second part, the paper deals with primary data from a cross-sectional survey of aspirations of the next generation of farmers in West Bengal. The paper engages with the issues like household dependency on agriculture, operable land possession, income from farming, productive investment, consumption expenditure, household debt condition, outcome versus satisfaction among farmers, and occupational continuance of the next generation from farmer households. The paper concludes by pointing out the gaps in understanding of sustainable agriculture where a cybernetic negotiation exists between the growers, food processors, distributors, retailers, consumers, and waste managers.

Keywords Agriculture · Farmers · Nutrition Security · Aspirations · Sustainability

P. Chattopadhyay (✉)
Department of Home Science, University of Calcutta, Kolkata, India
e-mail: purba25cu@gmail.com

© The Author(s), under exclusive license to Springer Nature Singapore Pte Ltd. 2021
M. K. Verma (ed.), *Environment, Development and Sustainability in India: Perspectives,
Issues and Alternatives*, https://doi.org/10.1007/978-981-33-6248-2_9

149

Introduction

India is one of the biggest economies of the world classified either through Purchasing Power Parity (PPP) or by nominal Gross Domestic Product (GDP). Besides, it is one of the most populous nations in the world, coming just after China. Farming or agriculture represents about 23% of its GDP and utilizing about sixty per cent of the nation's functional workforce. The enormity by its size in terms of both engagement and area under acreage vis-à-vis standing of the agriculture in the lives of the population can be gauged from the fact that about 70% of Indian families rely essentially upon it for their job and sustenance. With absolute foodgrain yield running to billions of tons, India is the biggest agrarian producer, importer and exporter in totality, on planet earth. India simultaneously also has the largest output of milk and jute, along with cradling world's second-biggest cattle populace. India holds the distinction of being the second highest producer of basic grains like wheat, rice, pulses and groundnut as well as other crops like, cotton, sugarcane, fruits and vegetables. The Green Revolution of 1960s showcased how mechanical advancement and the appropriate ventures and interventions could translate into increased efficiency and into productivity. This significant feat was achieved through intensive and innovative farming. By and large, upgrades in agricultural innovation including the advancement of 'high yielding variety' of improved seeds, chemical manures alongside rationalized administration and effective farm management techniques resulted in augmented agrarian productivity in India, in billion tons from early 1960s to 2000 (World Bank 2007; FAO 2011). Stimulation of the agro-dependent rural economy of the country cumulatively resulted in significant poverty reducing multiplier effects and malnutrition rates (Dalila and Dewbre 2010).

The composition of rural labour market in India reveals that more than 70% are agriculture-related workers, both cultivators and hired workers. The entire system thus banks heavily on the farmers who actually are instrumental to the future growth trajectory of agriculture. However, as per the survey conducted by National Sample Survey Organization (GOI 2005), 40% of the Indian farmers are willing to leave agriculture given a chance. Studies in the agrarian and rural sectors, in all the major continents of the world, have exhibited that youth's attitude towards farming is mostly negative (World Bank 2007). The pull factors of the urban service sector and the push factors of the rural primary sector have contributed towards formation of the present attitudes among the youth of the country. The disinclination and disenchantment with agrarian employment also revolve around profoundly disempowered position embedded in the agrarian structure and relations that limit youngsters' admittance to land and work—ultimately resulting in exodus of youth to urban areas (Sumberg et al. 2012). The matter has also been accentuated by the contemporary system of education that induces surreal material and consummatory aspirations among the youth. Diverse studies have come to conclude that education in general and secondary education in particular, as is practised in our country, resulted in 'deskilling' of rural youth, and dying out of sustainable farming practices. Consequentially, it has been observed that the outlook and approach towards learning of farming skills and engagement in agrarian occupation have become negative, and farming as an occupation is socio-culturally looked down upon (White 2012). Studies across states of India (Sharma 2007) show that there is a general decline in the agrarian population of the country. The trend is stronger in regions where there is a low value for agriculture production (Verma and Gupta 2015: 97–110) and in villages nearer to towns. Demographic profile indicates that this trend is stronger in youths who are better educated and those with nonfarm skills. This is supported by the findings of the sixth economic commission which found that the growth in nonfarm sector is outpacing the growth in the farming sector (Sharma and Bhaduri 2009). Studies further reveal that youth perceive farming as boring and stressing and as a profession which requires hard physical labour (Cotton et al. 2009). The biggest challenges of India in the coming years will be to retain its youth in agriculture (Swaminathan and Sambasivan 2011).

However, the rural sector of India depicts that for every thousand persons, 576 are employed in agriculture (Government of India 2012), which underlines the importance of agriculture, in the Indian context. Needless to say, being the second most populous nation housing the most populous democracy, the issue of food security for the people is the biggest challenge for achieving the ambitious goals of development of the nation. A deficit in the food scenario will negatively influence the Indian interests of becoming a key global powerhouse in all sectors, especially in political and economic terms. Moreover, India has the dubious distinction for accounting one-fourth of the world's under-fed masses since it is home of over 190 million malnourished people. More than 30% of people living in India can be categorized as extremely poor. As regards the data on stunting and wasting of under five children the prime indicator of malnourishment, India's performance has been dismal (Global Nutrition Report 2016).

Under these circumstances, there is no beating about the bush in understanding the primal importance of agriculture in shaping the future of the nation. Studies have revealed that there is an imminent need for innovations, diversification and significant increase in productivity so that the small farms do not become unsustainable or economically unviable (Jayne et al. 2010). This can counter uncertain and non-guaranteed returns, overcome the complications due to climate change, streamline market accessibility and replace segmented government policies with focused and contextual framework making the prospects of agriculture a profitable and remunerative undertaking.

Keeping the above-mentioned premise in view, here an attempt is being made to analyse the production pattern and consumption trends of agricultural produce on the basis of available secondary data. An exploration of viability and sustainability of the Indian agriculture would be attempted next. Based on it, an attempt will be made to locate the position of Indian farmers in terms of changing agricultural scenario. Finally, with the data of a primary cross-sectional survey of aspirations of the next generation of farmers in West Bengal, a critical engagement will be undertaken to examine the situation of farmers, their aspirations from agriculture as a means of livelihood and achievement of life goals.

Indian Agriculture Sector: Present Trends in Production and Consumption

Agriculture is regarded as the backbone of Indian Economy. The composition of the folks engaged in farming and allied activities is not homogenous in nature. There are farmers who are very rich to farmers who could barely maintain the minimum subsistence. Significantly, 82% of Indian farmers could be classified under small and marginal. Nevertheless, going by the production trends, it is observed that India's foodgrain production reached a stupendous 275.68 million tonnes in the year 2016–17. Going by the world rankings in volumes of productivity, India ranks second in producing wheat, rice, sugarcane, cotton and groundnuts. The total produce of fruits and vegetables comprises eleven per cent of the total world productivity. Agriculture-based exports constitute 20% of the total exports of the country (Economic Survey 2016–17). Rice and wheat are two traditionally grown main crops in the Rabi and Kharif[1] cropping seasons in India. As regards wheat and rice, post-independence, the productivity of these two stable crops in India has experienced a steady increase, especially after the introduction of the first wave of Green Revolution in 1960s. These crops are generally monsoon dependent and India had been witnessing a steady monsoon in recent past, which had contributed to the rise in productivity of these crops. Further, considering Jowar, which is the staple food of low-income groups and an essential food crop for cattle, as well as raw materials for some industries; it is seen that yield per hectare has increased despite the fact that Jowar cultivating land has been steadily falling. With regard to pulses, India contributes to one-fourth of the total produce of pulses worldwide and consumes about a matching amount, i.e. 27% of the total world production of pulses. Regarding fruits and vegetables, India's feet has been stupendous where it comes next only to China, the largest producer. However, the consumption pattern shows that there are interregional and spatial disparities in the intake of fruits and vegetable, mainly due to the prevalent socio-economic disparities. There are also other agricultural produces, like basmati rice, fruits, vegetables, cocoa, coffee, etc., that earn crucial foreign exchange. It can also be seen that increased production and sales of tractors are observed in the past decade, which of course is an indirect indicator of farm mechanization.

Regarding consumption, secondary data reveal wide disparities within the various socio-economic classes in India. Data show that topmost 5 percentile of the urban population spent about nine times more than that of the bottom 5 percentile of poorest of the rural population. The consumption pattern further reveals that the food composition basket of the rich is diverse and high-end than that of the poor. The rich can afford varieties of fresh fruit, milk, vegetables and pulses, which are considered a luxury for the poor. The food composition basket of the poorest lot contains items with low nutritional content like coarse pulses, Jowar, etc. The disparities in the aggregate consumption patterns are also evident from the rural–urban divide where the composition and quality of ingredients of the food basket of an average household in rural areas invariably lag behind their urban counterpart. Interestingly, this diversity in consumption patterns in India with wide inter- and intra-regional disparities is also undergoing a steady change. Rapid urbanization, rise in income and globalization have led to major transformation and vagaries in the lifestyle of average Indian household, having significant shift on the type of food being consumed. The emerging chains of supermarkets are stimulating their interest towards non-traditional food items. The picture is further complicated by the paucity of time leading to consumption of ready-to-eat or easy to cook food items, evident from the rise in sales of 'instant' or pre-cooked food items in India (FAOSTAT 2014; NSSO 2014; NCAER 2017; Economic Survey 2017–2018; Puri and Misra 2017).

We can reasonably then derive from above that the issue of food production and consumption is complex, multi-layered and dialectical in nature. It is multifarious as the production on the one hand is challenged by ever-growing concerns like climatic changes, resource depletion and overcrowding where upon on the other hand we have technological advancement leading to more productivity per hectare of land and increased land use efficiency. Further, we see that the demand and consumption trends and pattern of top layers of the socio-economic spectrum are very different from the bottom layers of the society.

The Issue of Sustainability

However, coming to the very basics of the economic functioning of the society, we find that resources are limited and these limited resources have to face the additional pressures of an ever-growing enormous population. India has millions of people living in poverty and a substantial percentage (about 80%) of them are residing in the rural areas. Of these people, most are either directly or indirectly dependent upon agriculture. Thus, efficient, meaningful and sustainable farming practices are vital for the agrarian sector to maintain its viability. This is extremely pertinent for India, if, on the macro-economic plane, she has to see more of her citizens coming out of the cyclical trap of poverty. It is seen that since independence, the number of farms has steadily declined over the years. On the other hand, agriculture has metamorphosed, in its both form and technique. Rise in productivity in food and fibre has been witnessed as a direct consequence of farm mechanization, advanced irrigation techniques, introduction of new technologies and increased use of chemical fertilizers. These changes have assured increased productivity, as well as the risks and uncertainties involved in agriculture because climate dependence and impact of calamities have been significantly mitigated. These undoubtedly have encouraging impact, and India surely needs more of it to root out poverty and secure food provisions for the population. However, the biggest hindrance to adoption of advanced farming techniques comes from its high costs, both direct and indirect, in nature. Procuring advanced technology, good seeds, chemicals, fertilizers, irrigation implements and facilities, electricity and mechanized organization by farmers requires direct-initial investments by them, vis-à-vis their lack of access to capital. The indirect cost comes in long run to upgrade, which as a consequence of over-exploitation of land and use of chemical fertilizer results in increased soil salinity, topsoil depletion and groundwater contamination.

Herein comes the question of sustainability! One needs to interrogate the socio-ethical core principle of fulfilling the wants of the present without exacerbating the limited ability of generations to come, to meet their own needs. For agriculture to be sustainable, it has to translate down to meeting society's needs for food, fodder, raw material in present times without exhausting and destroying the common-pool resources that would take away the ability of upcoming generations to meet such needs for themselves, again in a way integrated into their environment. Agriculture can be sustainable only if it seeks to integrate economic rationality with socio-economic equity and justice in a well-conserved and regenerating environment. Every individual involved in this complex system of food production—like the tillers and cultivators, the food processors, distributors and retailers, consumers, and waste managers—undertakes his role and job with the commitment towards guaranteeing to synergistically uphold a sustainable agricultural system that commensurate with the ecology of an area or region. Here, efficient economic organization *of both bio-geophysical resources and human resources* has key implications upon agriculture to flourish for it helps integrate well with the cultural patterns of a region having ecological knowledge. More precisely, ***human resources here*** include the sustenance and better conditions of agricultural labourers and cultivators at work as well as in their daily lives.

Looking at the big picture, it is evident that the agriculture system cannot be dismembered from the nature and natural resources. Thus, it is the functioning of this system within the economy which has far-reaching consequences upon productivity as well as human resources and environment. The main task is to seek a balance between humans with their natural environment, and also from the perspective of individual farm, to the indigenous ecosystem, to societies affected by the farming system, both locally and globally.

Sustainable agriculture is not only a collection of good and ethical practices, but also it is a dynamic process of arbitration between the competing interests of an individual farmer and of people in a larger community, while being the components of the same system. For these to work together synergistically means to resolve multifaceted issues involving in the process by which one grows food and fibre at the same time ensuring better living and working conditions for farm labourers. Thus, sustainability of agriculture requires farming in synchronization with resource base of the local or regional ecosystem. The basic motto of nurturing the ecology rather than exploiting it gets compromised when excessive mechanized tilling is done along with use of chemical fertilizers that makes the soil devoid of nutrients and in long run becomes arid. Sustainable farming requires convergence with the community so that farming is efficient enough to ensure the well-being of those who are carrying out the farm job. This can be achieved if and only if there are synergistic relationship of the farmer with the key players of the demand makers like the consumers, the retailors and other people involved, without forcing either one to fit some predefined norm defined for productivity. Thus, sustainable farming requires that farmers to reconnect with the community and vice versa. It is only through a two way causal relationship that agriculture can be sustainable. Invariably therefore, the issue of sustainability of the agriculture rests on its key player, i.e. the farmers, about whose condition we turn to discuss in the following section.

Locating the Indian Farmer

In this section, let us attempt to locate the Indian farmer in context of the socio-economic rubric of India and contextualize their multi-layered interaction with the local ecological system and consequent outcomes. Taking a closer look at the farmers or those involved directly with the growing crops reveals that they are not a homogenous group as they are highly segregated and widely dispersed in nature.

It is to be noticed here that though India has the distinction of being the country with second largest amount of arable land, yet 65% of her farming is rain dependent with only 35% of cultivable land being well irrigated. This goes on to show the heightened priority of ecological conservation and need for investment in creation of irrigational infrastructure. The rapidly degrading environment inducing fragility in the ecosystem has resulted in deep-rooted consequences on the poverty-stricken farmers due to their bondage to often erratic and elusive monsoon rains for irrigation, further pushing up their vulnerability. It is needless to reiterate that a common outcome of global warming and deforestation is flood and abnormal rains or drought or below average rainfalls. Both agricultural productivity and output, as well as the surety of cash flow and income from agriculture, suffer due to it.

To gauge the condition, one may do well to factually understand that the median annual wage of an average farmer in India dismally corresponds to approximately about sixty days of minimum wage a labourer earns in the city Mumbai. Meagre irrigation facilities and absent or inappropriate weather insurance schemes force farmers to sow only low-value crops, such as rice and wheat, instead of blue-chip, high-value fruits and vegetables making them more susceptible to poverty. At the macroeconomic level, it can be observed that per capita agricultural income is not only low, but also is volatile and unstable and at times even falls down to non-viable levels. Going by the agricultural output, it can be seen that it swings more often than the outputs of the industrial and service sectors. This has its consequence on income distribution with rural–urban wage gap rising to a dismal 45%. Moreover, there is lack of an operative government policy to regulate the food prices at which the agricultural producers buy inputs and the price at which they sell their output in the market. The unevenness of this price has serious repercussion on the socio-economic justice that an average farmer deserves. In India, if farmers want to sell their crop, they have only two formal options at their disposal. Abiding by the first option, the person has to sell directly to the government. The government in turn offers the Minimum Support Price (MSP). This is mainly done by government agencies such as Food Corporation of India (FCI). MSP is the lowest price assured for a given produce standardized by the government. It is implemented by making payments to food growers or farmers in case the market price of the produce falls below a definite quantifiable minimum. However, there is a problem with MSP. The trickiness lies in the fact that although MSP is generally higher than the market price, yet only a minuscule percentage of farmers have access to avail the facility. To be precise, there are very few villages which have a FCI window for facilitating purchase of the produce. Further obscurities arise when, even in presence of FCI, the government may not buy the produce at MSP if the farmers bring their harvest on some other date which may either fall before or after the notified procurement dates. The next best option left for the farmer is to carry their output to the nearby (state authorized) *mandi*[2] where in front of government regulatory officers they can auction or put their produce on sale to the brokers. There are about only 7500 such government-designated *mandis* spread across the length and breadth of the country, which by any conservative estimate is inadequate simply going by the size of the Indian agricultural sector and the population involved in it (Banik 2018). This restrictive age-old rickety structure of marketing of agricultural produce has led to exclusionary trends which the farmers at the lower strata fail to overcome and get within the available state-aided security net. The unfortunate offshoot of the overstretched structure is the ever-rising herd of intermediary broker-traders, and non-credible informal agencies that skew the market and deny the farmers of getting a just-income from their produce. Due to their lack of education and capital savings which could have allowed them to explore alternatives to get best price in the market, farmers most often than not fall in the trap of such intermediaries and informal agencies to sell off their perishable crops and minimize loss, despite knowing that they are being unfairly paid.

The farmer knows that availability of storage facilities for the crops is very little in comparison with requirement, and most of them are in a dilapidated state and are expensive. Since more than 80% of them are small landholders[3] so they can hardly hope to have the access or can afford to keep their produce in storage facilities. Even in this field, the intermediary broker-traders rule the roost. Inappropriate and outdated policies still guide the framework marketing practices of agricultural produce hindering the small farmer-producers to sell directly to exporters or supermarket chains or agro-processors, deriving better price and income. Rationality behind disallowing contract farming for keeping the country food secure, however, is not counter-balanced by policies that would ensure better price recognition for these farmers of their produce. The complex situation becomes more difficult and flagrant for the farmers due to the continued dearth of access to formal institutions of credit. It is no wonder then that the unscrupulous informal money lenders, despite their long legacy woe-adding capacity, continue to get the advantage of patronage of desperate-farmers.

The situation of the small farmers in India could be better assessed from the fact that in acute desperation and destitution these farmers are left with no option but to commit suicide. In fact, even within the farmers, a broad dualism is observed in India, where on one hand there are a small percentage of well-to-do farmers having command over land, labour, capital and human resources. These farmers use mechanized farming and have efficient and profit-making landholding. They have access to information and adequate risk mitigation insurance coverage. However, there is a much larger portion of farmers who have no or little access to the resources. And when the question of sustenance arises, it is these farmers at the bottom-most echelons of the socioeconomic framework on whose frail shoulders lies the burden of meeting the food needs of the gigantic national population. Insensitivity and thoughtlessness towards the agrarian sector of the country also root on the all-pervasive dualism in our society too, where the urban, educated, employed masses have a secured income and a set demand pattern, which is different from that of the poor and deprived classes located in the periphery. The consumption basket and pattern of those in periphery depend largely upon the produce of the small and marginal farmers consisting of subsistence items like rice, wheat or Jowar.

Recent studies, as cited before, reveal that the disillusioned farmers do not want to continue themselves as well their future generations in agriculture (Economic Survey 2015–2016; Banik 2017). Broadly speaking, we thus observe that with changed social norms and economic aspirations of the people on the one hand, and indifferent returns with growing environmental uncertainty on the other, the young farmer children are aware about the poor and insecure condition of this sector. The younger generation from the families of farmers would prefer to move away from agriculture and thus not invest, both materially and intellectually any further. Shift and disguised withdrawal of these skilled, knowledgeable and healthy youth from agriculture put a huge spanner in the policy trundle of the country that aims to reduce poverty and ensure food security simultaneously.

Aspirations of a Farmer: A Case Study

Keeping the above inferences drawn in view, a primary survey was undertaken in three districts of West Bengal (Burdwan, South Twenty-Four Parganas and Purulia). These districts were chosen purposively to represent farmers from all possible set-ups: like farmers working in farms with high soil salinity, farmers from fertile farms with adequate irrigation facility and farmers working in drought-prone/flood-risk areas. A total of 642 data were collected from farmers, that is, those who were between 20 and 40 years of age, and those who were more than 40 years of age, representing two generations. The mean age of the sample was 38.72 ± 11.07 years. Interestingly, despite age differences, 87% of the sample stated that given a viable alternative they would like to give up farming or agriculture. Moreover, it was observed in general that sampled farmers look upon agriculture as a lesser job, in terms of importance and status perceived to be endowed upon it by the greater community than employment in a desk-bound job, or being a teacher in school, etc. Their negative perception about their own job gets further skewed when placed in variation to the landholding status and/or close familial ties in the urban centres. For more specific inputs on these variations, a binominal-regression analysis[4] was carried out to look into the association (if any) between the aspirations levels and perceptions of those engaged directly in the farming activities. Allowing for variability within groups of farmers, the random effect model was considered appropriate, which in its mathematical form looks like:

$$\ln[x_{ij}/(1 - x_{ij})] = a_0 + a_1 * X_1^{ij} + \ldots + a_n * X_n^{ij} + u_j$$

Here, x is the likelihood of assessing the farming aspirations, a_0 is the constant value of the intercept, while the coefficients $(a_1 \ldots a_n)$ represent the change in the log-odds of the outcome variable which occurs due to change in each corresponding independent variable $(X_1^{ij} \ldots X_n^{ij})$ (this is done while we exercise control for other independent variables). Here, i and j are denoting groups and individual within the groups, respectively. The term u_j denotes the random-error term.

Here, there are two categories of the variables—*Dependent* and *Independent variables*. In this study, the farming aspirations are considered as the target variable. This is measured by the choice of career as farming or any other profession. The value of the variable is equal to 1 if any one of the other professions are chosen. Otherwise, 0 is given for stating farming as career aspiration. Thus, the variable is binary (0, 1 values) in nature. *Predictor* variables encompassed controls for demographic characteristics and other economic or environmental variables. Age was defined by two categories 20–40 and 40 and above. To measure education level which otherwise is a continuous variable, we have converted it into another categorical variable like age where illiterate farmers were the base category. The survey further categorized the factors or conditions which may have a facilitating effect on farming as a career choice, like type of farmer-small marginal or large—farming climate, irrigation facilities, alternative employment opportunities, size of the landholding, etc. To further validate the findings from the logistic regression calculated, odds ratios for some of the key variables were calculated to see the intergenerational differences in attitude towards farming as an occupation.

From the above study, it is seen that younger generation are unwilling to take up agricultural activities or farming as full-time occupation because as an occupation it offers little societal status/prestige and is perceived as dirty work. It was also observed by them that the returns or profits from farming generally get delayed and disparate. Table 9.1 shows that younger generation of farmers had significantly increased odds of having career aspirations other than farming (OR 1.98, 95% CI 1.35–2.41). Most of the samples across the geographical location, the young generation of farmers favoured urban life and perceived it as more secured and certain. The older generation of farmers also did not want their sons to get involved in agricultural activities in view of the risks on the returns and the amount of hardship it entailed. The odds against taking farming as a profession increased if the individual did not own the land (OR 2.17, 95% CI 1.07–3.33). This is an obvious consequence as without land ownership; returns from agriculture remain meagre and highly unpredictable. With increase in the years of education come greater awareness and changed outlooks. This was observed for both young people in their own context and their parent's expectations from their children. Further, the investments in education have often involved trade-offs where stable job and city life get a priority over getting engaged in agriculture. From the above table, it becomes clear that the more the educational attainment, the more are the odds (OR 2.27, OR 2.69) of choosing an alternative profession than taking up agriculture. Agriculture was found to be a feasible choice only for those who have command over land resources and could arrange material inputs required

Variables	Range	Other career aspirations of farmers
Age	20–40	1.98***
		(1.36–3.18)
	40 and above	1.00
Owner of land	Yes	1.00
	No	2.17***
		(1.07–3.33)
Education	Illiterate	1.00
	Primary level	2.27*
		(1.37–3.13)
	HS or above	2.69***
		(1.68–4.31)
Type of farmer	Large	1.00
	Small	1.74***
		(0.50–2.10)
	Marginal	2.02**
		(1.48–3.07)
Climate	Adverse	1.89***
		(1.37–2.71)
	Normal	1.00
Irrigation Facilities	Rainfed/monsoon dependent	1.80**
		(1.35–2.41)
	Available	1.00
Loan Liabilities	Yes	2.33**
		(1.35–2.91)
	No	1.00

Table 9.1 Odds ratios (OR) (95% confidence intervals for career aspirations of farmers)

$*\rho < 0.10, **\rho < 0.05, ***\rho < 0.01$ (levels of significance)

for carrying out steady agricultural activities. This is evident from the fact that the odds increased for the small (OR 1.74, 95% CI 0.05–2.10) and marginal farmers (OR 2.02, 95% CI 1.48–3.07). For many rural youths, the lack of land ownership, capital and other inputs (crucial for carrying out hands on agricultural activities) prevents them from considering agriculture as a full-time occupation. For some rural youth, farming could be considered a profitable activity only if there is necessary backup of capital for initial years, especially when the returns to agriculture are quite low at the initial phase.

In addition, to the uncertainty perceptions of farming as an occupational choice, it is further adversely affected by climatic and environmental uncertainties. Apart from natural disasters and extreme weather conditions (as cyclones, droughts and intense heatwave, hail storm), erratic change in climatic conditions is a consequence of such global warming more often causing destruction of livelihoods for many who are singularly reliant on agriculture. More and more, such cases are resulting in the need for such farmer families for shifting to alternative occupation at a different location as their primary mode of livelihood gets eliminated or destroyed. This can be observed in the (OR 1.89, 95% CI 1.37–2.71) table, which means that farmers who are facing the uncertainties of nature were more likely to give up farming in pursuance of a productive alternative. Here also difference was seen in the responses of the farmers whose lands had proper irrigation facilities in contrast to those who were dependent on the monsoon. The odds increased for farmers having to depend on the monsoon (OR 1.80, 95% CI 1.35–2.41). In the present study sample, the areas in Sundarbans and Purulia where the farmers were relatively poorer were found desperately wanting to move away to an alternative occupation from agriculture, for stability in livelihood. Further, the ever-hanging sword of loan liability acted as a great disincentive for the farmers to stay in agrarian profession (OR 2.33, 95% CI 1.35–2.91). Moreover, the survey revealed that those farmers who had loan liabilities, particularly the small farmers, were uncertain about the prospects of being able to service or guaranty return the debt liability within a reasonable timeframe.

The view that emerged from the respondents interviewed in this study was that engaging in agricultural activities, either as direct farming or involving in some form of farm labour, as not only physically strenuous but also financially unrewarding, more so for small farmers. Farming as a profession is viewed by most respondents as entailing economically insecure and socially of low esteem and prestige. However, in contrast, the urban metropolitan life is supposed to be hassle free, posh, simpler, stable and well remunerated. The arduous toil of farming combined with poor unsteady incomes makes it an unappealing prospect for many, particularly for young generation. Individuals perceived that the labour-intensive nature of work in farming, coupled with the apparent relatively low status of agriculture-based activities, is no match to stable jobs of urban factory workers or other professions particularly, in the services sectors where payments are stable and regular and can be used to acquire consumer goods and/or invested in savings.

The findings from the present study concur with the findings and conclusions of Leavy and Hossain (2014: 144) who state:

> agriculture's lack of appeal to young people reflects i) lack of effective public investment in small holder farming and the public infrastructure needed to link to markets; ii) constrained access to land and uncertain access to inputs among young people, including land fragmentation in many countries in past few decades; and iii) social change resulting from rapid increases in mass education provision but which have often resulted in a perceived decline in the status of agriculture.

Policy Implications

In addressing farmer's misery, and for improving their condition for better utilization of their potential, tactical re-thinking and grounded planning for better agricultural production and food security are required. Infusing synergy between agricultural practices in consonance with the location and diverse conditions of the farmers across the regions has become a necessity. The traditional policy responses mostly were positioned at *macro-level*, like raising minimum support prices, providing farm-loan waivers, etc., that did not build adaptive capacity of the farmers or augmented production. Such measures have not resulted therefore in providing a long-term sustainable solution to the farmers in alleviating their distress, risks and vulnerabilities. It also undermined the profitability of agrarian livelihoods by neglecting the majority marginal and small farmers or landless labourers. Moreover, shifting towards non-farm occupations have resulted in permanent drift, breaking them away from the traditional knowledge chain found suitable to a local condition of a region. Introducing changes in policymaking, prioritizing use of agriculture as a lever of growth and increasing the capabilities of farming households, it is expected to result in them staying put in agricultural farming without feeling deprived, vulnerable and humbled. Herein, policy framework needs to engage with promoting sentience in the greater society about the vital role of farmers, no less than our defence personnel protecting the sovereignty of the country. To do justice towards this end, the policymakers need to remove the blinders that force them to ignore the unique knowledge base and local challenges of farmers by heaping them into a homogenous category.

Conclusions

Following Gibson (2006), we can say that as the social, financial and ecological assets are harmed or drained with time, the communities face enormous challenges. What makes the problem more complicated is the interconnectedness of the resources—be it human resources or natural resources. Thus, there are no modest answers to the glitches that emerge in the society with its multidimensional causalities among its various argents. Based on these interdependencies and interconnectedness of the environmental, economic and social justice essentials, the present world requires systemic instead of symptomatic interventions, which will facilitate to form a future, where humans as apart of community or society, and nature can cohabit with shared benefit. Such mutually beneficial coexistence will be sustainable if poverty and human sufferings are eliminated, ensuring conservation of natural resources (Flint 2013).

The foregoing discussion, engaged with here, explains how it is important to have a nuanced understanding of agricultural distress where the farmers as the lynchpin in the gargantuan called agriculture have to be foregrounded in the system. To understand sustainable agriculture, one needs to keep in mind the cybernetic negotiation that exists between the growers of food at one end, distributors and retailers, and ultimately consumers and waste managers on another end. Herein, we need to theorize in a creative and constructive way about this interface between the significant key players. Leavy and Hossain (2014: 146) deciphered that:

four key determinants of the opportunity space in relation to agriculture within which aspirations and expectations were formulated and interpreted:

(a) the extent to which/how increased risks in the agro-food system out-weighed perceived benefits from price increases from the producer side;

(b) The extent to which land remains a significant obstacle, and where;

(c) How access to other inputs was being influenced by higher commodity prices, and whether younger smallholder farmers were seen to have access to agricultural input subsidies or supports;

(d) Experiences of being consumers; as cost of living rises spread beyond food staples and beyond food, for some, regular formal sector cash incomes seemed to become more desirable.

Thus, if we continue with the modernist tradition of intellectual representation of reality, expressed in the form of industrial slant for all developmental project, then we will fail to address the requirements of contemporary living conditions. After all, one cannot impose specialization, standardization and centralization of decision-making, as relevant for an industrial set-up, for agricultural production as well. Similarly, for example, if we continue to strategize following Smith (1887) who theorized that the equilibrium in any system is guided by a pair of invisible hands where all individuals work for their own profit or benefit, in the process a well-oiled system works out where everybody gets his/her dues. For Smith, even if there are temporary bottlenecks in the system in long run, the issues are sorted out and again the system gains equilibrium. So Smithian invisible hands can explain to a large extent the working of the economy.

Nevertheless, the problem lies in the fact that the economies at present have evolved and have become highly specialized and complicated than that of the simple economies at the dawn of industrial revolution. The system that worked on informality depending on interpersonal sharing of bygone era is now replaced by the economic system, especially in this post-globalization era that is highly complicated and dispersed. Therefore, with the issue of food production when seen through the prism of present economy, paucity of knowledge and understanding about the toil involved in food production and the social location of the person who is doing the job becomes clear. The picture of a farmer we have imprinted in our head is that of a utopian rural person, half-clad, stout and happy doing his job in field, for feeding us. The advent of supermarkets and malls has now changed the picture of food consumption by accentuating the disconnect of contemporary society. We as consumers remain oblivious to the plight of the farmer, how the vagaries of problem confronted by the small, marginal and large farmers vary, and thus, treat the entire group as a homogeneous lot with a 'mechanical' bonding. Elementary issue of such disconnectedness is that the broad concern of sustainability and climate change will remain in the domains of intellectual discussions with ideological polemic that are mostly deficient in empirical interrogation and internal critique and thus fail to gain sufficient authenticity to be part of policy-oriented actions. Further, in the process of this disconnect, the poorer farmers within the 'lot' will be pressurized by the system to look for other jobs perceived to more remunerative and less exploitative.

For skill-based jobs like agriculture, we need to remember that the chain of inter-generational transfer of wisdom will be broken in present circumstances, and in the process immense harm will be inflicted upon the society. In this context, things look exceedingly gloomy as India, as a country, is heavily banking on its farmers for feeding the ever-increasing sea of population, without realizing the plight in which the food-producing farmers are placed in. With high incidence of poverty, malnutrition and poor health indicators, the consequences of the stated problem become more complicated. Although persistence of poverty has been a common problem with most countries, the incidence of poverty and its changing nature in India is at an alarming state. As a developing country and as one of the leading nations of the third world, India in no way is free from the scourge of poverty. The issue of sustainability that rests on the provision of social justice gets compromised as we observe that cultivators and hired workers, as a section has to bear the brunt of feeding the masses, are laden with unreliable and non-guaranteed income, facing distressful uncertainties of climate changes, suffering deficient market accessibility and enduring piecemeal government policy. The farmers in India are denuded and disillusioned due to persistence with subsistence orientation. It is indeed frustrating that people do not understand how sustenance life in general is intricately connected to soil and its produce. The consumers on the other hand are just concerned about the prices of food, not delving deep as to how in the first place it reached there. The farmers, especially the marginalized and small farmers, are being repeatedly forced to place their worth at the economic bottom line well before the blue-chip companies and top-notch industrialists. Education, training and capability-enhancing socio-economic structural shifts can bring down the intense volatility in agricultural life and move towards economically and socially stable society. Maybe then, we can imagine to start repaying our debts to the hands that till and feed us.

Notes

1. In India, the cropping seasons are classified as—(i) Kharif and (ii) Rabi. This classification is done on the basis of monsoon. Months from July to October, which signifies the south-west monsoon is the Kharif cropping season. While the months from October to March, the predominantly winter months are the Rabi cropping season, (iii) the crops grown in the remaining months are summer crops.
2. *Mandis* are the markets or bazaar in smaller towns or cities where farmers come to sell their produce.
3. With less than 1 hectare of landholding.
4. Through the binomial logistic regression, the probability of the dichotomous-dependent variable (taking only 0 or 1 value) based on one or more independent variables (either continuous or categorical in nature) is predicted.

References

Banik, N. (2017). Farmer suicides in India and the weather god. *Procedia Computer Science, 122,* 10–16.

Banik, N. (2018). Benefit-cost analysis of farmer distress: Analysis of interventions addressing farmer distress in Rajasthan. *Woking Paper Copenhagen Consensus Centre.*

Cotton, et al. (2009). Broadening perspectives: Educating underrepresented youth about food agricultural sciences through experiential learning. *NACTA J, 53*(4), 23–29.

Dalila, G., & Dewbre, J. (2010). Economic importance of agriculture for poverty reduction. *OECD Food, agriculture and fisheries working papers*, No. 23, OECD Publishing.

Economic Survey 2015–16 Retrieved from https://www.indiabudget.gov.in/eBook_Economic_Sur vey2015/pdf/economic-survey.pdf on 8.1.2019.

Economic Survey 2017–2018. Ministry of Finance, GOI.

FAO. (2011). FAOSTAT. Food and Agriculture Organization of the United Nation.

FAO. (2014). Retrieved from http://faostat3.fao.org/download/FB/CL/E on 5.1.2019.

Flint, R. W. (2013). Basics of sustainable development. *Practice of sustainable community development* (pp. 25–54). New York, NY: Springer.

Gibson, R. B. (2006). Beyond the pillars: Sustainability assessment as a framework for effective integration of social, economic and ecological considerations in significant decision-making. *Journal of Environmental Assessment Policy and Management, 8*(03), 259–280.

GOI. (2007). *Report of the steering committee on agriculture for 11th five year plan.* New Delhi: Yojana Bhavan.

Jayne, et al. (2010). Principal challenges confronting smallholder agriculture in sub-Saharan Africa. *World Development, 38*(10), 1384–1398.

Leavy, J., & Hossain, N. (2014). Who wants to farm? Youth aspirations, opportunities and rising food prices. *IDS Working Papers, 439*(2014), 1–44.

National Sample Survey Organization GoI. (2005). Retrieved from http://www.mospi.gov.in/nat ional-sample-survey-office-nsso on 9.12.2018.

NCAER. (2017). *Kharif outlook report.* Report No. 2017-12-1.

NCAER Annual Reports 2015, 2016 and 2017 retrieved from Annual Reports http://www.ncaer.org/annual_reports.php as on 2.12.2018.

NSSO. (2014). *Household Consumption of Various Goods and Services in India 20ll-12*, NSS 68th Round, Ministry of Statistics and Programme Implementation, GOI.

Puri, V. K., & Misra, S. K. (2017). *Indian economy* (35th ed.). Mumbai: Himalaya Publishing House.

Sharma, A. (2007). The changing agricultural demography of India: Evidence from a rural youth perception survey. *International Journal of Rural Management, 3*(1), 27–41.

Sharma, A., & Bhaduri, A. (2009). The "Tipping Point" in Indian agriculture: Understanding the withdrawal of the Indian rural youth. *Asian Journal of Agriculture and Development, 6*(1362-2016-107622), 83.

Smith, A. (1887). *An inquiry into the nature and causes of the wealth of nations.* T. Nelson and Sons.

Sumberg, et al. (2012). Introduction: The young people and agriculture 'problem' in Africa. *IDS Bulletin, 43*(6), 1–8.

Swaminathan, M., & Sambasivan, (2011). *Production, income and equity—Key driver for sustainable development. Future agricultural extension.* New Delhi: Westville Publishing House.

The World Bank Annual Report. (2007). Retrieved from http://siteresources.worldbank.org/EXT ANNREP2K7/Resources/English.pdf as on 12.02.2018.

Verma, M. K., & Gupta, N. (2015). Development, food insecurity and health predicaments: A study of Bundelkhand region of Uttar Pradesh. In R. Garada (Ed.), *Communities and social development in India* (pp. 97–110). New Delhi: Avon Publication.

White, B. (2012). Agriculture and the generation problem: Rural youth, employment and the future of farming. *IDS Bulletin, 43*(6), 9–19.

Chapter 10
Environmental Migration: A Challenge for Sustainable Development

Clare Lizamit Samling

Abstract Anthropogeric factors as the cause of Climate Change and Global Warming is now being increasingly ascertained. The Intergovernmental Panel on Climate Change (IPCC) has reported alarming estimates and predictions that are expected to have greater impact on the poorest and most vulnerable people, especially those residing in the low lying island regions like the Sundarbans. Migration (voluntary or forced) of people has been identified as one of the major impact of global environmental changes. In this background, the paper attempts to focus on the aspect of human mobility due to environmental changes and the complexity that the subject arouses in the global as well as Indian scenario. Migration of people, voluntary or forced, has garnered attention in the Sustainable Development Goals (SDGs). It is one such area other than environmental change that is now being debated. However, the author underlines, India in its commitment towards Sustainable Development has far more demanding issues to tackle with, therefore, 'climate migration' is not featured high on its policy agenda. In a concluding remark, the chapter highlights that environmental migration is a burgeoning challenge for most of the developing countries, especially India, in their path towards achieving the SDGs and require solutions that are tangible as well as sustainable.

Keywords Displacement · Environment · Migration · SDGs · Sustainable development

Introduction

In recent decades, the phenomena of rising global temperatures and climatic changes have been increasingly ascertained to be a consequence of unsustainable anthropogenic activities, something that has been incessantly asserted by the Intergovernmental Panel on Climate Change (IPCC) and its much-acclaimed reports. The Panel concerns with the assessment of scientific, technical and the socio-economic data

C. L. Samling (✉)
Department of Sociology, University of Calcutta, Kolkata, India
e-mail: liza.samling@gmail.com

that are pertinent to comprehend the phenomenon of climate change. The reports also accentuate the impending impact, both ongoing and probable, and provide alternative responses. According to the IPCC Fourth Assessment Report (2007), Asia is among the list of regions that would face the major brunt of the impacts due to climate change and result in increasing their existing vulnerability (p. 11). The disruptions that are of particular concern are the increase in sea level rise, cyclones, storm surges, and riparian flooding. Such impacts would be immensely felt by those at the lower rungs of the society in the aforementioned regions (IPCC 2007: 9).

Human migration due to environmental changes is not a novel occurrence. However, it is only in the last few decades that climate change as a precursor to the emerging environmental changes, being a driver of population movements, has garnered attention. Climate change is predicted to directly impact the livelihood of the 'vulnerable' population and is further projected to undermine food security, sustained provision of fisheries productivity, intensifying competition of water resources be it surface or groundwater, due to their overall reduction and exacerbate existing health problems. Natural disasters be it rapid or slow-onset would affect the movement of people in diverse manners, oscillating between voluntary and forced. Climate change would magnify the risks of human and natural systems, a consequence of the integration of 'exposure' and 'vulnerability'. However, climate change may interact with other drivers, instigating human mobility and enhance the existing vulnerabilities of the people (IPCC 2014).

The IPCC fourth assessment report also indicated the developmental pathway adopted by both developed as well as developing countries as the plausible cause of the perceived global warming and climatic changes (IPCC 2007). Such a course of development was anticipated to be detrimental to man and environment in the long-run. To ameliorate this situation of impending doom, the member states of the United nations espoused the course towards responsible development with the signing of the Sustainable Development Goals (SDG) in 2015 (the Millennium Development Goals was signed in 2000, acting as a predecessor to SDG). This pathway mandates 17 goals and their subsequent targets to these member states that are to be achieved by the year 2030.

In the above context, the paper thus focuses on the subject of environmental migration and the debate that revolves around it. It further attempts to analyse the relevance of such migration within the context of Sustainable Development Goals (SDGs). Although human migration is a subject of focus in the SDGs, yet the question of 'environmental migration' is somewhat bypassed.

Climate Change and Environmental Migration: The Debate

The International Organization for Migration (IOM), defines 'environmental migrants' as *'persons or groups of persons who, for compelling reasons of sudden or progressive change in the environment that adversely affects their lives or living conditions, are obliged to leave their habitual homes, or choose to do so, either temporarily or permanently, and who move either within their country or abroad'* (IOM 2007: 1–2). This definition has become extremely debatable due to its varied definitions and the lack of consensus among scholars. To begin with, the issue arises with the terminology itself, whether the section of population who have migrated (forced or voluntary) due to environmental events should be labelled as 'environmental migrants', 'climate migrants', 'ecological migrants', 'environmentally-displaced people' or 'environmental refugees'.

Renaud et al. (2007) classified the environmental migrants as 'environmental emergency migrants', 'environmentally forced migrants' and 'environmentally motivated migrants', depending on the nature of the environmental event, duration of stay, option for choice or voluntary and the movement taking a national and international character. Barnett and Weber's (2010) categorization of migration that may be instigated by the vagaries of climate change include international labour migrants, internal labour migrants, internal displacement, international displacement, internal permanent migrants, international permanent migrants and relocation of communities. The end three groups experience significant risk to their lives and livelihood as these groups would involve a common process of permanent migration (pp. 18–19).

The nexus between migration and environment is convoluted. Considering the multi-causal nature of migration, isolating environment as the principle factor is a challenge. This is problematic due to the interconnection of socio-cultural, economic, political, demographic and environmental processes that influence the decision-making process of an individual (Foresight 2011: 44–45). Certain rapid onset disasters like hurricanes provide people with very little choice, but to move. Slow-onset environmental fluctuations allow movement to be more voluntary and interconnected with other economic, social and political factors. A problem emerges in delineating the dichotomy between forced and voluntary migration which are often referred to as two discrete categories. The movement in reality often oscillates between the two polars of voluntary and forced. Therefore, clear dichotomies between such human mobility are not always so simple. Thus, it is often portrayed as a continuum (IOM 2009:18; Foresight 2011: 41). The underlying difference between migration and displacement would be of 'choice', where in many situations these 'choices' of 'forced' and 'voluntary' intermix.

The concept of 'Environmental Refugees was first introduced by Lester Brown of the World Watch Institute in the 1970s (Black 2001: 1). But the concept was made popular by the work of El-Hinnawi (1985) who in the report for United Nations Environment Program (UNEP) has defined them as people *'who have been forced to leave their traditional habitat, temporarily or permanently, because of a marked environmental disruption (natural and/or triggered by people) that jeopardized their existence and/or seriously affected the quality of their life. By "environmental disruption" is meant any physical, chemical and/or biological changes in ecosystem'* (El Hinnawi 1985: 4).

Bates (2002), criticizing the vagueness of the typology provided by Hinnawi, categorizes the 'environmental refugees' grounded on three criteria—the factors that begets environmental disruptions, planned or unplanned migration as their ramifications and their duration. The first typology is the 'disaster refugees'—originating in acute events caused by natural or anthropogenic events that cause unplanned short-term migration producing either temporary or permanent refugees. The other is the 'expropriation refugees'—due to acute human activities like either for economic development or warfare, that purposefully dislocate people and usually permanently. Lastly, the 'deterioration refugees' are due to gradual anthropogenic events causing 'unintentional or unplanned migration'. Employing the classification of El Hinnawi (1985), Stojanov (2004) has classified the environmental disruptions into five categories, viz. natural disasters, progressive (slow-onset) changes, involuntarily caused accidents including industrial, developmental projects, hostility and warfare (pp. 78–79).

These contradictions in terminology and definition have largely emerged out of the problem in isolating climatic changes as the sole factor of migration from other drivers. Further, the label 'environmental refugees' has received considerable criticisms. The United Nations Commissioner for Refugees (UNHCR, n.d.) defines 'Refugees' as *'owing to well-founded fear of being persecuted for reasons of race, religion, nationality, membership of a particular social group or political opinion, is outside the country of his nationality and is unable or, owing to such fear, is unwilling to avail himself of the protection of that country; or who, not having a nationality and being outside the country of his former habitual residence as a result of such events, is unable or, owing to such fear, is unwilling to return to it'* (p.14). The term 'refugees' ignores the role of environmental disasters and degradation and only addresses international migrants and not internal migrants.

Black (2001) regards 'environmental refugees', as defined in the past studies, as more of a myth than a reality. According to the author, people migrating due to changes in climate is not a novelty which should be addressed differently. Among the variegated reactions to environmental change, migration may be just one of them, that too often interacting with the other drivers. In the same vein, Johnson (2009) observed that there is an inextricable link between 'environmental refugees' and 'economic migrants'. The author also agrees with UNHCR's stance that people who migrate due to economic reasons are more of 'migrants' than 'refugees', as an 'economic migrant' would voluntarily leave a country and would receive the protection of its government, in situation of his return. But the refugees have no scope of returning to their original homeland. Inclusion of these refugees into the international refugee cohort and expansion of the definition of UNHCR on refugees would *'blur the responsibilities of national governments and distract the international community from the root causes for environmentally forced migration'* (Johnson 2009: 245). According to the author, such inclusion would only work as a provisional solution to the problem rather than accosting the underlying cause. This is contradictory to other studies which demand the inclusion of such refugees into the international regime (Johnson 2009).

The terminology and definition which is now gaining ground is the 'Internally Displaced Persons' who are any *'persons or groups of persons who have been forced or obliged to flee or to leave their homes or places of habitual residence, in particular as a result of or in order to avoid the effects of armed conflict, situations of generalized violence, violations of human rights or natural or human-made disasters, and who have not crossed an internationally recognized State border'* (UNHCR 1998, para. 17). The people displaced due to environmental changes, within respective nations, are assumed to fall under the ambit of this definition. Furthermore, the problem in definition has led to variations in estimations of flows of environmental migration. These issues create problems in methodology, especially in quantifying the movements of people especially in the developing countries (ADB 2012: 10).

Climate Change and Environmental Migration: The Nexus

Giddens in his seminal work, 'The consequences of modernity' (1990) states that the *'world in which we live today is a fraught and dangerous one'* (Giddens 1990: 19). Although not discounting, the 'opportunity side' of modernity, Giddens also sheds light on what he refers to as the 'darker side of modernity'. Classical theorists on modernity in the likes of Marx, Durkheim and Weber, have overlooked the fact that the furthering economy would have large-scale destructive potential in relation to the material environment.

Foresight (2011) identifies three ecological regions play a significant role in this interaction between factors that induces migration and environmental change—low-elevation coastal zones, global dry lands and mountainous regions (p. 25). Further the interaction between climate change and population movement is influenced in different manner through escalation of rapid and slow-onset natural disasters, the pernicious repercussions of escalating and capriciousness nature of environmental changes on livelihoods, health, food security and availability of water, increasing sea level transmuting coastal areas into unlivable regions and contention over dwindling natural resources possibly germinating antagonism and, in further displacement (IOM 2009: 58). The contrasting repercussions of environmental stressors due to the variability in the vulnerability of the population tend to sprout different migration sequence on a regional and global level, considering similar environmental threats (Renaud et al. 2011: 15).

Myers (1993) in his study has made a projection of what he terms as 'environmental refugees' for the year 2050, using the baseline of 1990 as provided by the IPCC. The projected total number of environmental refugees by 2050 stands at 150 million people, of which China would have 30 million refugees, India-30 million, Bangladesh-15 million, Egypt-14 million, other delta regions and coastal areas-10 million, island countries-1 million and agriculturally dislocated areas-50 million. That is 1.5% of the total 10 billion world projection for 2050. Stern (2006) approximated a total of 200 million that would be 'climate refugees' by 2050. The United Nations Framework Convention on Climate Change in 2007 predicted an approximate number of 50 million environmentally displaced people by 2010. Bierman and Boas (2007) calculated the displaced population as 250 million by 2050. In 2011, the Internal Displacement Monitoring Centre (IDMC), reported that between the years 2009–2011, about 55.7 million have been displaced due to the dual forces of rapid-onset natural disasters and acute weather events, in Asia alone.

In the context of India, the country according to the Global Climate Risk Index of 2018 is the 12th most vulnerable country to climate change impacts (Eckstein et al. 2017). Myers (2001) analysed the current situation of selected countries of Bangladesh, India, Egypt, China and the Island States, their factors of endangerment in the epoch of climate change and their future projected scenarios of consequences. In the case of India, which out of a total coastal population of 180 million people in 1993, about 80 million inhabiting 120,000 km^2 of land, likely to be affected by rising sea levels. It has been specifically mentioned that they would be affected but not necessarily displaced. The question can be raised that if the place becomes inhabitable, then how would displacement not occur? The projected affected population is at 30 million by 2050. The flood zone refugee could be at 20 million to 60 million. In the Bay of Bengal region, inclusive of the states of West Bengal and Orissa, it is predicted that the impact would be felt on not less than 2000 km^2, and risking roughly about 1 million lives. It is opined that 07 large deltas in India are at risk. The cities of Calcutta, Madras and Bombay have also been featured as vulnerable and would equally contribute to producing refugees (Myers 2001).

The Energy and Resources Institute (TERI) in its 2009 report has demarcated 'Climate Change hotspots' bordering India. These areas include low lying deltaic regions which are susceptible to flash floods, like Bangladesh, Nepal, Bhutan, Tibet [China], prone to glacial lake outbursts like Nepal, Bhutan, Tibet [China], areas with volatile precipitation patterns in places of Nepal, Myanmar that results in loss of income source and areas susceptible to rising sea level like Sri Lanka, Myanmar, Bangladesh (TERI 2009: 3). These can produce migration trends of either long or short-term period, which could presumably lead to either an abrupt or a gradual inflow of migrants. Considering the flow of population within the country itself, the North-East and the Himalayan regions along with the flood plains of the major river basins of India would experience rapid movements (TERI 2009).

The IPCC, since its initial reports have projected a surge in the displacement of people. The reports reiterate the notion of the risk exposure to human and natural systems due to the impact and the character that climate change assumes. This risk materializes due to the interaction between the climate-related hazards, 'exposure' and 'vulnerability'. This is further enhanced by an interconnected but varied range of non-climatic factors (IPCC 2014). These factors amalgamated with multidimensional inequalities, a product of intersecting social processes like discrimination on the basis of gender, age, class, ethnicity, disability and caste in countries like India, results in differential but interrelated elements of 'exposure' and 'vulnerability' (Brauch 2011). This furthers the shaping of differential risks from climate change. Therefore, making it plausible that people who are socially, economically, culturally, politically or institutionally marginalized or at the lower rungs of the society, are more susceptible or 'vulnerable' to climate change. It is envisaged that the prevailing known risks be augmented by climate change but further generating new ones for both man and nature. The interconnectedness of climatic and non-climatic factors makes it arduous to determine an immediate interconnection between climate change and migration.

The IPCC in its Fourth Assessment Report (2007), has proclaimed that the regions of Asia and Africa will face the major brunt of climate change due to their higher vulnerability (p. 11). Furthermore, the shortfall in resources that assist in planned migration in the aforementioned regions results in prodigious subjection to extreme weather events.

The decision whether or not to pursue migration is generally dependent on the household, a pivotal driver to indemnify meagre earnings and livelihood. The environmental changes intervene with the socio-economic attributes and produces varied forms, extent and patterns of migration. It therefore sometimes also creates situations where the vulnerability of the people is heightened when they are unable to adopt a course of voluntary migration, especially those who are already susceptible to the vagaries of such changes. These are what is recognized as 'trapped population'. In contrast, migration can also act as source of supplementary earnings through remittances that may facilitate steadiness and resilience of the migrants and their households in their places of origin, thereby opting to remain behind (Foresight 2011: 13).

On June 2008, the Prime Minister's Council on Climate Change, Government of India, publicized the National Action Plan on Climate Change (NAPCC), as an initiative on the part of the country to combat the globally challenging phenomenon. The plan is to be executed through eight missions, visualized to be the mainspring of India's pathway to achieve development that is sustainable, both economically as well as environmentally. These national missions with long-term, integrated sustainable strategies included the sectors of solar, other forms of energy, habitat, water, Himalayan ecosystem, greening India, agriculture and strategic knowledge on Climate Change (GoI 2008). It was suggested that these missions would be implemented through various ministries working along with inter-sectoral groups and experts from various fields. Each of the above missions has stated their own objectives detailing further their timeline and strategies, which are supposed to be periodically monitored, evaluated and reported (GoI 2008). It is to be noted that only in the eight mission—'National Mission for strategic knowledge on Climate Change'— migration has been mentioned. The mission states in its agenda that research would be undertaken to comprehend the socio-economic ramifications of climate change on migration patterns apart from other domains like health, demography and livelihoods, specifically of the coastal communities. However, the implementation of any of the missions (apart from the National Solar Mission) has been sluggish or for that matter non-existent.

Environmental Migration and Sustainable Development

In 2015, heads of about 193 member states of the UN including, India pledged to transform the world by adopting '2030 Agenda for Sustainable Development', which is now operational. There are 17 Sustainable Development Goals (SDGs) and related 169 targets marked, to be achieved in the next 15 years i.e., 2030 (UNDP, n.d). Migration was a missing component in the erstwhile Millennium Development Goals (MDGs) of 2002 and so during the preparation of the Agenda there were increasing demands on its inclusion into the SDGs. The report 'Realizing the future we want for all' of the United Nations System Task Force recognized migration as one of the core enablers of development.

The Agenda 2030 with the principle of 'leave no man behind', declares that *'we will cooperate internationally to ensure safe, orderly and regular migration involving full respect for human rights and the humane treatment of migrants regardless of migration status, of refugees and of displaced persons'* (UN 2015: 8). The SDGs included a comprehensive category of migrants that included refugees, forced labour (especially child) and human trafficking. Migration has been specifically focused on Goals. 08—'Decent work and economic growth' and 10—'Reducing inequalities', and minutely in 07 other goals (Goals 3, 4, 5, 11, 13, 16 & 17). However, the 'migrants' that are focused in the SDGs appears to be taken as a composite whole and environmental migrant is not explicit.

The Bertelsmann Stiftung and the Sustainable Development Solutions Network (SDSN) (since 2016) produces an annual 'Sustainable Development Index' (SDI). According to the SDI of 2018, in terms of the overall achievement of the SDGs, India ranks 112 out of 157 countries, with a global index score of 59.1, a rise from 116th position in 2017. A quick analysis of the index shows that in terms of achieving Goal. 1—'No poverty' and Goal. 8—'Decent work and economic growth', India is 'on track', which means that the country has achieved much success; it is 'moderately increasing' in achieving Goal. 2—'Zero hunger', Goal. 3—'Good health and well-being', Goal. 6—'Clean water and sanitation', Goal. 9—'Industry, innovation and infrastructure' and Goal. 14—'Life below water' and 'stagnating' in terms of Goal. 5—'Gender equality', Goal. 7—'Affordable and clean energy', Goal. 11—'Sustainable cities and communities', Goal. 15—'Life on land', Goal. 16—'Peace, justice and strong institutions', Goal. 17—'Partnership for the goals'. Climate change has been dealt with in Goal. 13—'Climate action'. The index states that India is 'maintaining achievement' under this goal. The goal has set targets that largely pertain to 'adaptation' and 'mitigation' to climate change which is more generalized (Thomann 2016).

As a part of an initiative to assess or review the enactment of SDGs in across countries, the United Nations created a 'United Nations High-level Political Forum on Sustainable Development' (HLPF) in 2012, stipulated in the 'The future we want' document conceived during the United Nations Conference on Sustainable Development—Rio+20. The 2030 Agenda emboldens member nations to *'conduct regular and inclusive reviews of progress at the national and sub-national levels, which are country-led and country-driven'* (UN 2015: 33), which are however voluntary and thus named as the Voluntary National Reviews (VNRs). These VNRs would form the essence of the appraisal undertaken by the HLPF. The reviews should be country-specific, embarked upon by all countries despite the ranking in development, providing a foothold for partnerships and contributions by appropriate stakeholders.

In 2017, India along with other 39 countries volunteered to take part in the VNRs of the HLPF, where the emphasis was on 'Eradicating poverty and promoting prosperity in a changing world' (UNHPLF 2017, p. ix). Poverty, assumed as the root of all problems, where its obliteration is expected to overcome all deterrents in the pathway towards the achievement of sustainable development. This is emphasized as a national and an international agenda. India's VNR discloses making headway towards the achievement of Goals 1, 2, 3, 5, 9, 14 and 17 (UNHPLF 2017: x). It is claimed that the new economic policy of structural adjustment and liberalization undertaken post-1991 by India has stimulated faster economic improvements leading to the curtailment of poverty in varied sectors, to a certain extent. This unrelenting progress (calculated at 6.2% from 1993–1994 to 20032004 and 8.3% from 2004–2005 to 2011–2012) has produced lucrative employment and facilitated augmentation of wages therefore, empowering the poor. These new reforms also augmented proceeds for the government, aiding it to expend a greater portion on social sectors and therefore, multiplying the direct benefits of progress on poverty eradication. This has been made further possible by the enactments of livelihood schemes such as the Mahatma Gandhi National Rural Employment Guarantee Scheme (MGNREGA) (UNHPLF 2017: x).

Similar other progress has been made in achieving food security, promoting health and well-being, building resilient infrastructure, gender equality, conserving marine resources and revitalizing global partnership for achieving sustainable development (UNHPLF 2017: x). It is obvious that India has far more pressing issues and priorities to deal with for the timely achievement of the SDGs by 2030 that environmental migration, unfortunately does not fit in its agenda. It is therefore presumed that the advancement of one goal would percolate in the other goals and result in the simultaneous evolvement of other goals as well. Migration has potential benefits in terms of remittances on 'sending areas' which would act as a stimulant to poverty reduction. But the situation of the environmental migrants has not changed much despite numerous initiatives undertaken and policies formulated across globe to resolve their problems.

Policy Implication

Climate change may not be the sole driver of environmental migration (be it voluntary or forced) of people but it enhances the existing vulnerabilities (of the non-climatic kind) of the said population, working alongside multidimensional inequalities brought about by a variety of intersecting social processes like discrimination on the basis of gender, age, class, ethnicity, disability and caste in countries like India and most likely instigating some new ones. This causes differential but interrelated elements of 'exposure' and 'vulnerability'. Climate change can intersect with other 'conventional' drivers of migration and may further exacerbate the vulnerabilities. It is therefore essential to perceive the movement of people along the continuum of voluntary and forced. The policymakers should take into consideration while formulating policies, the fact that generally migration decision-making occurs at the household level, working alongside numerous socio-economic characteristics resulting in creating diverse situations, that allows people to decide on whether they should stay or migrate. It should especially examine the implications that such a formulation should have on people who do not choose to leave their place of origin. The remittances from migration may act as a source of resilience for the migrants and their households but the contrasting research (even though minute in number) that brings to the forefront the reverse impact that migration can pose on the environment, usually in a negative manner, needs also to be taken into account.

If viewed along the continuum of forced, environmental displacement is now considered to be one of the most probable impacts of climate change. The phenomenon is a burgeoning challenge for most of the developing countries, especially India. Although migration of people, voluntary or forced, has garnered attention in the SDGs but in the country's path towards achieving the SDGs, it is stated that climate change impacts can act as a hindrance to achieving these goals. India is still lagging on the thrust towards climate change and lacks effective implementation of climate action plans, where there appears to be a gap between policy and actual practice. Laws and policies like NAPCC, Disaster Managements Act(s) etc. that seem relevant for environmental displacement fail to take it under its ambit (even in the case of 'voluntary' migration). Thus, India requires solutions that are tangible as well as sustainable.

The strategy of the government in the pathway towards the achievement of the sustainable development goals, of obliterating poverty and relying on the trickle down-effect of such effects to other goals may be short-sighted. Economic progress is expected to decline in the long-run as an impact of climate change, making the eradication of poverty a strenuous task. Rather creating a predicament where the newly generated poverty encumbrances are affixed to the extant ones (IPCC 2014, p. 20). Dialogues in this respect should be enabled at varied levels of regional, national and international level as future migration trends would transcend national boundaries. The elucidation and employment of the perception of environmental migration is disposed exclusively at the discretion of the respective state policy creators, without any international guidelines, support nor synergy.

Conclusion

The problem is onerous, if not impossible to solve. The literature on the nexus between migration and environmental changes are still relatively few, and further their results are inconclusive. Human migration is of multi-causality nature. Climate change is increasingly being ascertained as a precursor of migration (whether forced or voluntary) that moves beyond the 'traditional' push, pull and intervening factors. The phenomenon of environmental migration still requires acknowledgement and inclusion in government policies and programmes, especially in India. Migration is now being rendered as a mitigation and 'adaptation' optional response to climate change.

Displacement, no matter the cause, threatens the social fabric of integration of society. It can be argued that development induced displacement and most recently even resettlement and rehabilitation of these victims are in a way being protected by the National and State Acts and Rules (even if only on paper). The question of environmental-induced displacement is left on the sidelines. The need of the hour is to develop strong and effective policies. However, the complexity of the problem of environmental-induced migration and displacement the debate in its terminology, definitions, predictions and intervening climatic and non-climatic factors, furthermore makes it extremely problematic to satisfy this need. Worldwide experience has shown that most people who undergo forced relocation be it due to natural or anthropogenic factors, are in a way ineffectual in wholesome restoration, resulting rather in more deplorable conditions than prior to displacement. The dismal state of displacement and resettlement in India is well known and dealt with. The world is now altering its pathway in the case of environmental displacement. Instead of debating on the terminology, a novel concept of 'Climate induced human mobility' is coming to the limelight. 'Planned relocation' or 'Planned resettlement' is now being considered as a viable option for the environmental displaced populations. Such relocations could be planned under Goal. 11—'Sustainable cities and communities', which aims at creating settlements that, are of course sustainable. It is extremely difficult to prioritize one goal over the other, in the case of SDGs; but at the same time there is also the need for India to at least dwell on the issues in a changing climatic scenario, so that the country can truly achieve sustainable development.

India, along with other member states of the United Nations, hopes to accomplish the SDGs by the year 2030. However, eradication of poverty, among others has received considerable attention, that focus on such instances of migration or displacement is minimal or ignored. Economic repercussion of climate change further paints a gloomy picture. Poverty reduction is expected to be extremely difficult due to the sluggishness of economic growth and creating new 'poverty traps', in the climate change era.

References

ADB. (2012). *Addressing climate change and migration in Asia and Pacific: Final report*. Philippines: Asian Development Bank. https://www.adb.org/sites/default/files/publication/29662/addressing-climate-change-migration.pdf.

Barnett, J., & Webber, M. (2010). Accommodating migration to promote adaptation to climate change. Policy research working paper 5270. Background paper to the 2010 World Development Report. The World Bank. http://documents1.worldbank.org/curated/en/765111468326385012/pdf/WPS5270.pdf.

Bates, D. C. (2002). Environmental refugees? Classifying human migrations caused by environmental change. *Population and Environment, 23*(5), 465–477.

Bierman, F., & Boas, I. (2007). Preparing for a Warmer world: Towards a global governance system to protect climate refugees. *Global Environmental Politics, 10*(1), 60–88.

Black, R. (2001). Environmental refugees: Myth or reality?. New Issues in Refugee Research, Working Paper No. 34. https://www.unhcr.org/research/working/3ae6a0d00/environmental-ref ugees-myth-reality-richard-black.html.

Brauch, H. G. (2011). Concepts of security threats, challenges, vulnerabilities and risks. In H. G. Brauch et al. (eds.), *Coping with global environmental change, disasters and security, hexagon series on human and environmental security and peace 5* (pp. 61–106). Berlin: Springer. https:// doi.org/10.1007/978-3-642-17776-7_2.

Eckstein, D., Kunzel, V., & Schafer, L. (2017). *Global climate risk index 2018.* Berlin: German Watch. http://re.indiaenvironmentportal.org.in/files/file/Global%20Climate%20Risk% 20Index%202018.pdf.

El-Hinnawi, E. (1985). *Environmental refugees.* Nairobi, Kenya: United Nations Environmental Programme.

Foresight (2011). *Migration and global environmental change: Future challenges and opportuni- ties.* Final project report. The Government office for science, London. https://assets.publishing. service.gov.uk/government/uploads/system/uploads/attachment_data/file/287717/11-1116-mig ration-and-global-environmental-change.pdf.

Giddens, A. (1990). *The consequences of modernity.* UK: Polity Press.

GoI. (2008). *National action plan on climate change, prime minister's council on climate change.* Government of India. http://moef.gov.in/wp-content/uploads/2018/04/Pg0152.pdf.

IOM. (2007). *Discussion note: Migration and the environment.* Ninety-Fourth Session, MC/INF/288. https://www.iom.int/jahia/webdav/site/myjahiasite/shared/shared/mainsite/micros ites/IDM/workshops/evolving_global_economy_2728112007/MC_INF_288_EN.pdf.

IOM. (2009). Migration, environment and climate change: Assessing the evidence. Interna- tional Organization for Migration, Geneva. https://publications.iom.int/system/files/pdf/migrat ion_and_environment.pdf.

IPCC. (2007). Climate change 2007: Synthesis report. In Core Writing Team, R. K. Pachauri & A. Reisinger (Eds.), *Contribution of working groups I, II and III to the fourth assessment report of the intergovernmental panel on climate change* (104 p.). Geneva, Switzerland: IPCC. https:// www.ipcc.ch/site/assets/uploads/2018/02/ar4_syr_full_report.pdf.

IPCC. (2014). Climate change 2014: Synthesis report. In Core Writing Team, R. K. Pachauri & L. A. Meyer (Eds.), *Contribution of working groups I, II and III to the fifth assessment report of the intergovernmental panel on climate change.* Geneva, Switzerland: IPCC. https://www.ipcc. ch/site/assets/uploads/2018/02/SYR_AR5_FINAL_full.pdf.

Johnson, L. S. (2009). Environment security and environmental refugees. *Journal of Animal and Environmental Law, 1,* 222–248.

Myers, N. (1993). Environmental refugees in a globally warmed world. *Bio Science, 43*(11), 752– 761.

Myers, N. (2001). Environmental refugees: A growing phenomenon of the 21st century. *Philosoph- ical Transactions of the Royal Society of London, 357,* 609–613.

Renaud, F., Bogardi, J., Dun, O., & Warner, K. (2007). Control, adapt or flee: How to face environmental migration. *InterSecTions, 5.* Bonn, UNU-EHS.

Renaud, F. G., Dun, O., Warner, K., & Bogardi, J. (2011). A decision framework for environmentally induced migration. *International Migration, 49*(S1), 5–29.

Stern, N. (2006). *The economics of climate change: The stern review.* Cambridge, UK: Cambridge University Press.

Stojanov, R. (2004). Environmental refugees—Introduction. *Geographica, 38,* 77–84.

TERI. (2009). *Climate change induced migration and its security implications for India's neighbourhood.* New Delhi: The Energy and Resource Institute.

Thomann, L. (2016). Organizational perspectives on environmental migration. In K. Rosenow- Williams & F. Gemenne (Eds.), *Organizational perspectives on environmental migration.* Routledge.

UN. (2015). Transforming our world: The 2030 agenda for sustainable development (p. 35). A/RES/70/1. United Nations. https://www.unfpa.org/sites/default/files/resource-pdf/Resolution_A_RES_70_1_EN.pdf.

UNHCR. (1998). Guiding principles on internal displacement. UN doc. E/CN.4/1998/53/Add.2 11 February 1998. http://un-documents.net/gpid.htm.

UNHLPF. (2017). On the implementation of sustainable development goals, voluntary national review report, a report submitted to the United Nations high-level political forum. http://niti.gov.in/writereaddata/files/India%20VNR_Final.pdf.

UNHCR. n.d. Convention and protocol relating to the status of refugees. https://www.unhcr.org/3b66c2aa10.

Chapter 11
Interface Between Tribes and Ecotourism: A Study on Sustainability and Development in Purulia, West Bengal

Biswajit Paul and Ramanuj Ganguly

Abstract Purulia is located in the western most part of West Bengal, where we find social backwardness, poor economic conditions, exotic environment and various ethnic groups. Thus, it makes the locale significant and ripe for the scholars of social sciences to understand the changing Indian scenario in present times. Despite several infrastructural problems, geographical specificity and social backwardness, the state government, in this district, has been promoting ecotourism steadily as a potential method and tool to exploit the exotic environment for attracting tourists who would pump in resources to the local economy. The major challenge that the government wishes to counter is that of the Maoist activities, often threatening the democratic fabric of the state. Here does the paradox lies. Since the motive and design of the state is to promote development of the people, whereas ecotourism is based on the principles of 'environment over development'. That means environment should be given more priority while taking up any developmental policy as the natural resources and environmental conditions are limited and non-renewable. If we ignore sustainability of the environment while taking up developmental programmes, then the development will jeopardize both environment and culture. In this context, the paper attempts to study: the ecotourism project in the Baranti area, a small tribal village of Santuri block under Raghunathpur subdivision within Purulia District to understand what these projects have achieved in tangible terms, like employment, housing, assets and skills; cultural impact of the project in terms of life style practices; land holding pattern of the place; nature of shifting identity; changing nature of aspirations of the local people; and possible social mobility among the people of the area. On the basis of the primary findings, the authors interrogate the paradoxical debate between 'environment over development' and 'development over environment', and attempt to draw conclusions for policy framing.

B. Paul (✉)
Department of Sociology, Sidho-Kanho-Birsha University, Purulia, West Bengal, India
e-mail: biswajit.paul.194@gmail.com

R. Ganguly
Department of Sociology, West Bengal State University, Barasat, West Bengal, India
e-mail: rg.wbsu@gmail.com

179

Keywords Ecotourism · Development · Sustainability · Environment · Purulia

Introduction

Development is a multidimensional concept. Many classical social scientist (economist, sociologist, historians, etc.) thought that economic growth is the ultimate development. That is why the developed country used natural resources arbitrarily and make sure of their development and the control over nature by using highly advance science and technology. This approach is also called positivistic development or mainstream development where all developmental policies are taken into consideration keeping in view the pace of industrialization and urbanization. This type of development is controlled centrally and did not consider balance between local and global needs which Elster called 'justice system' (Elster 1992). Gadgil and Guha, in this backdrop, distinguished between three types of people based on their relationship with the nature and its resources. Ecosystem people are one who resided in nature, for example, forest dwellers; ecological refugee are those who are displaced for various developmental programme like dams; and omnivores, people include those who gets direct benefits of economic development. All the benefits of economic and ecological development are ideally shared among the three but the reality is quite different; all the benefits of this positivistic development or mainstream development is shared among the omnivores people which include rich land owner, entrepreneurs, urban professionals, government and semi government employees (Gadgil and Guha 2004). The ecotourism projects aims for the ecosystem people, for their social and economic development. There is a close link between environment and culture in human society thus there seems a reciprocal relation between culture and environment but if any attempt done to control the environment externally the result would be devastating. In India such situation has reached due to two very significant reasons. In the words of Sukanta K. Chaudhury, these two reasons include (i) large-scale environmental destruction due to construction of dams, heavy industries and other developmental activities and (ii) maddening consumerism (Chaudhury 2014). Thus, to avoid the confrontation between environment and culture and to create a balance, the alternative way of development is very much in discussion among intelligentsia. Policymakers are giving more importance on balance between environment and development. To address the issue in India, we have an old debate between 'development over environment' and 'environment over development'. Professor Baviskar (2014) had discuss the debate between 'development over environment' and 'environment over development' by using the intellectual genealogy from Radhakamal Mukherjee to Ramachandra Guha in the context of environmental sociology and history in India. She also indicates that how environmental sociology had get rebirth from the crisis of its early phase. By discussing this debate she indicates that the class of people plays a significant role to understand the ecological concerns and thus until and unless poverty would eradicate from our country and the educational level increases this debate will go on and it will also create problems for policymakers (Baviskar

2014). Jayram Ramesh argues for the importance of space and time framework for this debate, and point out about the tension between those who espouse economic growth, which is consider the main stream development, and the second group who talked about the environmental protection. Moreover his main focus was on the reckless exploitation of nature (Ramesh 2010). Tourism is consider as mainstream developmental activity where the economic growth of the state or country remains the main focus of discussion and all the policymakers aim for that, but ecotourism opens the alternative way of tourism as well as development approach. By implementing this type of tourism, economic growth and environment protection both can be achieved. UNEP and WTO recently reported that a range of successful outcome is achieved by implementing ecotourism in proper way. They also stated that during IYE 2002 more than 50 countries made separate ecotourism policy and if it managed in right manner then it will be helpful for sustainable development, conserve biodiversity, it will be also helpful for elevating poverty from rural areas and it will became an economic driver (UNEP and WTO 2002). In this piece, we have tried to examine the outcome of ecotourism projects implemented in Baranti Village of Purulia District in West Bengal. In recent past, not more that 5 years, 16 resorts were constructed here for the visitors of nature, these are Muradih Baranti Ecotourism (A Government of West Bengal Project), Baranti Manbhum Resort, Baranti Wildlife and Nature Study Hut, Banabithi, Retreat, Palash Bari, Aakash Mani, Lake Hill Resort, Album, Baranti Gitanjali Resort, Aranayak, Ankhai Bari, Monpalash, Spangle Wings Resort Pvt. Limited, Muhul Bon, Tourist Point. These newly established resorts have an impact of the villager's social, cultural, political and economic life, and we tried to analyse these impacts in tangible forms like employment, shifting identity, occupation, land holding patterns. By analysing this complex impact we can understand the effectiveness or constrains of ecotourism projects in this area.

Towards Defining Ecotourism

Ecotourism is a multidimensional concept. It is a nature and culture based tourism which also consider 'as an alternative to mass tourism' (Weaver 2001). World Tourism Organisation (WTO) in their report on 2001 had defined tourism by several prefixes such as descriptive terms like adventure, nature; some value-based terms like ethics and sustainability, etc. All these terms were used to describe the different type of tourism like 'Nature Tourism', 'Culture Tourism', 'Farm tourism', 'Wildlife Tourism', 'Adventure Tourism', etc. The prefix 'eco' is used for ecotourism which has several dimension like ecological, developmental, economic and most importantly sociocultural. The concept ecotourism first introduced in 1960s then it was rigorously discussed by ecologist in 1970s era and after that in 1980s various tourism researchers accepted this issue and started conducting intensive research also it became the fastest growing segment in tourism industry thus the year 2002 was declared as International Year of Ecotourism by United Nations (Björk 2000). Many scholars gave several definition of ecotourism, the Mexican ecologist Hetzer (1965) coined the term 'ecotourism' and identified 'four building blocks' of this concept:

1. "Minimum environmental impact;
2. Minimum impact on and maximum respect for host cultures;
3. Maximum economic benefits to the host countries or grassroots; and
4. Maximum recreational satisfaction to participating tourists".

The above definition gives emphasis on the preservation of environment and respect of host culture. From another perspective, Ceballos-Lascurain (1987) gave a definition of ecotourism where the issue of preservation of the nature was missing, on the contrary, argues that

> ecotourism is travelling to relatively undisturbed or uncontaminated natural areas with the specific objective of studying, admiring, and enjoying the scenery and its wild plants and animals, as well as any existing cultural manifestation found in this areas.

However, later he modified this definition by adopting the preservation dimension to ecotourism (Ceballos-Lascurain 1996). Gössling (1999), Fennell (2001) and Das (2011) have progressively defined the term ecotourism in the context of several key factors, like a responsible travel to nature, appreciate the nature, conservation of nature, local people engagement of tourism and their socio-economic development. Suchismita Das also critically examine the paradigm of ecotourism by placing it within the context and discourse of sustainable development by evaluating several central government policies, which were taken or implemented in the name of ecotourism in several places in the country. Another definition in these line was proposed by Boo (1990), where he states ecotourism can be understood as 'Travelling to relatively undisturbed and uncontaminated natural areas with the specific objective of studying, admiring and enjoying the scenery and its wild plants and animals, as well as any existing cultural manifestations (both past and present) found in these areas'.

Björk (2000) offers, by critically analysing the trajectory of emergence of the concept of ecotourism.

First, that gave emphasis on comprehensive multidimensionality of the concept, like as defined by Ziffer (1989), as follows:

> A form of tourism inspired primarily by the natural history of an area, including its indigenous cultures. The ecotourist visits relatively undeveloped areas in the spirit of appreciation, participation and sensitivity. The ecotourist practices a non-consumptive use of wildlife and natural resources and contributes to the visited area through labor or financial means aimed at directly benefiting the conservation of the site and the economic well-being of the local residents. The visit should strengthen the ecotourist's appreciation and dedication to conservation issues in general, and to the specific needs of the locale. Ecotourism also implies a management approach by the host country or region which commits itself to establishing and maintaining the sites with the participation of local residents, marketing them appropriately, enforcing regulations, and using the proceeds of the enterprise to fund the area's land management as well as community development.

Second, that gave functionality of the concept more emphasis, as likewise given by the International Ecotourism Society that stated ecotourism as:

Responsible travel to natural areas that conserves the environment and improves the well-being of local people.

However, it the International Ecotourism Society in 1993 delineated the following value principles towards understanding the concept of Ecotourism, which are:

(a) "Minimizes impact.
(b) Builds environmental and cultural awareness and respect.
(c) Provides positive experiences for both visitors and hosts.
(d) Provides direct financial benefits for conservation.
(e) Provides financial benefits and empowerment for local people.
(f) Raises sensitivity to the host country's political, environmental, and social climate.
(g) Supports international human rights and labour agreements".

While considering the dimension of nature to ecotourism one can distinguish between embedded active and passive nature of participants in the definition. The definition given by Ziffer is consider as active ecotourism where he concentrated on the protection and improvement of natural resources whereas the definition given by Ceballos-Lascurain is considered as passive ecotourism because it seeks for minimum damage of environment (Björk 2000). Weaver (1999) proposed a 'liberal ecotourism model' which is 'sustainable and fit' into ecological as well as human-altered landscapes. Another definitional approach that we come across talks about 'deep and shallow ecotourism' (Acott et al. 1998), wherein former approach focuses on the sustainability of environment and the latter approach considers degradation of environment as a natural corollary to consumptive functionality of nature to humans.

The above discussion undertaken exhibit the dynamic fluidity of the concept of Ecotourism changing its shade with the greater understanding of capitalist forces and sustainability concerns that have mired our world for last five decades. The major shift in the idea embodying this concept is to protect the environment as well as use tourism as a means and a tool to bring about change in the socio-occupational conditions of the local folk encircling the area where ecotourism is being proposed (Hall 1992). Needless to say such change in analysis also expose the pull and push dimensions of ecotourism. The inner intentions and intrinsic motives of the tourists to pursue events to lessen their desires and requirements can be considered as push factors whereas pull factors are features and uniqueness of destination places whose awareness causes patronage from thee tourists.

Most push factors are intrinsic motivators, such as the desire for escape, rest and relaxation, prestige, health and fitness, adventure and social interaction. Pull factors emerge due to the attractiveness of a destination, including beaches, recreation facilities and cultural attractions. Traditionally, push factors are considered important in initiating travel desire, while pull factors are considered more decisive in explaining destination choice. (Mohammad and Som 2010: 283)

Following this line of argument we can say that the pull factor works emerging out of awareness programmes undertaken at the state administrative level for potential tourists to realize the necessity to behold, protect and preserve the precarious balance of the nature, and the push factor emanates from the changes in socio-economic condition of the local people in and around ecotourist spots that is brought about by the resource inflow for the promotion of ecotourism and the patronage received from the tourists.

Ecotourism is consider as an industry today, it has slow but steady growth in all around the world. Thus, it has several interest groups and aims or goals to fulfil. These interest groups are (1) ecotourists, (2) organizations, (3) host population, (4) nature and culture, (5) tourism businessman (Cater 1993, Björk 2000). Apart from these, the said interest groups do keep several other aims in their purview, which are:

1. Preservation of indigenous physical and cultural resources on which ecotourism is mainly depended.
2. Ecotourism essentially has to make available open access to public with ease, and create such kind of infrastructure and ambience that ecotourism firm must be survived for long time (Hvenegaard and Dearden 1998).
3. Weaver (2001) and Fennell (2003) added one more goal that is the social dimension linked to the host community by dint of which the host community can get involved with the various projects related to development of the place, and such participatory-engagement can improve their skill, potential and employability.

The International Ecotourism Society in 1993 published a set of guideline for operating or establishing an ecotourism centre, which are as follows:

- "**Prepare travellers**: One reason consumers chose operator rather travel independently is to receive guidance.
- **Minimise visitor impact**: Prevent degradation of the environment and the local culture. To minimize accumulative impacts, use adequate leadership and maintain small groups to ensure minimum group impacts on destination. Avoid areas that are under-managed and over-visited.
- **Minimise nature tour company impacts**: Ensure managers, staffs and contractual employees know and participate in all aspect of company policy that prevent impacts on the environment and local culture.
- **Provide training**: Give managers, staffs and contractual employee's access to programme that will be very helpful to upgrade their ability and that they could manage better the client and the local culture.
- **Contribute to conservation**: Fund conservation programme in the region being visited.
- **Provide competitive local employment**: Employ local people in all aspects of business operations.
- **Offer site sensitive accommodation**: Ensure that the facilities are not destructive to the natural environment and particularly they do not waste local natural resources. Design structures that offer ample opportunity for learning about the environment and that encourage sensitive interchanges with local communities" (Source: UNEP 2002).

The United Nations Environment Programme (UNEP) highlighted in 'Ecotourism: Principles, Practices and Policies for Sustainability' (2002) certain key components for ecotourism:

- "contributes to conservation of bio-diversity,
- sustains the wellbeing of local poor people,
- includes an interpretation or learning experience about nature,
- involves responsible action on the part of tourist and tourism industry,
- it helps the small groups to build the small scale industry or they getting a chance to show their skills into the arena of global platform and by the time according to merit they can flourish,
- stresses more and more on local participation and involvement,
- more importantly it also promote lowest possible consumption of the non-renewable resources". (ibid)

In a way, this endorsement of ecotourism by United Nations partly internationalized the approach of self-serving vision of capitalist class which ignore that 'there is no single nature, only natures' (Macnaghten and Urry 1998; Vivanco 2002). In this way, one can see that ecotourism, as promoted, as a concept is an occidental social construct that encompasses the spheres of cultural, economic and political life allowing global power knowledge with the backing of global capital to invade the pristine and unchartered territories.

It is this context, now we turn to engage with the passage of ecotourism in India, as is being witnessed in the district of Purulia in West Bengal. The purposive choice of the field is reasoned by factors like underdevelopment and backwardness of the region, pristine natural environment, and tribal and non-tribal community residing side by side. It is considered that in such a locale, to test the applicability of a western-bred concept like ecotourism would be most appropriate in terms of viability and sustainability.

Purulia and Nature of Its Ecotourism

Purulia is the western most district of West Bengal, sharing its border with the state of Jharkhand. The district is known for its social backwardness due to its infertile land settlement, extreme weather, absence of industry, low employability among the inhabitant, and Maoist insurgency. The state government of West Bengal, in its attempt to counter insurgency, has adopted various measures to mainstream developmental process in this district, as part of which centres around ecotourism projects. Like elsewhere in the state, simultaneously it is being observed that a sense of urbanity and new form of consumerism is growing in the district town of Purulia, propelled by the establishment of various outlets and Malls that give access to products of top brands produced by multinational corporations. Moreover, a super speciality Hospital and Medical college, a University, a Government General Degree College and a Government Engineering College have been established within a very short period, and they are functioning well, serving the needs of the people. However, these have in no way brought any significant change in the employment of locals in the district, which could have happened had there been entry of new production units or industry taken place in the district by providing diverse range of regular engagement from highly skilled jobs to unskilled manual labour as per the capability of the populace. Herein ecotourism came as a handy tool for the state government since by exploiting the exotic natural beauty and pristine landscape of the district there was a possibility of promoting tourism, in particular ecotourism in various parts of this district.

The study was conducted at Baranti, a small tribal village within Santuri block under Raghunathpur subdivision. The total number of household in this village is fifty-five with total population of 267 (Census 2011). The population is divided between two communities: Santhals (tribe—40%) and the Mals (Scheduled caste—60%). The area surrounded by Panchokot hill, Biharinath hill (part of Araku valley), a water reservoir Ramchandrapur Medium irrigation project (a state Govt. project), Muradi lake and lots of beautiful *Palash* flower trees (*Butea monosperma)* whose brilliant flaming colour in its bloom transcends the place to another level. There are sixteen resorts that have come up in last five years in this area, and most of which are under private ownership or have been leased out by government. The data were collected from the villagers of Bara Baranti, and from those local inhabitant who works at the resorts and shoppers of local market.

It was found that the gradual development of ecotourism broadly has affected the life of the local people both at macro- and micro-level. This becomes clear as we examine the operating principles and strategies used by these tourist resorts for their business end. During the high tourist-traffic season, which starts from July and ends somewhere in February, about eight months duration every year, the resorts for their residents organize functions of tribal cultural performance, mostly constituting music and dance programme. Such functions, on the one hand, give the local tribal people to earn by portrait their cultural unique practices to the outsiders (the tourists), and on the other hand it is also resulting in commodification of their cultural traits, taking away the cultural essence embodied in such practices. For the requirements of resort administration, and the taste of particular type of tourists (specifically depending upon the place and culture from where they hail from) visiting, such performances are tailored and fitted to suit the purpose resulting in an exploitative, mechanistic and mass performed system. The matter gets further complicated by the fact that since the tribal are not oriented to designer production of their cultural practices keeping in mind the aesthetic values of the tourists. Therefore, invariably a group of middle men have emerged, from within, who work as conduit between them and the resorts. These middle men within the tribes are those who either enjoy relatively better normative status or have primary level educational attainment. They solicit with the resorts as per the changing set of requirements of the day, which, they then fix upon particular tribal groups, found suitable, to execute. Thus, it is these middle men who have emerged from within have become key persons with a controlling handle on the income generated from such activity. Paradoxically, when we observe the Mal caste group, we find that they have not been beneficiary like the tribes simply because they do not have any specific or typical cultural practice that they can offer or sell. Thus, they are found to crib about the lack of opportunity and benefit for them due to the growth of ecotourism in the area.

Next issue of contention that has emerged is regarding land holding and agriculture in this area. Baranti village is in the edge of Gorongi Hill encircled by the Panchokot Hill and the Biharinath Hill, close to it is the Medium Irrigation Project water reservoir under Ramchandrapur area. Evidently, therefore, this area is hilly with forest cover that come under the 'backward hill areas' under the ownership of state with common access and usage. In these areas, all arable lands were used for monocrop by the landless villagers who were mostly under below poverty line. However, as this area started to develop as an ecotourism centre, the land was given away to investors by the state for development. In the process, the villagers, especially the Mal caste group, lost their possibility and option for agriculture and food production. In addition, to the loss of the community of resource users, the movement of locals is also getting restricted due to gated-ecotourist properties, alienating them from the common property resources as well as from their natural setting.

The alternative source of contractual employment provided to few villagers, by the ecotourism ventures, were of low wage support staff who were engaged seasonally in various centres. They were mostly engaged as contingent-paid staff for work of casual nature subject to the seasonal pressure of tourist visit with a view to making rational use of the personnel as per requirement. The better paid middle-level managerial jobs and hospitality (food) jobs in ecotourism centres are mostly from the urban centres of West Bengal with the logic that taste and sensibilities of the tourists could be better taken care of. What is observed that ecotourism in and around the village has resulted in a form of underemployment with unscheduled and unpredictable working hours, or days, or weeks, contributing to income instability and constraining household consumption. This has also created a situation where simultaneously there is an experience of involuntary part-time employment, on the one hand, and phases of longer working hours, on the other. Poor condition of villagers does not permit them to offer 'home-stay' facility for earning either by accommodating in their family home, or in a separate adjacent quarters. They have not been able to make any investment in ancillary sector, like transport facility in the area. Villagers were found to be stressed out under such situation that gets reflected in their family life as well since they were often found to be irresponsible in fulfilling their personal responsibilities towards the household. Work role–family role conflict that was unknown before to these villagers has become common now and at times acute also.

The efforts to conserve the forest and the hills in this area is no more an obligation for the villagers which they had assumed would provide them with remedies to escape poverty and hunger. With income instability becoming a permanent outcome of ecotourism now for the host villagers in Baranti, possibility of poaching of flora, fauna, birds and small animals is becoming more and more probable.

Thus, it is observed that the host villagers are not particularly alleviated from their poverty as proponents of ecotourism would like us to believe. This also jeopardizes the long-term possibility of conservation of nature. Due to multiple tires and level of involvement of people from the village in diverse activities, ranging from agriculture, to handicraft manufacturers, to casual workers, there exists a structure that is unaffected by the thriving ecotourism ventures in the area. The gap between local and global justice of consuming resources is increasing as at the time of planning and policy formation form ecotourism in the area, decentralized inclusive approach appropriate for the locale was missing. The sloppy implementation of the guidelines of the International Ecotourism Society and the United Nations Environment Programme, both at the administrative and investor's level has created a situation that is exploitative in nature though as a business project they can be termed as successful ventures.

Policy Implications

Ecotourism project, as discussed here, presents potential opportunity to the host communities for betterment of their living condition by providing uplifting opportunities of employment and skill development that will ultimately empower them to be autonomous and self-sufficient. However, such project unless overcomes the limitations of the existing approach and strategies adopted by the developers and state, and the differential priorities of the host communities of the region are not addressed, cannot bring about well-being of the tribals. The suggested substitute action plans, preferences, operating principles, policy tools and guidelines to hasten the enabling and emancipatory strategies and procedures adopted contingent to the local conditions. For framing meaningful and productive methods to confront and withstand with the predicaments and circumstances of the host population necessarily we may have to think in terms of best-fit programmes of development. Further, ecotourism in places like Baranti can offer a stable platform to sustain and uphold folk cultural tradition.

The present work exhibits that due a standardized and all-embracing approach in policy framing, there is a lack in efficiency and utility in the existing institutional arrangements directed towards welfare programmes relating to protection and development of host tribal population. Taking these into account, policy improvements can be initiated. In pursuing the physical and financial targets for empowering the tribes, a better proactive role of the Panchayati Raj Institutions, and/or Tribal Councils needs to be evolved in the policy framework so that self-governance and planning becomes stronger.

Conclusion

It is important to remember that, in India, tourism and ecotourism are becoming fastest expending and maturing industry. It aims to promote economic development that is sustainable albeit if it is planned and executed in a way commensurate with the local conditions. Unless ecotourism uphold the cause of protection to environment, initiate the spread of awareness among the people from all strata and empower the host population by meaningful engagement in the process so that their commitment is unwavering.

Further, at the level of policy framing the following key factors, not exhaustive though, can be recommended for consideration for sustainable and equitable ecotourism projects:

1. The resources generated from such projects for the State need to be invested back to the area for the development of the community through education and skill development;
2. Prioritize the employment from the host population in such projects who have traditional knowledge about the ecology;
3. Tourist management and projection of nature requires strategic planning and intensive investment that those who are promoting such ecotourism projects need to do;
4. Promotion of social forestry and plantation consistently, keeping the topography and local flora-fauna in mind;
5. Facilitation in a smaller scale of farming and cultivation of economic crops by preparing pieces of arable land under community responsibility;
6. Apportion certain welfare schemes, like assistance in house building or health programs or environment friendly garbage recycling projects, specifically for host population in and around ecotourism projects that will ensure smooth assimilation of the community with such projects.

Ecotourism has the potential to bring about sustainable change in society for its wholesome betterment. It also gives the folk cultural tradition of a place, like Chou dance, Chou Mask, Tusu, Vadu and unnamed Tribal dance patterns of Purulia in this case, a global forum to present, expand and grow. Ecotourism project of Baranti also presents an opportunity to diversify beyond the 3S of tourism, i.e. Sun, Sea and Sand. It would be pertinent at this age for us to make ecotourism a potential winner for a better future that will be possible only when all stakeholders, especially the host community, are involved from inception in every possible activity and decision making.

End Note

1. Some NGO like Third World Network, opposed this declaration of IYE, because they were concerning about the accelerated exploitation of nature and natural resources by insensible tourist and also they were worried about the capitalist mode of exploitation of nature by capitalist classes.

References

Acott, T. G., Trobe, H. L., & Howard, S. H. (1998). An evaluation of deep ecotourism and shallow ecotourism. *Journal of Sustainable Tourism, 6*(3), 238–253.

Baviskar, A. (2014). Ecology and development in India: A field and its future. In S. K. Chaudhury (Ed.), *Sociology of environment* (XVI–XLII). New Delhi: Sage Publications.

Björk, P. (2000). Ecotourism from a conceptual perspective, an extended definition of a unique tourism form. *International journal of tourism research, 2*(3), 189–202.

Boo, E. (1990). *Ecotourism: Potentials and pitfalls.* Longman Pub.

Cater, E. (1993). Ecotourism in the third world: Problems for sustainable tourism development. *Tourism Management, 14*(2), 85–90.

Ceballos-Lascurain, H. (1987). The future of ecotourism. *Mexico Journal* (January), 13–14.

Ceballos-Lascurain, H. (1996). Tourism, ecotourism and protected areas. International union for conservation of nature and natural resources, Gland, Switzerland.

Chaudhury, S. K. (2014). Introduction. In S. K. Chaudhury (Ed.), *Sociology of environment* (XVI–XLII). New Delhi: Sage Publications.

Das, S. (2011). Eco tourism, sustainable development and the Indian state. *Economic and Political Weekly, 46*(37), 60–67.

Elster, J. (1992). *Local justice: How institutions allocates the scarce goods and necessary burdens.* New York: Russel Sage Foundation.

Fennell, D. A. (2001). A content analysis of ecotourism definitions. *Current Issues in Tourism, 4*(5), 403–421.

Fennell, D. (2003). *Ecotourism.* London: Routledge.

Gadgil, M., & Guha, R. (2004). *The use and abuse of nature.* New Delhi: Oxford University Press.

Gössling, S. (1999). Ecotourism: A means to safeguard biodiversity and ecosystem functions?. *Ecological economics, 29*(2), 303–320.

Hall, M. C. (1992). Ecotourism in Australia New Zealand and the South Pacific: Appropriate tourism or a new form of ecological imperialism. In E. A. Cater & G. A. Lowman (Eds.), *Eco tourism—A sustainable option.* London: Royal Geographical Society.

Hetzer, N. D. (1965). Environment, tourism, culture. *Ecosphere, 1*(2), 1–3.

Hvenegaard, G. Y., & Dearden, P. (1998). Ecotourism versus tourism in a Thai national park. *Annals of Tourism Research, 25*(3), 700–720.

Macnaghten, P., & Urry, J. (1998). *Contested natures.* London: Sage.

Mohammad B. A., & Mat Som, A. P. (2010). An examination of satisfaction on tourism facilities and services in Jordan. *Anatolia, 21*(2), 388–392.

Ramesh, J. (2010). Two cultures revisited: The environment-development debate in India. *Economic and Political Weekly, 45*(42), 13–16.

UNEP Publication. (2002). The international eco tourism society report 1993.

Vivanco, L. (2002). Seeing the dangers lurking behind the international year of ecotourism. *The Ecologist, 32*(2), 26.

Weaver, D. B. (1999). Magnitude of ecotourism in Costa Rica and Kenya. *Annals of Tourism Research, 26*(4), 792–816.

Weaver, D. (2001). *Ecotourism.* Milton: John Wiley & Sons Australia Ltd.

www.prurulia.gov.nic.in.

Ziffer, K. (1989). *Ecotourism: The uneasy alliance. Conservation international.* Washington, D.C.: Ernst & Young.

Chapter 12
People's Movement Against Ecological Conservation Policies in the Western Ghats: Reflections from Kerala, India

S. Gurusamy and P. V. Basil

Only after the last tree has been cut down,
Only after the last river has been poisoned,
Only after the last fish has been caught,
Only then will you find that money cannot be eaten.
—Native American saying

Abstract This chapter is an attempt to articulate the need for ensuring environmental sustainability of Western Ghats and reflections on people's resistance over ecological conservation in Idukki District, Kerala, India. Western Ghats has been conferred by UNESCO with 'World Heritage Status' and it is one among the eight 'biodiversity hotspots' in the world. Idukki District, a part of Western Ghats, the major population here is of migrant farmers from lower lands of Kerala settled through state-promoted schemes for poverty eradication. The diverse moves initiated in the Western Ghats with regard to conservation of biodiversity created insecurity among people and led to Farmers Movements in Idukki. Misunderstandings and disinformation campaigns on Madhav Gadgil Report and Kasthuri Rangan Report regarding conservation of Western Ghats added fuel to fire and converted the movements into Social Action form. People in Idukki received information regarding Environment Sustainability and Conservation Policies mainly from the discussions of local tea shops, church, Election Campaigns of Political Parties, controversial news and media discussions which act as catalyst to the movement. The author stresses that land rights and environmental sustainability are complimentary to each other which deserve proper address from stakeholder's point of view. And hence, the paper concludes that proper environment education and participatory environment conservation is need of the hour to tackle the barriers in conservation and ensuring sustainability of Western Ghats.

S. Gurusamy (✉) · P. V. Basil
Centre for Studies in Sociology, Gandhigram Rural Institute, Deemed to be University, Gandhigram, Tamil Nadu, India
e-mail: sellagurusamy@gmail.com

P. V. Basil
e-mail: postforbasil@gmail.com

Keywords Ecological conservation · Western Ghats · People's movements · Environment education

Abbreviations

CHR	Cardamom Hill Reserves
CSE	Centre for Science and Environment
ERRC	Environmental Resources Research Centre
ESAs	Ecologically Sensitive Areas
ESZ	Ecological Sensitive Zone
IRTC	Integrated Rural Technology Centre
KSEB	Kerala State Electricity Board
KSSP	Kerala Sastra Sahitya Parishad
MoEF	Ministry of Environment and Forest
NSDP	Net State Domestic Product
SEEK	Society for Environment Education Kerala
SPEK	Society for the Protection of Environment- Kerala
SWGM	The Save the Western Ghats Movement
UNESCO	United Nations Educational, Scientific and Cultural Organization
WGEEP	Western Ghats Ecology Experts Panel
WWF-India	World Wide Fund for Nature-India

Introduction

The Western Ghats region in India which covers 129,037 km^2 is one among the wealthiest biodiversity 'hotspots' conferred with 'World Heritage Status' of UNESCO. The region sheds around 25 crore population in six states between Gujarat and Tamil Nadu. Located on the southern end of peninsular India, the Kerala state is too well known even in the colonial era for its spices export and trade to the rest of the world from the fourteenth century onwards. The timber teak wood and tea are significant resources produced in Kerala. Besides, the spices which have international demand are largely grown in the hill stations say Idukki, Wayanad and Silent Valley in the Western Ghats of Kerala exemplify the biodiversity richness of the region. One of the prominent high range districts in Kerala named Idukki is having the lowest population density in the whole of the state. It is known for their mountainous hill which lies in Western Ghats. The population other than tribes are migrant farmers and bonded plantation labourers who were brought into Idukki from the greater mainland of Kerala, Tamil Nadu and Coorg region of Karnataka. In order to address the great famine after the First World War, the Travancore province of Kerala which was an ally of the British rule in India initiated a number of schemes to accelerate its domestic production in the agricultural sector. One among the many schemes introduced by the Divan (Governor) of Travancore, Sir C. P. Ramaswami Iyyer was occupying the high lands of Idukki region to grow more food. So, migrations towards the highlands were largely promoted and supported by the state. During the period 1920–1990, land use and the land cover pattern was considerably changed in the Western Ghats, especially in Kerala, Karnataka and Tamil Nadu states. This trend resulted in mass encroachments and destruction of forest land in Idukki. Encroachments of forests as well as Cardamom Hill Reserves (CHR) took place mainly during the post-industrialization period. Destruction of the natural environment initiated through industrialization further accelerated environmental problems in the region.

The first World Earth Summit, 1992, convention on environment opened the eyes over the ecological degradation and discourses came in prominence. The global attention on ecological conservation led to identification of remaining biodiversity rich areas and policy formulation for conservation of the same. With the purpose to comprehend the state of the environment of Western Ghats, Ministry of Environment and Forest (MoEF) formed a committee named 'Western Ghats Ecology Expert Panel' (WGEEP) later known as Gadgil Committee. The report given by the panel emphasized that in the neoliberal era, thoughtless exploitation of scarce non-renewable resources by corporate houses steered to environmental degradation. That led to ecological conservation of Western Ghats became a hot topic in discourses in Idukki District. The report advocated for environmental conservation in the area. Following it, to see the controversy surrounding it, another committee was constituted by Government to review the situation under the chairmanship of Kasturirangan. Both the reports highly criticized for being extravagantly environment focused and not fit the social reality. People living in these areas had lot of apprehensions regarding its effect on their lives, especially those who hold land without tenure rights (Title Deed). Different stakeholder movements started in Kerala related to conservation policies which was followed by widespread rebel movements embarked especially in Idukki. While addressing the problems of environment and formulating policies on ecological conservation, equal emphasis should have been given to the land rights of the individuals which is enshrined in the constitution of India along with protection of the source of livelihood of the locals which is mainly plantation agriculture. In Idukki's context as a result, the reports on ecological conservation triggered a state of panic among the settler farmers. Disinformation created regarding these reports increased the threat of eviction in the minds of settler farmers who were holding land without Title Deed. This created an existential crisis among the people and led to widespread Social Action Movements initiated by political parties, religious groups and land mafias. The misguiding was purposefully done by the mafia groups to protect their vested interest in the nature. It indicated that a bottom-up approach where grievances of stakeholders could have been redressed with genuine concern to the people was missing there. That led the working and peasantry class to address the environmental conservation policies with supreme conviction and challenge it with utmost seriousness which converted it into an environmental movement.

Theoretical Aspects

Environment is defined as everything that surrounds us. It essentially includes two parts: (a) physical conditions such as air, water and landforms which are interlinked with the survival of ecosystem and development of an individual or a community and (b) social and cultural conditions such as ethics, aesthetic and economics on which the behaviour of an individual or a community is dependent (Sundar and Muthukumar 2006). Environment conservation is a systematic approach to finding practical ways for saving water, flora and fauna, energy, and materials, and reducing negative environmental impacts. Efficient environment conservation policies with people's participation are needed for ensuring sustainability. To this end, proper environment education has to be spread in order to augment livelihood security of stake holders. The ecological conservation is a part of Environment Conservation policy. The word 'ecology' emphasizes the stable and sustainable human-environmental relationship with the physical surroundings. Even the term ecology reflects the apolitical nature of the interaction; the critique fostered a burgeoning interest in the association between politics and environmental change in the third world which resulted as a political process. The connection between politics and ecology is vital in shaping ecological conservation policies.

Jurgen Habermas proposes a theory of 'communicative action' which lays 'Between Facts and Norms' in society, he calls 'lifeworld', the basic conception of society, to be amended or supplemented only for cause. Systems of economic and political action in the course of social evolution arise whereby action is coordinated by the consequences of self-interested action, rather than consensual understanding. The Habermas's idea of such 'systems' is based on his reading of Talcott Parsons and its application in resistance movement taken place in Western Ghats. Critical model developed by Habermas in Theory of Communicative Action is more functionalist than straightforwardly normative (Habermas 1972).

Habermas's (1972) theory of communicative action serves to analyse deeply, how social movements counter 'instrumental' interests of the government through communicative claims making. The society which uncritically accepts the information provided through any social interaction or discourses which involve opposing or altering perspectives pointed at achieving new rational consensus is defined as Communicative Action.

Here, Habermasian framework explains how the resistance movement over ecological conservation policies in Western Ghats has formulated two new domains of communicative action. Firstly, the ecological movement took place to conserve Western Ghats through collaborations with experts and associations. The second is the emergence of resistance movement over ecological conservation of Western Ghats, and how it set up to a Social Action Movement through Communicative Action. The self-reflection among people through communicative action followed with public communicative claim as the people in the ecologically sensitive area has to face several kinds of restrictions led to the mass movement against ecological conservation policies (McCormick 2006).

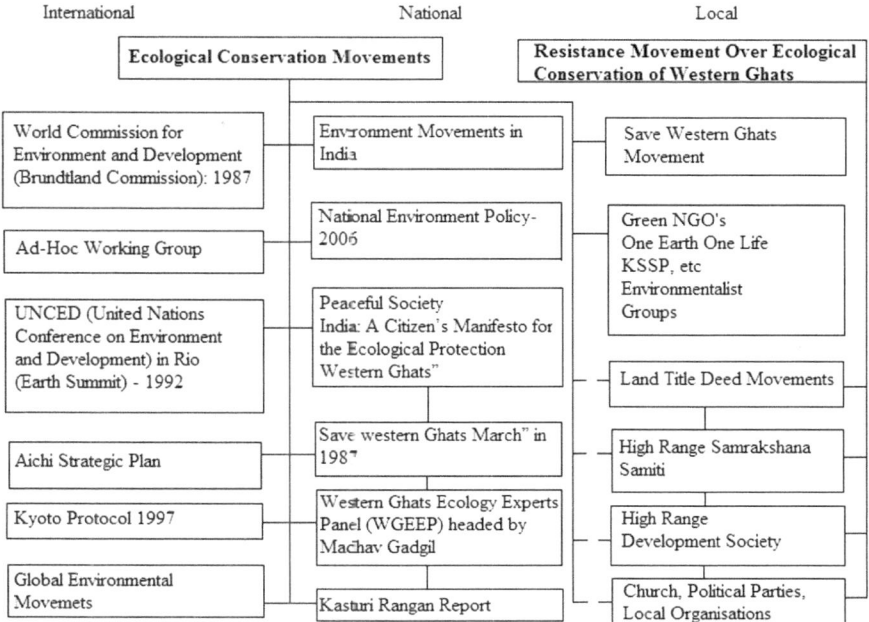

Fig. 12.1 Organisational chart of ecological conservation and its resistance. *Source* Author

As it shows in Fig. 12.1, the ecological conservation came into prominence through international discourses. International environment movements and policy formulations reflected in national level as well. It led to environment movements in Western Ghats. The Save Western Ghats Movement constitutes communicative action to argue scientific orientation of conservation policy. The Save Western Ghats March resulted in appointing Dr. Madhav Gadgil headed Western Ghats Ecology Expert Panel (WGEEP) to study the ecology of Western Ghats. The criticism over WGEEP Report resulted in revising it by Kasturirangan. Therefore, the academic work and scientific theories help to explain the need for research and knowledge development as a source of argument for ecological movements.

As part of resistance movement over ecological conservation policies in the Western Ghats, the first move formed with the support of legal experts in developing new research and followed with the formation of network with the collaboration of church, political parties, NGOs and farmers associations.

Biodiversity of Western Ghats

The Western Ghats, one among the eight-biodiversity hotspot, running along the entire west coast of India from Maharashtra to Tamil Nadu with 1600 km length comprises about 27% of the country's total species and over 4000 species of flowering plants. Evergreen trees are another essential part of biodiversity, and 56% of 645 varieties of evergreen trees are endemic to the Western Ghats. The lower plants groups consist of 850–1000 bryophytes species are located in the Ghats which includes mosses and liverworts with 28% and 43% of endemics.

Location map (Fig. 12.2) of the Western Ghats running in the peninsular region of India and the location map of Idukki District (Fig. 12.2) are provided for overview.

The invertebrate groups in Western Ghats consist of large amount of different species like butterflies (330), ants (350), odonates (174) and mollusks (269), freshwater fish fauna (288), amphibian fauna (220), reptiles (225), birds (500), mammals (120) are found in Western Ghats and most of them are endemic. The Ghats enrich

western ghat mountains

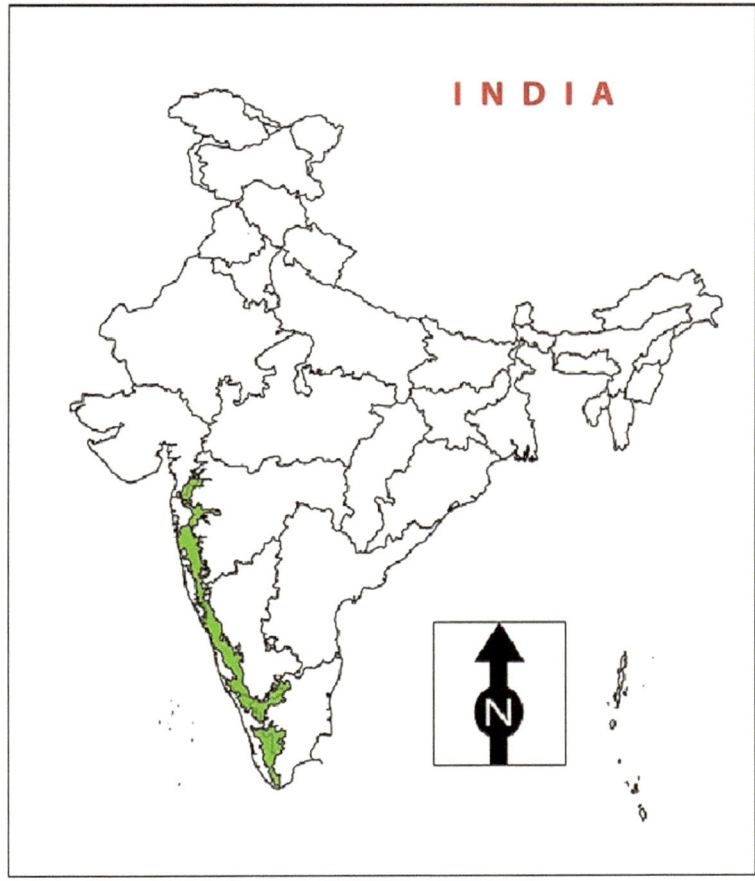

Fig. 12.2 Western Ghats's location in India. *Source* Wiki Commons 2018

its importance for ensuring sustainability by having lion share of country's 20 known species which are being endemic. The Western Ghats biological wealth enriches by having several globally threatened species. Furthermore, the hilly tracts were well known for its wild produce of cardamom, pepper, sandal and ivory (Gadgil 2014).

Human Settlement and Deforestation in Western Ghats

Traces of human settlement in Idukki dates back to the Stone Age. Migration and settlement in Idukki after Stone Age is of five traditional Tamil groups settled in Marayur, Kanthallur, Keezhamthur, Karayur and Kottakkudy. Later, British invasion initiated in 1860 in Idukki through formation of Tiford Estate in Peermade. Travancore King gifted Peermade Hills to Hendry Baker Jr. for coffee plantation. The emergence of comparatively cheaper Brazilian Coffee in the world market led to decline of price of Indian Coffee and Brazil occupied the prominent position in the market. The decline of coffee and abandonment of estates in late eighteenth century, Travancore Kingdom decided to support the planters for Tea cultivation. As a substitute to coffee, tea became the major crop in the hilly areas. In 1877, the Poonjar Kingdom gave Kannan Devan Hills (Munnar Region) on lease to Daniel Monroe for Tea Plantation. During coffee era, the labour requirements was partially met by local labour from Travancore. When it shifted to Tea Plantations, the planters had to recruit Tamil labourers from districts like Trichinapally, Salem, etc. and later they settled in Idukki (Tharian and Tharakan 1985). East India Company had a monopoly in Tea Production and the increased demand for tea in international market led to widespread Tea Plantation in hilly tracts of Western Ghats.

Migration of people from their usual habitat is backed by some push and pulls factors. Particularly in agricultural communities, pressure of increasing population on mainland became a major push factor. But the population pressure may not always lead to migration, since migration is one among the various method of adjustment to tackle the pressure of increasing population. By the year 1911, the encroached area in Travancore's high ranges, especially in Idukki's ecologically sensitive areas (Fig. 12.3), by the settler farmers covered 73.7% of the total land excluding the plantations; later by 1951, it became 98.1%. This includes the cultivable land and forest land. As these statistics show the pressure on land precipitated, the migration of farmers who settled in Idukki District was for the purpose of agriculture and allied activities. Another point to be noted here is that during 1921–1931, migration took place from the Travancore Taluks (revenue sub-districts) of Meenachil, Thodupuzha and Muvattupuzha to the neighbouring high range tract of Devikulam and Peermade caused significant increase in density of population. The decennial rate of growth density of migrated population in Idukki District since 1901 is shown in Table 12.1 (Fig. 12.4).

The land exchange and ownership agreements held during the early 1900s in the high range region were not properly documented as per the Travancore land laws. It was largely based on verbal agreements and unregistered documents, so the

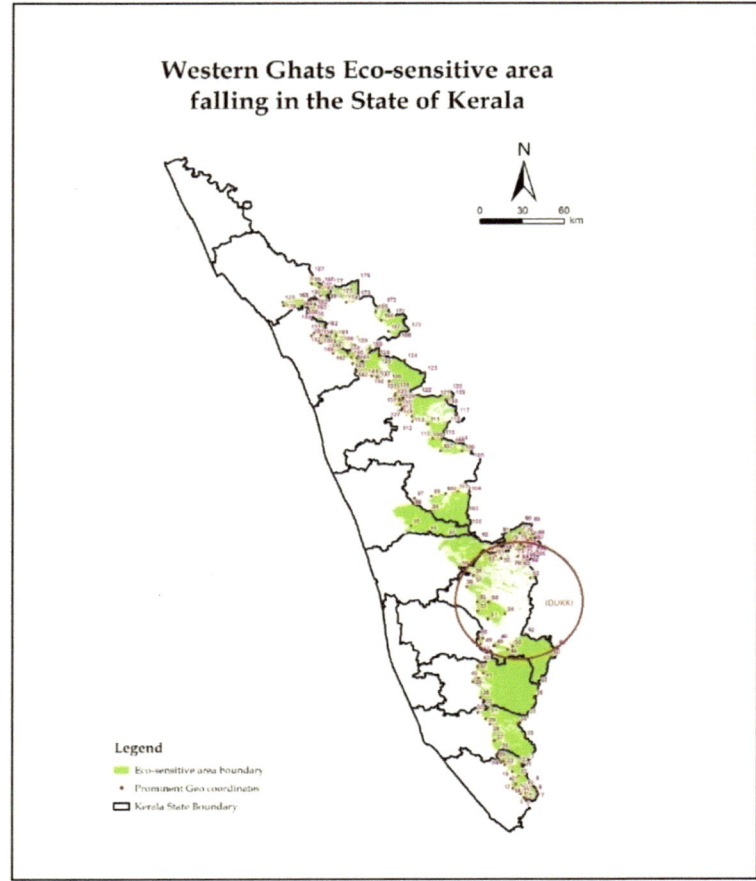

Fig. 12.3 Idukki District, Kerala State. *Source* Wiki Commons 2018

number of migrants from Travancore to Idukki cannot be estimated. The decrease in forest cover is the only available clue regarding estimating agriculture and habitation encroachments in Western Ghats in that decade.

To address the great famine after the First World War, the Travancore province promoted migration to hilly tracts of Western Ghats. These migrants (Settler farmers) mainly focused on agriculture especially the cash crops (cardamom, pepper, ginger, etc.) which were then having a lion share of state's export revenue. In 1940s, Travancorean settlements were developed in hilly tracts of Western Ghats which initially concentrated in Upputhara and Moonnamkandom region of Idukki District. The wetlands and bent forests which Government auctioned were early centres of migration and the document of that auction held in 1931 is still available in records. Ghats's roads and forest roads to Idukki built by the British made migration easier. The migration during 1933–1947 was on a smaller extent as part of Pallivasal Hydroelectric

Table 12.1 Growth of population in Idukki District since 1901

Year	Persons	Decadal variation	Percentage decadal variation (%)
1901	47,666	0	
1911	99,564	51,898	109
1921	108,751	9187	9
1931	187,680	78,929	73
1941	244,296	56,616	30
1951	331,422	87,126	36
1961	579,071	247,649	75
1971	765,392	186,321	32
1981	969,292	203,900	27
1991	1,055,023	85,731	9
2001	1,129,221	74,198	7
2011	1,108,974	− 20,247	− 2

Source District Handbook, Idukki, Census of India (Kumar 2014)

Project, which was first of its kind in Travancore, made for the industrial benefit of the tea companies. The post-Second World War era witnessed the transfer of political power to natives. The state allocated forestland to two thousand individuals in the year 1946 while Colonies were established for Ex-servicemen in 1950. To confront the famine affected by the Second World War, Government initiated 'Grow More Food Scheme' in the 1940s. 10,000 acres of land was allotted to Settler farmers in Ayyappankoil and Adimali region and further 3000 acres of land was allotted in Kattappana region for agriculture purpose to address food shortage. During 1954–1955, the ruling Thiru-Cochi Minister 'Pattom Thanupillai' declared 'High Range Colonization Scheme' and formed colonies in Kallar, Pattom, Marayur, Kanthallur and Deviyur by allotting more than 7500 acres of land to settlers. In the period of 1950–1970, the Travancorean settlement took place in Idukki and in 1962–1963, 15,000 acres of land was allotted to farmers in Kalkoonthal Village itself. Encroachments also took place during that time. Another catalyst for migration was during 1960–1970 as part of construction of Arch Dam at Idukki, on account of demand for labour. As a whole, during 1901–1971, the massive migration resulted in population growth on a higher scale which was nearly 16 times compared to the rate growth of population of the State (Suneesh 2016).

In discussions of migration, the intra-regional differences in agrarian systems have to be confined to the pre-1950 period. The Travancorean settlements developed in hilly tracts of Idukki District particularly in the land suited for commercial crops in which the migrant farmers were interested and were comparatively available in plenty. In 1956, State of Kerala was formed by integrating Travancore, Cochin and Malabar, and subsequent legislations made laws governing land tenure uniform for the whole state, thereby changing substantially the conditions under which migration took place before 1950.

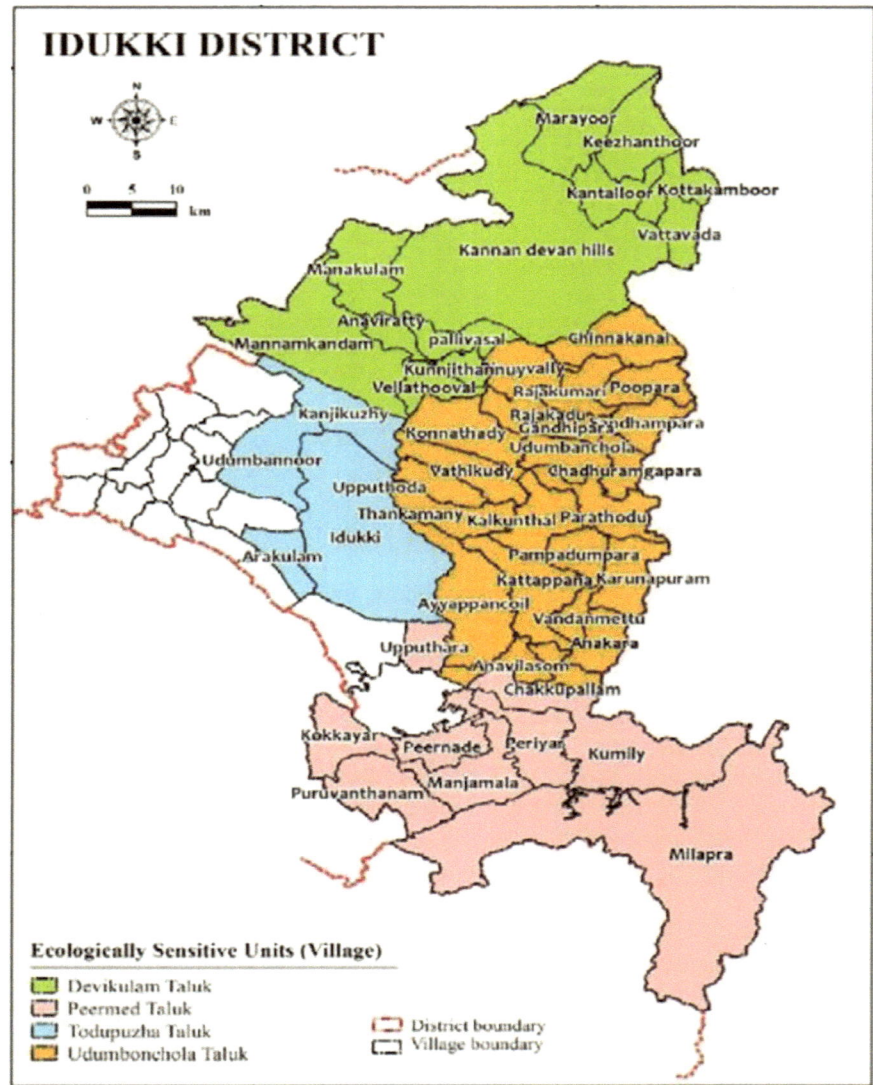

Fig. 12.4 ESA, Idukki District, Kerala State. *Source* Wiki Commons 2018

In the early stage of twentieth century, Travancore Government declared policies and programmes to promote paddy growing to address the food shortage. In the areas of Wandanmedu, Anakkara, Chakkuppallam and Kalthotty, Government exempted 1500 acres of land from Government reserves and allotted to farmers in 'Puthuval Rule' and as per Land Reform Act 1963 they got tenurial rights. In 1940s, the proclamation of 'Kuthakappatta' Government allotted 24,000 acres of grasslands and wetlands owned by Government for growing food crops and in 1949 Thiru-Cochi Government allotted 3000 acres of land for 'Kuthakappatta' in Kunthalampara and Kalthotty region. In 1950s as per High Range Reclamation Scheme, Government allotted 4000 acres of land for 'Kuthakappatta' besides of Kumali Devikulam road.

In 1949, 30,714 acres of forest land was transferred from Forest Department to Revenue Department. In 1954, 6860 acres of land exempted from Forest Reserve and in 1968 formed High Range Colonization Scheme to allot land tenure for these lands. In 1954, the land subject to migration grouped as land convertible and non-convertible to forest. As per this grouping, 34,100 acres land without dispute including Cardamom Hill Reserves transferred to revenue department and declared as agriculture land. In 1970 as per Arable Forest Land Assignment Rule, 56.73 acres of land was exempted from forest land and transferred to revenue department.

The plantation agriculture and settlements of migrant farmers tremendously altered the topography of Idukki District which was historically transformed by land use and land cover changes and the vegetation areas were converted to different forms in as a result of human activities at different scale (Kumar 2014). The increase in human migration accelerated the conversion of forest into agricultural land which resulted in environment degradation of Western Ghats.

During the 1980s, the sustainability and the need for ecological conservation came into prominence in discourses. The mountain range is considered as one of the world's richest biodiversity hotspots and received the World Heritage Site tag from UNESCO. To address problems of environmental degradation, state compelled to undertake a comprehensive study on biodiversity of Western Ghats and Ministry of Environment and Forest (MoEF) appointed different committees for the same.

Ecological Conservation Policies

To study the biodiversity of Western Ghats, MoEF appointed Western Ghats Ecology Experts Panel (WGEEP) headed by Dr. Madhav Gadgil and later Dr. Kasturirangan which made recommendations for policy formulation.

Madhav Gadgil Report

Gadgil Report recommended for sector-wise activities to ensure sustainability of Western Ghats. Most of the settlers in Idukki region of Western Ghats are farmers and Gadgil provided few suggestions in agriculture practice like promotion of organic agricultural practices widely and cultivation of perennial crops on slopes. The report suggests introducing incentive payments for conservation initiatives as stakeholders took part in it and encourage participatory breeding programmes to improve productivity of traditional cultivars; encourage scientific agricultural practices, phase out all use of chemical pesticides and weedicides within five years. The sustainable development projects which help to enrich the biodiversity through subsidising the construction of biogas plants, restoration of community grasslands as well as grasslands in tea estates and forest grazing lands outside the Protected Areas are few of the ideas recommended by Prof. Gadgil and team. Genetically modified crops will be restricted in the region to protect traditional varieties, changes in land use and land cover pattern like filling of wetland should not be permitted in the region.

The report discourages road and other infrastructural expansion and restricts new hill stations as well as conversion of public land into private in Western Ghats. Mining and quarrying as well as polluting industry have to be restricted as per the report and existing polluting industry has to ensure zero pollution within 5 years with strict regulation and has to be socially audited. Decentralized renewable energy production and minimal impact tourism has to be promoted in the Western Ghats area with proper waste management, traffic and water-use measures. Different kinds of educational programmes emphasizing on conservation of Western Ghats for children, youth and other stakeholders has been proposed in this report. The report advocates to prepare People's Biodiversity Registers, Student's and River Clubs and Biodiversity Management Committees (BMC). Gadgil report insists on participatory way of ecological conservation by recommending GramSabhas and grass-roots level decision making (Gadgil 2014).

Kasthurirangan Report

As a result of widespread allegation and protest movements against Gadgil Report, state appointed Dr. Kasturirangan to revisit the report of Gadgil Panel. The Ministry of Environment and Forests issued a notification on 19 October 2013 accepting the recommendations of the Kasturirangan panel. The panel omitted 3100 km^2 of agricultural land, plantation and habitations from Ecological Sensitive Zone (ESZ) notification. The Kasturirangan panel recommends ban mining and quarrying activities, highly polluting industries, construction of thermal power plants, buildings bigger than 2000 m^2 and townships larger than 50 hectares (ha) in the 123 villages across 12 districts in Kerala listed as ecologically sensitive areas (ESAs) comes under Western Ghats region (Suchitra 2015).

Western Ghats Ecology Experts Panel (WGEEP) suggested Ecological Sensitive Zone 1 and 2 has to be physically verified according to Kasthuri Rangan Report. The report gave relaxation to conservation in populated areas, farm and plantations within village by excluding those from ESAs. The activists found it as favoring to the big plantations rather than small farmers and criticized it (Suchitra and Kumar 2015).

Challenges in Policy Implementation

Conserving biodiversity by excluding indigenous/local populations may threaten not only the survival of the Western Ghats but also the sustainability of the region itself. Such policies result in the suffering of the local population as well as exploitation of natural resources and biodiversity. The problem of Land Right versus Ecological Conservation has been debated for long time in Western Ghats region. Nevertheless, not much academic work has been carried out in this aspect. During the 1920s and 1930s, migration was on account of poverty and the world war but later it was for business interest. Large-scale forest encroachment accompanied the 'High Range Colonization Scheme'. For many decades, people have been living in Idukki without title deed. Along with this, statistics of Human-animal conflicts are reported very high in Western Ghats area.

While addressing a function, People's Campaign for Western Ghats Conservation, organized by Salim Ali Foundation and the Paristhiti Aikyavedi at Thrissur, Madhav Gadgil opinioned that, 'there is widespread misconception about the report. The farmers are confused. The report insists on equitable development and protection of livelihood. But the biggest allegation that has been leveled against the report is that it violates the rights of the people and ignores the livelihood of the common man, which is totally wrong' (The Hindu 2013). Even though the report insists conservation in a participatory way, the preparation of report was less participatory. The people in Idukki got threatened by disinformation campaigns and the lack of participatory approach during report preparation.

Due to the widespread allegations on Madhav Gadgil Report, MoEF appointed K. Kasturirangan panel to revise the report. In April 2013, the panel recommended, 37% area of the Western Ghats represent a band of contiguous vegetation and polluting industries should not be allowed (Suchitra and Kumar 2015). This kind of recommendations generated state of insecurity among people withholding land without Title Deed in Idukki District, who created resistance in implementing conservation programmes and policies. The settler farmers residing in Idukki have experienced development induced displacement earlier also. The Ayyappankoil eviction for the purpose of Idukki Hydroelectric Project launched in 1961 wherein most of the migrants were Christians, and hence, church has control over the settler farmers (Suneesh 2016). The row over the protection of the Western Ghats reveals the extent and manner in which politically resourceful groups with short-term goals decide on the plight of a common natural heritage and its residents by engaging in agitations and disinformation campaigns, resulting in information asymmetries. The church, major political parties, and those with economic interests in the Ghats helped to reinforce the political uncertainty at the cost of environmental issues (Nair and Moolakkattu 2017). The conflict among Environmentalists and Settler Farmers on the issue of eco-restoration, therefore, has to be scientifically studied to explore a Participatory Environment Conservation Mechanism.

Environmental Movements in Kerala

The first environmental agitation in contemporary Kerala was in the year 1964, led by Mr. K. A. Rahman. It was against environmental pollution caused by Grasim industries situated in Mavoor, Kozhikode. Traditional workers of that area took part in the protest to save Chaliyar River (Russel 2015).

Another prominent movement was in the Silent Valley forest reserve which is also situated in the Western Ghats. During the early 1960s, the Government of Kerala started planning to construct a dam reservoir at the river named Kunthipuzha in Palakkad District as a river basin for a hydroelectric project. The state government favored the project while environmental groups resisted the state's move and protested to protect the undisturbed rainforest in the proposed region. Kerala People's Science Movement (Kerala Sastra Sahitya Parishad—KSSP) which promotes environmental education programmes in the villages of Kerala was frontrunner of the protest against the hydroelectrical project (Karan 2019).

Another example of people's resistance was in Plachimada, located in Western Ghats to resist the groundwater exploitation of multinational giant Coca-Cola. The Plachimada struggle was spearheaded by the 'Coca-Cola Virudha Janakeeya Samara Samithy' (Anti Coca-Cola Peoples Struggle Committee) on 22 April 2002 demanding that the cola plant to be shut down. Mayilamma, the leader of this people's resistance later became an icon of the environmental movements in the state. The destruction of the environment and livelihood resources by the cola company led to the movement. While the KSSP upheld the positive aspects of Kerala's achievements in development; it also raised concerns about the environmental damages. Environmental agitations became its major area of activism during the 1970. The organization gained wider attention not only in the state but also at the national level.

Another notable agitation was against the construction of 'Pooyamkutty Project'. It was another hydroelectric project in the Western Ghats region located in the Ernakulum District of Kerala. The project was temporarily halted because of strong agitation.

KSSP was always at the forefront of struggles against deforestation. A 'Forest Unity Procession' was organized in 1986 to protect the Mundery Forest (KSSP 1984: 34). The movement organized a state-wide campaign on Forest Protection Convention. KSSP also firmly opposed Rudravanam Project which intended to carve out a township in the thick forests for the pilgrims (KSSP 1984: 34). It also led agitations against encroachment of freshwater resources in several parts of the state, for e.g., Chalakkudy River by Maduri Coats Factory (1978), Kalladayar Anti-pollution Struggle (1980) and agitations against polluting Muvattupuzha river.

Role of Non-governmental Organizations

It is estimated that there are 57 non-governmental organizations (NGOs) with environmental interest functioning in Kerala along with 30 NGOs are listed by World Wide Fund for Nature-India (WWF-India). Apart from that, Integrated Rural Technology Centre (IRTC), a research institute of Kerala Sastra Sahitya Parishad (KSSP), One Earth-One Life, Zoological Club, Thanal, Kottayam Social Service Society, Mithranikethan, Friends of Trees, Society for the Protection of Environment Kerala (SPEK), Friends of Periyar, Environmental Resources Research Centre (ERRC), Bio-Watch, Kottayam Nature Society, Kerala Gandhi Smaraka Nidhi, Jananeethi and Society for Environment Education Kerala (SEEK) is also functioning in the region. IRTC have conducted numerous studies related to various environment issues in Kerala. The campaigns lead by KSSP and SEEK gathered public opinion against the Silent Valley Project. At the same time, the Centre for Science and Environment (CSE) also supported the movement by providing scientific knowledge to the movement against coca-cola in Plachimada. The NGO named Thanal has a prominent role in shaping the history of anti Endosulfan agitation (Nirmala 2010).

The environmental movements in Kerala have a rich history of local participation by people owned initiatives and subaltern sections aroused for their control over the resource. 'Environmental movements in Kerala encompass the enhancement in the quality of life through recognition of people's right over their natural resources, their right to live with dignity and their participation in the decision making' (Russel 2015: 73–78). From above, it becomes evident that people's participation in environment decision making was intended to bring sustainability in the region.

Environment Movements in Western Ghats

In October 1986, Peaceful Society, a voluntary organization based in Goa, has coordinated a national consultation on environment. The programme acknowledged an urgent need to address environmental degradation issues resulting due to irreparable damage in Western Ghats's biodiversity and decided to organize a march along the Western Ghats region to draw attention. 'India: A Citizen's Manifesto for the Ecological Protection of Western Ghats' published by Save Western Ghats Campaign to earn attention on the issue. The year 1980 was marked by a number of environmental movements in India. 150 NGOs and over 9000 people from the region took part in it. The above said movements took place in Western Ghats region of Kerala and the Save the Western Ghats Movement (SWGM) was one of the prominent environmental activisms which became the model for numerous campaigns all over India. On the other hand, Kerala Sastra Sahitya Parishad supported the Gadgil report. For them, it involved democratic space for decision making at the grass-roots level. Some Tribal Groups in Western Ghats region also backed the conservation policies based on Kasthurirangan and Gadgil Report as it was needed for the human existence in the region (Chengappa 1988).

Land Right and Sustainability, Conflicts in Idukki

Due to the lack of Title, deed farmers could not prove the ownership of their land, and thus, they were denied bank loan of any kind along with government subsidies. Subsequently, the farmers were approaching only Co-operative Societies for monitory needs since nationalized banks do not offer any financial assistance without proper document.

The people's protest to get Title deed started before 2000. After the cancellation of Perinjakutty Hydroelectric Project of Kerala State Electricity Board (KSEB), Settler farmers conducted movements by picketing different Government Institutions in Kerala. People in Idukki were facing social issues of various kinds after the environmental controversies plugged in. The land titled under CHR (Cardamom hill reserve) and Kuthakappattom (monopolistic lease contracts) was another hurdle in availing Title deed. Estimated catchment area of the proposed Perinjakutty dam also became a barrier to avail the Title deed. Settlers who encroached this catchment area of the proposed project were forcefully evicted without any financial compensation.

The current predominant strategy for the conservation of the environment has managed to be centralized. It always neglected people's concern in the decision making. These approaches are executed by bureaucracies who are often hierarchically organized, and hence, the involvement and participation from the part of the public are excluded. In addition, these forms of initiatives are primarily backed by a conservative scientific standard that ignores the socio-environmental outcomes of the exploitation of the environment in which they live. Despite, there exists an evolving body of literature that promotes a participatory procedure which is decentralized, people and their society oriented and holistic in its perception of the environment. The central goal of which is to make environmental decision making sustainable and socially inclusive (Kapoor 2001).

Kerala is popular all around the globe for its different development patters with comparatively leading social and individual development, and hence, it holds the first position in the UNDP's ranking. In per capita consumption, Kerala stands at fifth in per capita NSDP (Net State Domestic Product) (Sundaresan and Kumar 2011: 483–486). The ecosystem of the state is sensitive and fragile, which creates obstacles and possibilities for development which is sustainable. Population density is a massive issue in Kerala because it has the third-highest population density in India (Kerala: 819, India: 324). Although Severe scarcity of food and need for land for housing and livelihood is a significant issue that contributed to the invasion of forests and low-lying wetlands, the intensity is further increased by eco-tourism and uncontrolled backwater tourism in the ecologically sensitive areas. The ecology of Western Ghats which is a habitat of a wide range of flora and fauna is seriously affected by the degradation of forests and loss of biodiversity. This, in turn, results in a frightful drop in availability of water in essential resources such as wells, streams and rivers. Other impacts involve landslides, heavy sedimentation of reservoirs, etc. Thus, this region demands sustainable and remedial conservation policies and practices. However, the organizations that work for the conservation of environment are locally functioning and fragmented.

On 4th March, 2010, the Ministry of Environment and Forest (MoEF) decreed the Western Ghats Ecology Expert Panel (WGEEP) consists of 14 members through a G. O. (Government Order). In one and a half years, the panel led by environmentalist Dr. Madhav Gadgil conducted 14 meetings in al over the state and submitted the report on 31st August 2011. In a sense, it can be called a participatory approach. Later on March 2012, the report of Western Ghats Ecology Expert Panel (WGEEP) published by the Government and several peasant organizations started sociopolitical movements, and governments of all the states from Maharashtra to Tamil Nadu backed the extensive critique that various recommendations for ecological conservation are against the fundamental rights and livelihood of the people reside there. Kerala State Environment Policy 2009 and Kerala local development. The massive opposition against the Madhav Gadgil report (Gadgil 2011) resulted to form the High-Level Working Group (HLWG) on 17th August 2012 headed by Dr. Kasturirangan to revise it. The HLWG conducted 10 meetings and four field visits during the preparation of the report and submitted it on 15th April 2013.

Under the pressure of the Supreme Court and National Green Tribunal, the MoEF forced to approve the report before further verification about the concerns raised by the local people. To impose the Indian Environment (Protection) Act, began moves to perform the HLWG directions and notified 4,156 villages in all the six states as Ecologically Sensitive Areas (ESA). The decision of the government led to widespread resistance in Kerala especially in the areas listed under ESA. Movements initiated against Gadgil and Kasthurirangan report (Rangan 2013) were a result of information asymmetries, disinformation campaigns led by the church, political parties, and other vested interest groups to mislead settler–farmers and created a panic situation. These reports specifically targeted 'ecologically sensitive areas' and efforts of Kasthurirangan report to omit some areas from it failed to address a broader perspective on sustainability of environment (Nair and Moolakkattu 2017).

People's View on Conservation

A study conducted in Idukki District regarding comprehending political ecology of Western Ghats reveals that the global debates of ozone depletion and problems related to it have little significance in the lives of the people living there, they have more inclination and craving for protecting the natural diversity of the Western Ghats. The awareness and attitude towards adverse effect of mono-crops like eucalypts, on the environment and activities like rock quarrying, sand mining, filling paddy fields, etc. have failed to fetch popular attention in a correct way. However, it is interesting to note that some people in Idukki observe environmental agenda as a pessimistic intrusion on their peaceful life. When examined closely, the people's understanding of various reports published on the conservation of the Western Ghats, it revealed that their general awareness about the content of the reports are inappropriate and inadequate. It is unfortunate that people raise protest against the two environmental reports without understanding the recommendations in them (Kumar 2015).

Policy Implications

People-centric approach with environmental education can be the way to address the lack of participation among stakeholders. Policy-level implications have to focus more on it. Western Ghats Conservation has to be a prime concern of conservation policy implementation in India. The problem of ownership security is one of the significant threats faced by people in Idukki. To tackle this dilemma of livelihood insecurity, ownership rights have to provide to the migrant farmers in Kerala, especially in Idukki District through policy formulation and implementation.

Proper environmental education and root-level interventions have to be the focus to ensure the environmental sustainability policy in the Western Ghats. The state has to develop an Environmental Education Policy which is suitable for the Grama Panchayats to implement widely. Biodiversity Management Committees which belong to Grama Panchayats in Kerala is a model for decentralized environmental planning. Biodiversity Register prepared with the support of Biodiversity Management Committees in Hrama Panchayat level demand more participatory and conscious effort which can help the members to create environment consciousness among the public. Along with this, the policy is to address the problems faced by farmers in High Ranges. Agrarian problems like human-animal conflict, restrictions in agricultural crops have to transfer into beneficial projects which promote organic mixed farming among farmers in the Western Ghats. National, as well as State policies, has to provide attention to this since Western Ghats Conservation is a global concern.

Conclusion

Title Deed Movements led by Settler Farmers and Social Movements against ecological conservation reports are the major rebel movements in relation to Western Ghats Conservation in Kerala. The main premises of the movement centred around the viewpoint that the land we live does not belong to us. We are trustees of nature. Human beings are also an organic component of the environment. We must handover it as a better place to live in for the future generations to come.

To tackle these rebel movements, 'Save Western Ghats Movement' based on the report of Gadgil Panel introduced as an environmentalist's initiative for the implementation of conservation policy. Due to the widespread protest, Kasturi-rangan Committee was appointed to review Madhav Gadgil Report. Kasturi Rangan Committee excluded highly populated areas from the Ecological Sensitive Zone and liberalized certain norms mentioned in Madhav Gadgil Report. Misunderstanding over these reports created conflict among people and it further aggravated in the form of mass movement. The threat of eviction fuelled rebel movements against eco-conservation. Therefore, solution must focus on reducing confusions about the situation among those who are immediately affected. Further, to streamline the situation, people centered conservation practices such as community forest management and decentralized planning strategies need to be strengthened further more.

People's movements against ecological conservation policies in Western Ghats arise due to the threat of survival of the inhabitants over there. This threat of survival is exploited by certain vested interest groups to protect their ulterior motives. The policymakers have to be aware about this social condition during policy formulation and programme implementation. Proper Environment Education and awareness creation regarding the ecological conservation policies would ensure sustainable environment in the region wherein people, environmentalist groups, Non-Governmental Organizations and Governments are the partners in the successful implementation of the policies.

References

Chengappa, R. (1988 February 29). Ecologists encounter mindless devastation in Western Ghats. *India Today*.

Gadgil, M. (2011). Report of the Western Ghats ecology expert panel. Ministry of Environment and Forests Government of India.

Gadgil, M. (2014). Western Ghats ecology expert panel—A play in five acts. *Economic and Political Weekly, 49*(3), 18.

Habermas, J. (1972). Science and technology as ideology. In B. Barnes (Ed.), *Sociology of science: Selected readings*. Canada: Penguin Books.

Kapoor, I. (2001). Towards participatory environmental management. *Journal of Environmental Management, 63*(3), 2001.

Karan, (2019). Environmental movements in India. *American Geographical Society, 84*(1), 32–41.

Kerala State Environment Policy. (2009). Department of Environment, Government of Kerala.

Kerala State Environment Policy KSSP Annual Report 1984:34.

Kerala State Environment Policy KSSP Annual Report 1996:38.

Kumar, A. (2014). *Land degradation in high lands: A study in Idukki District of Kerala*. Thiruvananthapuram: Kerala University Library.

Kumar, M. (2015). *The politics of environmental governance—A case study of Western Ghats Region of Idukki District*. Kottayam: Mahatma Gandhi University, Kerala.

McCormick, S. (2006). The Brazilian anti-dam movement—Knowledge contestation as communicative action. *Organization & Environment, Sage Publications, 19*(3), 321–346.

Nair, V. N., & Moolakkattu, S. J. (2017). The Western Ghats imbroglio in Kerala. *Economic and Political Weekly, 52*(34), 2.

Nirmala. (2010). Political left and new social movements in civil society: A study of CPI (M) and Shastra Sahitya Parishad in Kerala. *The Indian Journal of Political Science, 71*(1), 241–262.

Rangan, K. (2013). Report of the high-level working group on Western Ghats. Ministry of environment and forests Government of India 15 April 2013.

Russel, O. (2015). New environmentalism of Kerala for sustainability. *OIDA International Journal of Sustainable Development, 8*(6), 73–78.

Suchitra. (2015 July). Western Ghats conservation: Kerala panel seeks dilution of Kasturirangan report, stirs controversy, *Down to Earth.*

Suchitra, M., & Kumar, S. S. (2015 August). Curb on mining in Western Ghats. *Down to Earth.*

Sundar, I., & Muthukumar, P. K. (2006). *Environmental sociology.* New Delhi: Sarup & Sons.

Sundaresan, J., & Kumar, L. (2011). Climate change impact—A novel, initiative for Kerala. *Indian Journal of Geo-Marine Sciences, 40*(4), 483–486.

Suneesh, K. K. (2016). *The marginalised in revolt: Capital, migration and tenurial rights in Idukki 1961–1972.* Sree Sankaracharya University of Sanskrit, Kerala: Kalady.

Tharian, G., & Tharakan, M. (1985). *Development of tea plantations in Kerala: A historical perspective.* Working Paper, Centre for Development Studies, Thiruvananthapuram.

Chapter 13
Balancing Growth with Livelihood Sustainability: Social and Political Action Over Land Acquisition in India

Siddhartha Mukerji

Abstract This paper articulates the issue of land acquisition in India with respect to growth and livelihood sustainability. In the present times, the agenda of growth informed by the precepts of market has given rise to livelihood insecurities to millions of those who stay in the margin. Land acquisition is one such phenomenon that has caused tremendous displacement and unsettlement without adequate compensation. Additionally, it has led to exploitation of natural resources like forests, water and other resources that remained preserved under a sustainable agrarian economy. The necessity of land acquisition emerges from the relentless capitalist expansion process that is underway, especially since the emergence of SEZs and the booming real estate sector in India. Legislations in recent times have attempted at smoothening the process of land acquisition to give a boost to the commercial drive. However, it has faced resistance at both societal and institutional levels. The dispossessed have resisted big land grabs that have taken place after independence. However, the social and political agitations have intensified with greater civil society activism and consolidation of opposition forces. Democratic politics has provided both the strength and motivation to not only resist land acquisition but also express public dissatisfaction over the unregulated drive towards corporatization that has led millions of people dispossessed, displaced, unemployed and unsettled.

Keywords Livelihood · Displacement · Sustainability · Land acquisition · Resistance

The growth agenda that became central to India's economic policy since the initiation of economic reforms of 1991 has been a major driver of land acquisition drive in recent times. It resulted in a closer alliance between state and capital leading to a retreat from public sector-driven industrialization. The new industrial projects that have come up since 1991 have either been in the domain of private enterprises or under public–private partnerships. To run the projects, business enterprises needed large stretches of land which was, though abundantly available, difficult to procure due to

S. Mukerji (✉)
Department of Political Science, Babasaheb Bhimrao Ambedkar University, Lucknow, India
e-mail: butku9@gmail.com

© The Author(s), under exclusive license to Springer Nature Singapore Pte Ltd. 2021   211
M. K. Verma (ed.), *Environment, Development and Sustainability in India: Perspectives, Issues and Alternatives*, https://doi.org/10.1007/978-981-33-6248-2_13

social opposition and cumbersome land acquisition laws. Land was also required to establish SEZs (special economic zones), an institutional model of economic development that was borrowed from China. The SEZ Act 2005 permits the government to acquire land for public purpose and transfer it to private enterprises to meet its developmental objectives (Verma 2016: 23–48). As per the 2016 data, a total of 3591.56 ha of land was notified for SEZs in which Gujarat secured the highest with 1912.29 ha for two SEZs.[1]

The land acquisition drive therefore had to be smoothened to meet the goals of commercialization with greater involvement of private sector including multinational companies in the economy. However, it has led to displacement of millions of farmers and other rural communities for whom land is the basic source of their livelihood. Causing displacement without adequate compensation, land acquisition has added to their misery and vulnerability. More so, it has threatened sustainability that remained secure under an agrarian economy. This emerges from the linkage between land use and sustainability of livelihoods of the farmers and several other rural communities. Sustainability entailing ecological limits may only present a partial representation of land use; the concept is ethically linked with the notions of justice, equality and democracy. It is therefore a multidimensional conception with its social, political, cultural and environmental contours. As Attfield et al. in their study of South Africa point out that:

> The challenge of sustainable development includes, but is not restricted to, living within ecological limits; it also entails addressing issues of justice and equality, as well as eradication of poverty and ensuring sustainable livelihoods.[2]

The misery of land acquisition continues to weaken the social justice system and appears irresolute about balancing the growth agenda with the social promise of ensuring sustainable livelihoods to all. So a development agenda secured through new land legislations makes it difficult to balance growth necessitating corporatization with social equality and sustainability. The larger question that the paper raises is whether growth embodies sustainability or not? Can the present legal developments on land acquisition assure sustainability in terms of both human well-being and ecological protection? Thus, state's role to promote sustainability becomes critical in the light of the growing need or aspiration for rapid urbanization and commercialization. State itself gets encumbered by the society, when social pressures mount over a critical subject like land acquisition and its sustainability implications. Social pressures then churn into political action leading to policy debates and dissenting opinions within the state machinery.

Growth and Sustainability: Are They Contradictory?

Growth and sustainability are distinguished more by the ideology from which they emerge as agendas of economic governance. As a core objective of market economies, growth came to be conceptualized in terms of rise in per capita income of an economy.

It found its ideological basis in neoliberalism with Milton and Rose Friedman, and Jagdish Bhagwati as its major proponents and was institutionalized by World Bank and IMF as the remedy to all social and economic ills in developing societies. The theme of economic growth in neoliberalism stems from the idea of individual freedom that became the foundational principle as well as a goal of modern, democratic society. Drawing upon the economic philosophy of Adam Smith, Friedman views economic freedom as a prerequisite of political freedom and observes that the combination of two produced the golden age in Great Britain and the USA.[3] Economic freedom embodies the spirit of enterprising individuals in contributing towards the growth of commerce and industry. Second, growth is believed to be achievable in market economies that combined the freedom of people with extensive collaboration and cooperation. Market economies by promoting such voluntary exchange facilitate enterprising individuals and thereby drive economic growth. Friedman builds upon classical political economy to advocate greater opportunities for private investment and minimization of state intervention in the economy. The theory clear voices the perspectives of Adam Smith and David Ricardo that views the progress of civilizations in the rising spirit of enterprising individuals and the rapid growth of economies.

Yet another advocacy for economic growth through privatization and globalization is made by Jagdish Bhagwati. Comparing protectionist economies with trade liberalizing ones, Bhagwati justifies the necessity of an open economy with greater involvement of private players in promoting economic growth. As he comments that:

> The contrast in success with industrialization has been so enormous between trade liberalizing and protectionist countries that the old fashioned view that protection favours manufacturing in developing countries has lost its appeal.[4]

Bhagwati also cites the examples of East Asian growth-promoting economies like South Korea, Japan, Singapore and Taiwan for underscoring the significance of state-assisted privatization. State intervention in the economies of the developing world has determined the growth outcomes and to conceptualize the variable patterns of political intervention Bhagwati has given the nomenclature of prescriptive and proscriptive governments.[5] Prescriptive governments are growth-seeking regimes that encourage privatization by following open economies. They support their enterprising businessmen with relevant incentives and therefore play a positive role in promoting economic growth. In contrast, proscriptive governments inhibit private entrepreneurship with policies favouring public sector monopoly. This bias against privatization deprives the economy of the vital energy and spirit that is needed to push exports and promote growth. As a result, such economies stagnate as the energies get diverted to all unproductive channels.

These theoretical positions necessitate privatization for attaining economic growth. All in all, government's task remains confined to creating a conducive environment for private investment. This means that in a market-driven economy of which economic growth is the central goal, is there an orientation for livelihood sustainability? Neoliberalism maintains a Stoic silence on this critical question.

Taking a cue from the aforesaid radical liberal thought, Amartya Sen puts livelihood security at the centre of social and economic governance. Notwithstanding the necessity of market economy, Sen highlights the need to ensure human security as the real and substantive goal of development. The larger goal is still human freedom which is not narrowly conceived as freedom of enterprise, investment and trade but as social and economic securities needed for sustained livelihoods. Civil and political rights are found to be vital elements of this broad conception of freedom. Without them, citizens would fail to claim their social and economic entitlements. As Sen aptly puts it:

> Political and civil freedoms such as liberty of participation, dissent and opportunities to receive basic education and health care are constitutive concepts of development and effective for economic progress. So, freedom here refers to social and economic arrangements like health and education as well as civil and political rights.[6]

So, sustainable livelihood appears to be a citizen's right in the given theoretical framework. Market-driven growth only emerges as means to achieve progress. It is relevant to the extent that it contributes towards sustainable development. Here, sustainable development goals include: no poverty, zero hunger, gender equality, decent work, industry and innovation and reduced inequality. As long as growth remains a political goal, it will narrowly serve private interests and remain apathetic towards substantive and sustainable development. It will promote sustainability only when the state reorders its economic priorities to serve the larger public interest. Often private interests equated with the growth agenda tend to undermine such public interest and thereby divert the state from the inclusive goals of sustainable development. In such situation, will it be possible to establish a complementarity between growth and sustainability embodying inclusiveness? In other words, is inclusive growth that will balance growth with livelihood sustainability a possibility? If so, under what political conditions can the state pursue the agenda of inclusive growth? These questions may also lead us to resolve the political dilemma of land acquisition in India where the state finds itself trapped between its narrowly conceived growth agenda and the democratic compulsion of ensuring livelihood security to the displaced communities. As long as the state pursues growth as emerging from the vested interests of capitalist forces, it may continue to undermine inclusive development. In this wake, relentless land acquisition for commercialization may continue without a concern for livelihood security and sustainability. The thrust for capitalism and the political necessity of preserving capitalist interests has caused land dispossession, an observation that comes prominently in Michael Levine's study of land dispossession in neoliberal India. Michael Levine's study of land dispossession in India clearly contextualizes the big land grabs in India as emerging from the necessities of capitalism when he points out that:

> The practice of states forcibly taking land from rural populations for economic purposes has a long variegated history, but it takes on particular importance in the history of capitalism.[7]

Inclusive growth, on the other hand, is grounded in an ethical domain of public action. It divests growth of its agenda to serve the particularistic interests in the society. Growth then becomes a means to achieve the egalitarian goals of sustainable development. It focuses on poverty reduction, livelihood security and human development. The distributional effects of growth become relevant to policies pursued by the state and corporatist initiatives under corporate social responsibility. Growth then believes in 'giving at large' rather than 'taking away'. Ideology of the ruling elites becomes relevant in pursuing inclusive growth. This may either emerge from the logic of path dependency, wherein the ideology of the governing elites is historically grounded in the democratic ideals of social equality, distributive justice and sustainable development. The historical path of democratic socialism may reinforce the institutional path followed to ensure inclusive development. In such situation, new ideas and models of development may get moderated to meet the terms of an ideology that is historically grounded. Democratic socialism followed for almost three decades after independence was difficult to break in the light of the economic transformation of 1991. This meant that the growth-seeking agenda had to be moderated or even mitigated by relevant equity and sustainability initiatives at times for gaining legitimacy. So growth continues with a social concern for egalitarian and sustainable development. This may be the political stance of parties whose legitimacy was grounded in the democratic ideals of the past. Their own ideology gets reinforced to gain legitimacy even when new agendas in the light of changing social and economic circumstances prevail. Such parties then pursue inclusive growth as a top-down model of development by virtue of the organization's ideological goals of the past.

Parties that emerge powerful in a transformed political and economic environment may care more about newly-constructed ideological goals emerging in a neoliberal economy. BJP offered an alternative developmental model that was grounded more in the ideals of capitalism. In doing this, it was constrained by the compulsions of democratic politics that demanded moderation of land acquisition to secure livelihood sustainability of the affected communities. The pattern of party politics that emerged in view of rising social tide against land acquisition is accounted for following a brief summary of land acquisition in India.

Steps Towards Land Acquisition

The story of land acquisition goes back to the colonial period with the advancement of the 1894 Land Acquisition Act that made the legal provision for land acquisition for public purposes and companies and for determining the amount of compensation given to the land losers. This Act stipulated that the compensation amount was to be determined by the Collector, and in case of any dispute, it is he who would be the final arbiter. It left little scope for fair dealing as the compensation amount was arbitrarily decided without the consent of the land looser.

After independence, the government commenced the land acquisition drive for its large developmental projects. These projects were undertaken by the public sector and included the building of dams, highways, airports and roads. Many industrial and mining projects were also started by the government under capital-intensive industrialization. According to a study by ICRIER (Indian Council for Research on International Economic Relations), the use of land for non-agricultural purposes like industrialization, infrastructure development and urbanization increased from 9.36 million hectare in 1950–51 to 26.88 million hectares in 2014–15.[8] Land acquired for the massive developmental projects led to large-scale displacement of people from their lands. Some of these have been resisted by people and have drawn nationwide attention like the Damodar River Valley Project, the Hirakud Dam Project, and the BSF training camp in Hazaribagh. Asif (1999) observes that:

> For the Hirakud dam in Orissa, 112038.59 acres of cultivable land was acquired in 1950s and today 50 years later, according to the Orissa government, approximately Rs. 6 crores out of the assessed compensation of about Rs. 9 crores has not reached the 3098 affected families.[9]

The process of land acquisition was triggered with the advent of globalization and fast-track liberalization after 1991. With a thrust for commercialization and privatization, the government was required to smoothen the process of land acquisition. While land use in the previous decades was meant for public sector projects, it now came to be utilized by the expanding private economy involving both local and multinational companies. Government acquired land to be transferred to private companies for their business projects. It also created EPZs and SEZs (Verma 2016: 23–48), large commercial estates meant for the prospective investors in the rapidly expanding market. This made the earlier justification of public purpose questionable.

A new regime emerged to support a globalized economy. Its developmental initiatives prioritized economic growth not only for its desirable internalities but also the externality of exalting India's global status as an emerging economic superpower. A new generation of policy elites replaced the earlier ones that had corroborated a socialist pattern of economy based on the industrial strategy of import-substitution. The ideals of self-reliance and socialism had guided the policies of industrialization that sought to establish the monopoly of public sector. As Kohli observes,

> Advisors with pro-liberal proclivities and World Bank or MNC backgrounds like Arun Nehru, Arun Singh, Montek Ahluwalia, Abid Hussain, Bimal Jalan and Manmohan Singh were in sharp contrast with the advisors of Left-leaning in earlier regimes like Pitambar Pant, P. C. Mahlanobis, I. G. Patel, Vishnu Sahay, Tarlok Singh, Ashok Mehta, and V. T. Krishnamachari (Kohli 2006: 201).

New land legislations initiated since the mid-1990s can thus be contextualized in the transformed policy and economic environment. The process of institutional change supporting the land acquisition drive for commercialization had already begun in the 1980s more specifically under Rajiv Gandhi's regime and reached its climax, the tipping point in 1991 (Mukherji 2014).[10] It took a little while to streamline land acquisition laws for meeting the spatial needs of the prospective investors in the private economy. At the same time, the regime that had experienced socialism in the past could not really digress from its egalitarian goals. As land acquisition was ridden with social and political risks, especially when it was primarily meant to serve private, commercial interests at the cost of livelihood insecurity of the displaced families, the government was required to follow a balanced approach within this complex permutation and combination mathematics of conflicting interests. Sustainability in terms of both livelihood security and ecological balance had to be maintained while acquiring land for commercial purposes. This social concern remained implicit in the legislative action for two reasons:

1. Reinforcement of ideals defining the policy goals in the past.
2. Democratic logical of promoting citizen's welfare by a rights-based regime.

Leadership in the socialist regime has often been acclaimed for its commitment to the egalitarian goals of which sustainability is an important component. While pursuing the new institutional change, the new leadership with a neoliberal orientation was required to moderate its ideology to adjust with the ideals upheld by the predecessors. At the same time, the leaders could feel the frustration of masses with privatization and globalization, especially when it was done at the cost of their welfare. In a democracy where leaders need to constantly adjust their policies by the democratic logic of gaining larger public support, an impactful measure like land acquisition is bound to create political complexities.

Legislative Measures Under Recent Regimes

Legislative changes were initiated by the UPA government in 2013 LARR Act to provide necessary livelihood security to the displaced population. It placed a capping on land acquisition by mandating the consent of 80% affected families in case of private projects and 70% in case of PPP projects. This was a necessary step towards carrying out the process of land acquisition consensually. The second important measure to seek popular will was the introduction of social impact assessment mechanism to determine the possible effects of land acquisition on the livelihoods of the affected families in the concerned region. The communities covered under SIA included: artisans, labourers, sharecroppers, tenant farmers, fishermen and small traders. Also, a compensation amount relevant to the needs of the displaced people was fixed. It amounted to 4 times the market rate in rural areas and twice the same in urban areas. The central objective of the Act was to balance the growth agenda with the egalitarian goal of livelihood sustainability.

BJP's rise to power in the 2014 national elections under the leadership of Narendra Modi brought in new possibilities for easing the process of land acquisition to support the drive for a higher order of privatization and globalization. The prior motivation for the legislative action in this regard came from the political aspiration to showcase India as a comely investment destination. For this, the government launched the Make in India campaign to draw prospective investors particularly NRIs from different parts of the world. It goes with the caption of 'ease of doing business' to highlight India as the most-favoured investment destination. The prime minister took individual initiative in advertising the new programme by paying diplomatic visits to many countries. NRIs were addressed and appealed during his diplomatic visits in countries like USA, Germany and Australia where the population of Indian settlers are high. In fact, series of foreign visits that the prime minister paid was just after the launching the Make in India campaign. NRI industrialists also responded positively to the Make in India initiative by showing willingness to invest in India. In this regard, UK's Parliamentary Undersecretary of State for Energy and Climate Change Baroness Sandip Verma was reported to have said that:

> We are pleased to hear Prime Minister on Make in India that is an absolute an excellent direction forward, because we investors in Britain want to come to India and support Make-in-India moves.[11]

He led a business delegation to Rajasthan and responding to the chief minister's offerings for foreign investment, pitched for long-term investment in the state's agro-industry, skill development and other sustainable areas like renewable energy like solar and winds. This shows that in line with centre's agenda, states ruled by BJP also took individual initiatives to attract NRI and foreign investment. On a similar note, another Indian enterprise in UK Lord Swaraj Paul's Caparo Group decided to set up its mattress manufacturing plant at Bawal in Haryana.[12]

The step towards the new land legislation under BJP was therefore temporally linked with its Make in India initiative. It was believed that spatial requirements of the prospective investors could be met by further relaxing the procurement of land for business purposes. Given the thrust for manufacturing with greater involvement of private investors from across the globe, the government went ahead to amend the LARR Act passed by the previous regime. The incumbent government found the Act a bit too rigid for the investors. Land was a prior requirement for them and therefore had to be procured through desirable governmental support. This drove the government towards introducing a new Land Ordinance in 2014.

The new Land Ordinance of 2014 was brought under the pressing need to provide a boost to the emerging entrepreneurial spirit in India. It was promulgated just nine months after the new government came into power at the Centre. To broaden the scope of land acquisition, it substituted the term 'private company' with 'private entity'. This was mainly done to cater to the spatial needs of the growing number of private hospitals and educational institutions.[13] The term was further clarified and elaborated by the April 2015 Ordinance as including 'an entity other than a government entity or an undertaking and includes a proprietorship, partnership, company, corporation, non-profit organization or any other entity under any law for the time being in force'.[14]

Most importantly, five areas of investment were exempted from social impact assessment and the consent clauses pertaining to rural and urban areas introduced under LARR Act, 2013. These were: defence projects, rural infrastructure including electrification, affordable housing or housing for the poor people, industrial corridors and infrastructure and social infrastructure projects including projects under public–private partnership where the ownership of land continues to vest with the government.[15] Induction of these amendments expanded the scope of land acquisition for a variety of private enterprises and almost all types of public–private partnership projects under the social infrastructure arena. With regard to undertaking lands for building industrial corridors, two critical observations are pertinent: one that the industrial corridors appeared to be a major encroachment on the agricultural land and second that it disturbed the ecological balance of the occupied land affecting its overall sustainability that was preserved under agriculture.

Democratic Assertion and Political Action

Resistance within the limits of law and order is inbuilt in our democratic politics. Also democracy offers several avenues for opposition. The land ordinance also had to qualify the test of such resistance. With the partial dilution of social security and sustainable development measures for the displaced communities, the dissenting voices within society and politics surfaced in. Opposing forces used both societal and institutional platforms to denounce the new ordinance. Protests have taken place over land grabs for commercial use. In his study of land dispossession, Michael Levine points out that:

> India is arguably the epicentre of land grab protests…In the mid-2000s, just as India's economy launched into an extraordinary boom, farmer protests against land dispossession—for real estate, infrastructure, factories, and most controversially special economic zones.[16]

Land acquisition movements are not a new phenomenon. History is replete with instances of agitation by farmers and tribal communities. Most recent among these is the tribal movement on Niyamgiri hills in Odisha. Niyamgiri hills rich in bauxite and mineral resources covers 12 villages in the region. These villages are inhabited by the Dongria Kondh tribal group. The government allocated land for bauxite mining to the Vedanta Group to meet the company's requirement of its aluminium refinery in neighbouring Lanjigarh. The project also affected the livelihood of another tribal community called Kutia Kondh, next to the Dongrias in the region. In the face of these agitations, the Odisha Mining Cooperation, the major state enterprise, filed a petition in the Supreme Court. The apex court quashed the petition and restored the rights of the indigenous communities over their land and its natural resources. The Forest Rights Act of 2006 that conferred the tribals and forest dwellers the right to cultivate their land was being referred. The Act also requires the consent of Gram Sabha of the concerned region in which land is to be acquired. The Gram Sabhas of the region unanimously disapproved the project and the company had to withdraw following the verdict of the Supreme Court. As many as meetings of 12 g Sabhas were held in the monsoon months of year 2013 in the presence of a district judge, nominated by the Odisha High Court. The Niyamgiri story is a good illustration of how democratic assertion by indigenous communities can serve as a powerful tool for not only claiming justice for the marginalized but also protecting nature and environment from the relentless drive for commercialization. The resistance was an attempt to not only secure livelihood but also preserve indigenous culture that the tribals intricately linked with their land and its natural resources. It therefore signifies a trinity of land, nature and culture that the indigenous communities associate with their livelihood. The democratic power of the commoners was not only felt nationwide but also sent ripples abroad as the Canadians in Toronto held placards to support the social cause. The movement was therefore internationally acclaimed and grounded in a larger attempt at opposing the particularistic growth agenda that was being pursued at the political level.

The social mobilization drive to oppose land acquisition has mainly taken two forms: first, the consolidation of civil society organizations with diverse yet complementary social concerns and perspectives and second, political mobilization of the displaced communities by the opposition forces. In the first case, the common concern for social justice and ecology emanates from a range of interrelated social issues that the organizations address and bring into public discussions. The latter subscribes to the logic of politics that motivates the opposition forces to consolidate the masses to denigrate the incumbent government. The leaders play a vital role in getting the social cause rightly communicated and then appropriately engineering the strategy of mobilization. This includes identification of target communities, selecting a politically appropriate location for protest and ties-ups with like-minded, ideologically motivated organizations.

The social drive for mobilization by civil society organizations is well exemplified by the state-wide ecological movements against land acquisition. When the issue of livelihood security also figured in the public debates over amendments made by the 2014 Land Ordinance, the movement assumed a national character. The movement followed a holistic approach addressing every bit of loss that the indigenous communities experienced, from their land to its natural resources including water, forests, fauna and minerals. Large-scale agitations were carried out by organizations all over the country to defend the rights of tribals, Dalits and other indigenous communities by giving a call to repel the ordinance. Several organizations came together to lead a nationwide protest on the 24th of February, 2015 at New Delhi. The mahadharna was joined by NAPM, All Indian Union of Forest Working People, Akhil Bhartiya Kisan Sabha, Narmada Bachao Andolan, Jan Pahal, Kisan Sangharsh Samiti, Jan Sangharsh Samanvay Samiti, Delhi Solidarity Group, Ghar Bachao Ghar Banao Andolan, National Fishworkers' Forum and Rashtriya Mazdoor Kisan Sangathan. Simultaneously, state-wide protests were carried out in Patna, Lucknow and Ludhiana. So, the movement was both local and national in its character and unique in its way of connecting the local with the national.

A parallel social movement has been carried out in the form of political mobilization by the opposition forces. Land acquisition being a critical mass concern could not have possibly remained an issue of civil society organizations. It entered the political domain as the major opposition leader Rahul Gandhi organized a nationwide farmer's rally assembling 7000 farmers from different states of India on 19 April 2015. The choice of Ramlila Maidan as the site of the rally was strategically appropriate. For having served as the site for the nationwide anti-corruption movement led by Anna Hazare in the recent past, it was bound to provide greater impetus to the movement. The leader while denouncing the land ordinance highlighted many other hardships that the farmers faced like reduction of the budget for Rashtriya Krishi Vikas Yojana, reduction in MSP for wheat and inability to waive loans. The mobilization drive therefore assumed a holistic approach towards addressing livelihood insecurities that the farmers faced due to relentless drive towards commercialization. The public address delivered by Rahul Gandhi had an anti-incumbency overtone and reminiscent of Congress Party's past efforts at securing sustainable livelihood to the farmers and other marginalized communities.

The second level of resistance was carried out at the institutional level. Needless to say that our constitution provides multiples sites of institutional check and balance. Within the parliamentary structure, the institutional complex created has often brought institutions at loggerheads with each other. The tussle between legislature and judiciary to define the basic structure of constitution with reference to the unsettled right to property until it got repealed under 44th amendment is a case in hand. In a similar manner, the two houses of parliaments with two opposing political majorities diverged on the issue of land bill. Even within the Lok Sabha, where BJP held a majority, the opposition parties including few allies of the ruling party vehemently hindered the passage of the bill which was tabled before it in the March session of 2015. Shiv Sena, an ally of BJP, abstained from voting. Given the feeble presence of opposition forces in Lok Sabha, the ruling party could get its way through and the bill got passed without much hassle. However, it could not see the light of the day in Rajya Sabha with Congress using its effective majority in the house to oppose it. The upper house was thus employed as an institutional avenue to deter the possibility of the new enactment. It shows how by providing multiple pressures points in the institutional complex, our constitutions offer wider possibilities for democratic resistance.

Policy Implications

Relevant policy intervention can restore livelihood securities in projects involving land acquisition. Firstly, social impact assessment must be carried out for every possible group that is found to face displacement and livelihood insecurity. All commercial projects including public, private and PPP must be brought within the ambit of social impact assessment. This may be necessary because the distinction in the nature of project does not in any way determine the magnitude of insecurity. Public sector projects can be as devastating as private projects when it comes to displacement. Selective elimination of SIA, as the present policy ordains, may require reconsideration. Secondly, in all cases of land acquisition, decision regarding compensation may be taken in consultation with all the affected communities. The Gram Panchayat's consent is imperative but only after full satisfaction of the displaced people. The present framework of decision making on compensation lacks such participatory thrust. In the present scenario, when the large commercial projects are being undertaken and the rural-urban gap is getting bridged, the farmland is shrinking. This is adversely affecting agricultural production. In view of this, a special policy focus is required to check excessive land use for commercial production. Fourthly, as noticed, relentless land use has serious ecological repercussions like deterioration of soil quality with release of chemicals and depletion of groundwater use for irrigation purposes. Land acquisition policy must account for these ecological implications to restrict business enterprises from exploiting and depleting the natural resources. The Panchayats while deciding their consent for a commercial project must pay due attention to its ecological impact in addition to the livelihood implications. Lastly, a policy on employment of displaced communities needs separate attention for both long-term security and a sustainable future. As land acquisition is rampant on tribal land, employment guarantee scheme for displaced tribal communities must be separately framed. Therefore, the paper provides possible thrust areas where government interventions are solicited to resolve the issue of involuntary displacement and livelihood securities. The initiative can come from the Ministry of Tribal Affairs.

Conclusion

Livelihood insecurities have loomed large on the face of the growth-centric agenda of governance creating an imbalance in the developmental process. Growth that was believed to be inclusive is bereft of social concerns. Clearly, politics has driven the growth agenda that is central to the phenomenon of land acquisition. This has been conceptualized by Atul Kohli as the politics of growth in India. As Kohli opines, 'What may be a good approach for promoting growth may not always be a popular and just ruling strategy. When a democratic state is narrowly committed to growth and business groups, not only is the quality of democracy likely to suffer, but it is also likely to create distributional and political problems'.[17]

The narrowly conceived growth agenda informing the land acquisition drive has three important sociopolitical implications. First, there is little space for negotiation between the governing authorities and the affected communities. Growth being an ideologically-driven goal prioritizes commercialization over social development. It necessitates greater proximity between the state and the corporate sector. More flexibility in land acquisition through business-friendly laws is a logical outcome of the close collaboration. Second, the thrust for resistance comes through a process of social churning, whereby civil society organizations consolidate themselves to defend the rights of the marginalized communities. The nationwide movement to stand against the land bill is not only a mark of social action against land acquisition but a collective effort at securing citizen's welfare with a special thrust for ecological preservation. The latter implies that many organizations that saw land acquisition as a move towards exploitation of natural resources like forest, water and minerals also joined the movement to resist it. The movement was thus holistic in its approach informed by the need to secure livelihood sustainability by raising both human and ecological concerns. Lastly, the social action translated into political action as opposition forces resisted the move on ground and at the institutional level. It goes without saying that democracy offers multiple pressure points leaving ample scope for political resistance. To resist the land bill, the opposition resorted to social mobilization of farmers and other affected communities and at the same time used the Rajya Sabha platform at the institutional level. Thus, it can be concluded that democratic politics that has driven the movement opposing relentless acquisition and exploitation of natural resources is serving to strike a balance between growth and livelihood sustainability as two critical yet contradicting developmental agendas in India.

End Notes

1. http://pib.nic.in/newsite/PrintRelease.aspx?relid=142161, last accessed on 5 November 2018, Press Information Bureau, Ministry of Commerce and Industry, Government of India, 25 April 2016.
2. Attfield et al. (2004)
3. Freidman (1980, p. 3)
4. Bhagwati (1988)
5. Ibid.
6. Sen (1999)
7. Levine (2018, p. 1)
8. Hoda (2018)
9. Asif (1999, p. 1564)
10. Mukherji (2014)
11. *The Economic Times*, 12 November 2014
12. *The Economic Times*, 13 November 2014
13. http://dolr.gov.in/acts-rules-policiesacts/acts, Official Website of Department of Land Resources, Ministry of Rural Development, Government of India, "The Gazette of India", Part II, Section 1, Legislative Department, Ministry of Law and Justice, Government of India, "The Right to Fair Compensation and Transparency in Land acquisition, Rehabilitation and Resettlement (Amendment) Ordinance 2014", 31 December, 2014, p. 2
14. http://dolr.gov.in/acts-rules-policiesacts/acts, Official Website of Department of Land Resources, Ministry of Rural Development, Government of India "The Gazette of India", Part II, Section 1, Legislative Department, Ministry of Law and Justice, Government of India, "The Right to Fair Compensation and Transparency in Land acquisition, Rehabilitation and Resettlement (Amendment) Ordinance 2015", 3 April 2015, p. 2
15. Op. cit., p. 3 (reference to endnote number 11)
16. Levine (2018, pp. 1–2)
17. Kohli (2006, p. 1252)

References

Asif, M. (1999). Land acquisition act: Need for an alternative paradigm. *Economic and Political Weekly, 34*(25).

Attfield, R., Hattingh, J., & Matshabaphala, M. (2004). Sustainable development, sustainable livelihoods and land reforms in South Africa: A conceptual and ethical enquiry. *Third World Quarterly, 25*(2), 405–406.

Bhagwati, J. (1988). *Protectionism*. Massachusetts: MIT Press.

Freidman, M. (1980). *Free to choose: A personal statement*. New York: Harcourt Publishing Company.

Hoda, A. (2018). Land use and land acquisition laws in India. *Working Paper no. 361*. Indian Council for Research on International Economic Relations, July 2018.

http://dolr.gov.in/acts-rules-policiesacts/acts, Official Website of Department of Land Resources, Ministry of Rural Development, Government of India "The Gazette of India", Part II, Section 1, Legislative Department, Ministry of Law and Justice, Government of India, "The Right to Fair Compensation and Transparency in Land acquisition, Rehabilitation and Resettlement (Amendment) Ordinance 2015, 3 April 2015.

http://dolr.gov.in/acts-rules-policiesacts/acts, Official Website of Department of Land Resources, Ministry of Rural Development, Government of India, "The Gazette of India", Part II, Section 1, Legislative Department, Ministry of Law and Justice, Government of India, The Right to Fair Compensation and Transparency in Land acquisition, Rehabilitation and Resettlement (Amendment) Ordinance 2014, 31 December, 2014.

http://pib.nic.in/newsite/PrintRelease.aspx?relid=142161, last accessed on November 5, 2018, Press Information Bureau, Ministry of Commerce and Industry, Government of India, April 25, 2016.

Kohli, A. (2006). Politics of economic growth in India: 1980–2005. *Economic and Political Weekly, 41*(13), April 1–7, 2006.

Levine, M. (2018). *Dispossession without development: Land grabs in neo-liberal India*. New York: Oxford University Press.

Mukherji, R. (2014). *Political economy of reforms in India*. New Delhi: Oxford University Press.

Sen, A. (1999). *Development as freedom*. Oxford: Oxford University Press.

The Economic Times, 12 November 2014.

The Economic Times, 13 November 2014.

Verma, M. K. (2016). Development induced displacement, SEZs and the state of farmers in India: Some insights from the recent experiences. *Journal of the Human Rights Commission, 15*, 23–48.

Part III
Alternatives for Environmental Sustainability

Chapter 14
Constructed Wetland: A Sustainable Approach for Wastewater Treatment

R. R. Patil, Karuna N. Pohekar, and Neetu Rani

Abstract Disposal of wastewater in urban area relegates to sewage treatment plant located at remote places. On the contrary, in rural area wastewater disposed off into rivers and lakes through open drainage without pre-treatment which ultimately causes public health issues in the urban and rural areas. As the volume of wastewater is high and the number and capacity of treatment plant are low, it leads to water pollution and reduces the quality of water for daily use. Further conventional method for the treatment of wastewater is expensive and requires trained personnel for the maintenance and operation. Hence, other alternate easy to build, inexpensive methods for the treatment of wastewater are essential. Similarly, as there is increase in scarcity of freshwater, the benefits of wastewater reuse are equally necessary to reduce the pressure on the use of water from the natural resources. It is in this context, other than conventional method, new emerging constructed wetland technologies using various locally available macrophytes for the treatment of wastewater are receiving greater attention due to their cost-effective and environment sustainable approach. This paper mainly focused on the types of wetland, its use for the treatment of wastewater and its reuse for gardening, irrigational and pond reclamation purpose, so as to reduce the pressure on the use of freshwater from the natural resources. Moreover, for a comprehensive understanding, the paper highlights brief history of constructed wetland, its types, uses and importance for the treatment of various wastewaters by using various macrophytes and model/design to treat different types of wastewater.

R. R. Patil (✉)
Department of Social Work, Faculty of Social Sciences, Jamia Millia Islamia, New Delhi, India
e-mail: ravi_patil72@yahoo.com

K. N. Pohekar
Department of Biosciences, Jamia Millia Islamia, New Delhi, India
e-mail: kpatil76@gmail.com

N. Rani
University School of Environment Management, Guru Gobind Singh Indraprastha University, New Delhi, India
e-mail: neetu_rani@ipu.ac.in

© The Author(s), under exclusive license to Springer Nature Singapore Pte Ltd. 2021   227
M. K. Verma (ed.), *Environment, Development and Sustainability in India: Perspectives, Issues and Alternatives*, https://doi.org/10.1007/978-981-33-6248-2_14

Keywords Constructed wetland · Macrophyte · Treatment of wastewater · Water pollution

Introduction

Our dream of sustainable development cannot be achieved unless urban and rural communities provided with potable water and safe sanitation facilities. The major cause of poor sanitary, unhygienic and inferior ecological condition is due to perennial increase in population and industrialization which puts monumental pressure on the ecology and natural resources. The studies have found that most of the developing countries are facing moderate to high water crises, and by the year 2025–2030, more than half of the population in the world will face severe water scarcity problem (UNEP 2002; WWAP 2012; WHO 2019). The water is a vital need of human society but the scarcity of water in daily life not only affects the human health of deprived population but also interrupts socio-economic development of society and hampers the overall progressive growth of the nation.

The World Bank Organization in 2019 has revealed that the population of India has been growing rapidly and many densely populated pockets are severely affected by scarcity of water in India. The variation in rainfall pattern, overuse of groundwater, persistent droughts, poor water infrastructure, regional underdevelopment, etc., are the major reasons for water scarcity in rural India. However, the overcrowdedness and population explosion due to rapid large-scale migration from the rural, remote and underdeveloped region to urban areas are major factors putting extra-pressure on existing water resources and infrastructure resulting poor access and uneven distribution of water in urban areas contributing to the ever-growing problem of water-deprived communities in India (CPHEEO 2013).

Moreover, apart from scarcity and uneven distribution of water, the major factor of global water crisis is increasing level of water pollution in urban and rural areas. The water pollution is contamination of water usually by human/anthropogenic activities pollutes water and makes it unusable for utility and consumption. However, water pollution not only impacts accessibility of water for human consumption but has many disastrous effects on human health, animals and deeply damaging the sustainability of water bodies such river, lakes, ponds, streams, wells and wetlands. The Center for Public Health and Environmental Engineering Organization (CPHEEO) report has revealed that increase in population, rapid urbanization and massive industrialization contaminated water bodies by disposing persistent organic pollutants (POPs), domestic and industrial waste, toxic and chemical pollutant, runoff of sewage and medical waste, leakages of sewer lines, etc. (CPHEEO 2013). Similarly, in the rural areas, runoff of sewage, agricultural waste, river dumping and washing and bathing of livestock are the major factors of water pollution largely affecting human health and accessibility of water for daily consumption. In addition, some of the rural communities adjacent to the urban areas are converted into peri-urban areas that have less capacity to handle the demand and supply of water creating severe shortage of water in the urban hinterland in India. Moreover, less availability of freshwater resources and inadequate planning of water conservation and distribution by the government further worse the water needs of the future generations.

It is a well-known fact that the demand for water for multiple uses is continuously increasing in India. As per estimate, the total water requirement across all sectors in India will rise to 1093 billion cubic metres (BCM) in 2025 and 1447 BCM in 2050, while the actual usable water resources available in India were estimated to be 1123 BCM in 2015 (Sahasranaman and Ganguly 2018). Moreover, the water availability is not uniform across the country, and some states and regions are always facing an acute shortage of water for the consumption and productivity purposes (Wintgens et al. 2016).

Above state of scarcity of water along with the rise in water pollution problem in India and around the world is a serious threat to human existence and sustainability of existing water resources. This state of water crisis demands innovative practices of water conservation and comprehensive policy mechanism towards minimizing water pollution and increasing accessibility of water to all. It compels us to think about the reuse of wastewater that would be in the long run a sustainable secondary source of water to fulfil necessities of the present and future generations.

To resolve water-related problems needs effective planning for the management of water keeping in mind a long-term sustainability of water supply. The major focus should be on the rainwater harvesting, watershed management and water infrastructure. Apart from the water harvesting and watershed management, the strengthening of water infrastructure is an urgent and essential task for resolving the problem of water scarcity.

Water infrastructure, especially wastewater treatment plants, should be constructed at the site of generation, and treated water should be used for non-potable uses like flushing, washing, irrigation, agriculture and also for potable uses after treatment through latest treatment technologies (Panwar and Antil 2015). Further, the treated water can also help to recharge underground water, restore the water cycle and protect the natural ecological environment.

Thus, there is an urgent need to store and generate water from all available resources including wastewater by recycling, reuse and storage to fulfil the increasing demand for water.

Status of Wastewater Treatment in India

According to the estimate, sewage generated in the major part of India is 61,754 million litres per day (MLD), but only 22,963 MLD has a treatment capacity and remaining 38,791 (63%) MLD is disposed off without a treatment in the water bodies (Roy 2020). It is also found out that small-scale industries contribute polluted water almost around 40% and agricultural runoff containing residue of fertilizers and pesticide increases nitrogen content in wastewater leading to eutrophication in water bodies. The wastewater disposal of rural India is highest, as 70% population of

India lives in rural areas. In rural areas, disposal of wastewater from the houses (i.e., kitchen, wash water and bath) is highly unregulated and polluting the water bodies. Moreover, there are 63.3% households in rural and 18.2% in urban areas that have no drainage systems to carry wastewater into sewer (Suganya 2020). Apart from this, most of the Indian villages are facing water scarcity and water-borne diseases such as skeletal fluorosis and dental fluorosis due to high fluoride and arsenic content in groundwater. The skeletal fluorosis is a major health problem in India and nearly 6 million people are disabled because of fluorosis (Kurdi 2016). The skeletal fluorosis is more common among poor and vulnerable sections and widely prevalent in the states such as Andhra Pradesh, Bihar, Gujarat, Haryana, Karnataka, Kerala, Maharashtra, Punjab, Rajasthan, Tamil Nadu and Uttar Pradesh (Kurdi 2016). Similarly, arsenic-contaminated water is poisonous to human health and the consumption of arsenic water causes diseases such as diffused melanosis, spotted melanosis, diffused keratosis, hyperkeratosis and gangrene. The arsenic-related diseases are more common among tribals, poor and other vulnerable communities. The states such as Assam, Bihar, Chhattisgarh, Jharkhand, Manipur, Uttar Pradesh and West Bengal are most affected by arsenic-contaminated groundwater.

Above state of untreated wastewater and groundwater is highly hazardous for consumption and human health in India. Hence, there is an urgent need to wisely maintain available water resources, and proper water treatment should be conducted to make water usable for human consumption. Similarly, there is also need to focus on reuse of treated wastewater and conservation of water to increase its accessibility to the people. This reuse of treated wastewater is not only a most possible alternative option to overcome the scarcity of water, but it will also protect natural water resources, improve surrounding environment and most importantly safeguard human health.

The treatment of wastewater is usually done by two methods: first is by conventional method that comprises of physico-chemical and biological methods, whereas the second is by natural methods such as septic tank, natural wetland or constructed wetland. The conventional method is an expensive treatment method and apart from it is also a complex in functioning that requires skilled technical person to operate on a regular basis, whereas the constructed wetland is an easy and affordable wastewater treatment method, but due to lack of awareness and unskilled personnel the constructed wetland method is rare in India.

Mostly, the disposal of wastewater is unsystematic in India. In rural areas, domestic wastewater is directly disposed off in the nearby water bodies without any treatment. The situation is more serious in urban areas due to massive waste generation. In some of the urban areas, there are community sewers, which disposed off wastewater directly into nearest water bodies or terminate into a municipal treatment plant. All municipal treatment plants are not always in a working condition. Moreover, operation and maintenance of the plants are not adequate due to the lack of power supply and unavailability of skilled labour. It is also found that the numbers of sewage treatment plant are far less in number than required to treat the amount of wastewater generated.

In India, Ministry of Environment and Forest is putting an additional effort for the treatment of municipal wastewater but still there is a huge gap between the wastewater generation and its treatment.

All these facts force us to think about some other alternatives, easy to build, inexpensive and sustainable method for treatment of wastewater. The alternate treatment method should be useful to treat domestic as well as industrial wastewater in urban as well as in rural areas to resolve the existing problem of water pollution and water conservation.

Thus, the constructed wetland is an effective natural way of treatment of wastewater that can be constructed easily without use of complicated machinery and requires less energy. It is operated by gravitational energy for the passage of wastewater from one system to another.

Constructed Wetland: Emergence and Concept

The term 'Wetland' characterizes wetland as a place where water comes in contact with the land and land surface remains wet for most of the period in a year. It is a transitional zone between terrestrial and aquatic systems (Kadlec and Knight 1996). The importance of wetland system is an immense to the ecology and human society. It helps in conserving/storing water for longer period, checking floods, preventing soil erosion, increasing biological productivity, improving water quality, recharging groundwater quality, etc. (Ramsar 2020).

Wetland is not a new concept used for water conservation and improving the quality of water. It has been in practice for a longer time. According to the sources, wetlands like ponds and ditches were used for the disposal of wastewater (U.S. EPA 1999) as a traditional practice of wastewater treatment. However, over the period of time, the disposing of highly contaminated wastewater into a natural wetland resulting pollution of surface and groundwater and harming the natural environment by affecting living organisms, aquatic plants and animal life.

It is in this context that the constructed wetland emerged as an alternate model for treating the wastewater and improving the quality of water. It is one of the low-cost sustainable approaches to wastewater treatment and water conservation in the rural and urban areas. The properly designed, functional and maintained constructed wetlands can effectively remove various pollutants from the municipal, industrial as well as other wastewater sources.

In 1953, Dr. Keithe Seidel was the first who experimented on the treatment of over-fertilized inland waterway, sewage wastewater as well as siltation by planting appropriate plant spices in wetlands. Till 1956, she carried out various experiments for the treatment of different types of wastewater like phenol, dairy and livestock wastewater by using wetland plants. In the early 1960s, she augmented her research by using macrophytes (emerging plants) in wastewater and mud of different origins and also tried to implement the experiment to improve the inefficient septic tanks or pond systems in rural areas of Germany. Earlier, she planted different macrophytes in a narrow natural and artificial tray like ditches called hydro-botanical method. This hydro-botanical method is upgraded by using sandy soils in place of mud with high hydraulic conductivity. Further to overcome the oxygen-deficient (anaerobic) condition in the wetland system that reduces the treatment efficiency, she introduced filtration medium using primary sludge along with soil and planted with *Phragmites australis* following horizontal flow bed (Seidel 1965). However, the concept of growing macrophytes in the treatment plant was not acceptable to the sewage treatment engineers, so for many years plants were uprooted by them at the treatment site (Borner et al. 1998). Hence, it was difficult to construct full-scale constructed wetlands in Germany. This natural method of treatment spread all over the world from Germany in the late twentieth century. In the early twenty-first century, different wastewater was also tried to treat with constructed wetland method.

In India, the constructed wetland has been studied since the late nineteenth century. Initially due to insufficient information available, many laboratory-scale studies were carried out (Juwarkar et al. 1992). India's first constructed wetland was built in Sainik School in Bhubaneswar, Odisha, for the treatment of domestic wastewater. Later, many field-scale studies had been conducted in various part of the country for the treatment of different types of wastewater such as in Ujjain for the treatment of distillery effluent and in Bhopal around the Kshipra River for the treatment of domestic and storm wastewater. Along with it, the mechanism involved in grey water, sewage wastewater, industrial wastewater, heavy metal removal, spent wash, medical wastewater, textile and dye effluent, dairy wastewater, leachate, etc., has also been studied comprehensively, and efficient treatment mechanism is evolved by the intensive use of constructed wetland.

Constructed wetlands (CWs) are designed that are similar to the natural wetland that includes wetland vegetation, soil, sand, gravels as sediments for filtration and sedimentation and also provides space for the growth of microorganism in between the space of the sediments and on the surface of the roots to treat contaminants from surface water, groundwater or waste streams. CWs evolved during the last three decades of the twentieth century are environment-friendly treatment technology for the treatment of various wastewaters around the world. In the 1970s and 1980s, CWs had been improved to treat domestic wastewater or municipal sewage by using various naturally available sediments as adsorbents for the removal of nutrients and heavy metals. Since then, it has been used for the treatment of other kinds of wastewater such as landfill leachate, runoff from the urban wastewater, highway, airport and agriculture wastewater, food processing industries includes winery, cheese, dairy, chemical industries, paper mill and oil refineries wastewater.

Constructed Wetlands and Its Types

Constructed wetland system is broadly categorized into two types: (1) free water surface system and (2) subsurface flow system and combination of both free water surface flow and subsurface flow wetland system formed as hybrid constructed wetland. The classification of constructed wetland is as follows, in Figs. 14.1–14.4.

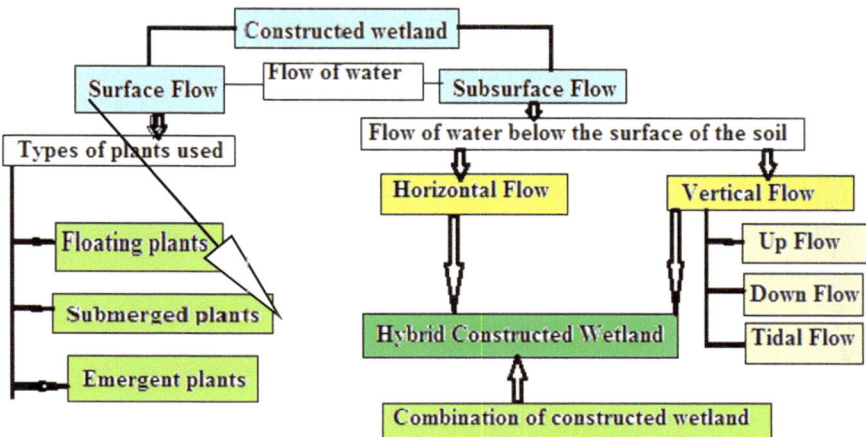

Fig. 14.1 Classification of constructed wetland

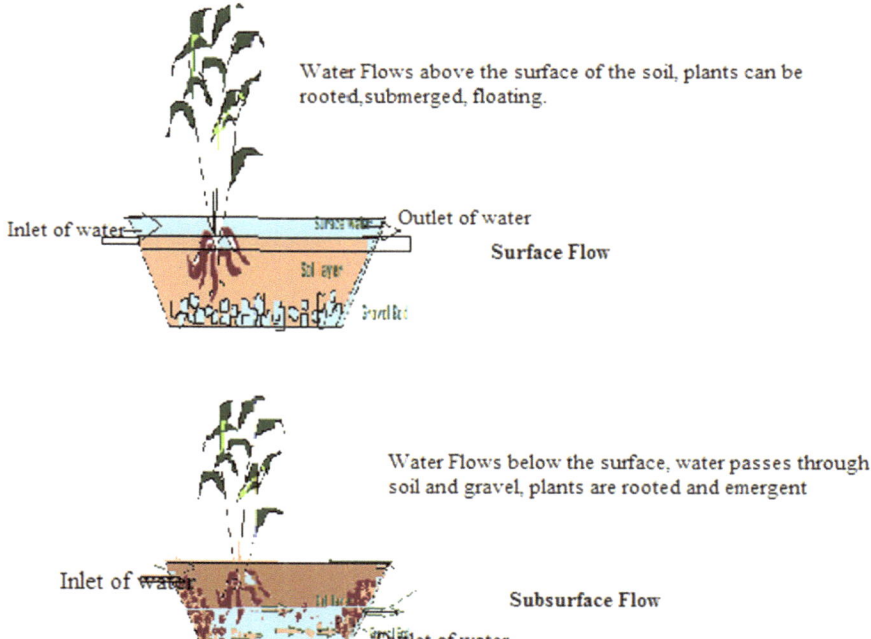

Fig. 14.2 Free surface flow and subsurface flow constructed wetlands

Free Surface Flow Constructed Wetland (FSFCW)

It is just like natural wetland consisting of the shallow basin, soil or other substrates such as large and small-size gravels to support roots of the plants. Flow of the water in this system is above the surface of the soil layer similar to natural wetland. The primary use of the free water surface wetland is for secondary and tertiary treatment of different types of wastewater such as storm-water runoff, municipal sewage and mine drainage (Kadlec and Knight 1996; Kadlec and Wallace 2008). Although construction and operation cost of FSFCW is low and maintenance is easy, its main disadvantage is that it normally requires a larger area of land for construction than other systems and highly prone to mosquito in breeding and nuisance odour in the neighbouring areas.

Subsurface Flow Constructed Wetland (SSFCW)

Subsurface flow constructed wetland is a covered basin with a porous medium that may include different sizes of soil, sand and gravels. The water that flows below the soil surface in the subsurface system is completely free from mosquito breeding and provides more scope for scenic beauty unlike the FSFCW system. Similarly, the flow of the water in subsurface constructed wetland can be made horizontal as well as vertical for aerobic and anaerobic condition and space management. Moreover, the advantages of water flow below the soil surface are many such as tolerance to extreme climatic condition, which reduces pest and odour problem, and improves aesthetic value. Further, the permeable substrate in the subsurface flow system provides greater surface area for the decomposition. The main disadvantage of subsurface system is the problem of clogging and unintended surface flows.

Types of Subsurface Flow Constructed Wetland

Subsurface flow system is of two types:

(a) Horizontal subsurface flow (HSSF) is the system in which water flows horizontally below the surface through the substrate (soil sand gravels layers) along the root zone area and aerobic and anaerobic microorganisms (Vyamazal 2009, 2014). It has also observed that anaerobic microbes are more in the substrate of HSSF and it is found more efficient in treatment of nitrate-nitrogen (Tunçsiper 2009; Zhang et al. 2014). HSSF-constructed wetland has been used for the treatment of onsite domestic wastewater, municipal sewage from villages, industrial and agricultural wastewater. Compared to FSFCW, it requires less space but construction cost is more (Tsihrintzis et al. 2007) and hydraulic conductivity is slower than the VSSF (Kadlec and Wallace 2009).

(b) Vertical subsurface flow (VSSF) is the system in which water flows vertically downward in the substrate leading to a high rate of oxygen transfer in the system. It removes biochemical oxygen demand as well as suspended solids from the wastewater. Nitrification in vertical flow system occurs easily, as it contains oxygen and microbial growth in the root zone area but denitrification is insufficient in the VSSF system (Zhang et al. 2014). This type of wetland is allowed to treat a large amount of wastewater, which is beneficial for the agricultural sector. The flow of the wastewater in each of above system can be continuous, intermittent or fill and drain mode.

From the above description, it can be proved that the VSSF system having the characteristics of aerobic condition is good for nitrification, while the HSSF system having anaerobic conditions is good for denitrification. The combination of different systems of constructed wetland known as hybrid constructed wetland system is the combination of free surface flow, HSSF or VSSF, or mixture of the system according to the requirement (Vyamazal 2007). It compensates the disadvantages of each other and has a potential to perform better for improving the treatment capacity of contaminated water and achieve higher removal rate, especially nutrient-containing wastewater.

Earlier treatment of domestic wastewater has been widely treated by constructed wetland, especially in rural and remote areas where municipal treatment facility was unavailable. Nowadays, most of the hybrid system combined with FSFCW other than the combination of VSSF and HSSF found useful in the removal of total solids and nitrogen compound from many types of wastewater (Vyamazal 2005; Ayaz et al. 2012). Similarly, the different types of wastewater have also been treated with hybrid constructed wetland such as winery wastewater (Serrano et al. 2011), cheese dairy (Reeb and Werckmann 2005), landfill leachate (Bulc 2006) and hospital wastewater (Laber et al. 1999). Thus, to improve the treatment capacity and also to treat diverse type of wastewater, especially nutrient-containing wastewater, the application of hybrid system had been experimented in the varied areas and sectors.

During 1970s and early 1980s, various combinations of hybrid wetland systems were constructed in many countries for wastewater treatment (Lienard et al. 1990). However, it has been found that hybrid wetland systems are more prevalent in developed countries like Europe, France, Italy, Norway, etc., than the developing countries (Masi et al. 2002; Vyamazal 2005). Similarly, there are not much studies and experiments available on hybrid constructed wetland in the developing countries. This may be due to lack of awareness about concept, design and importance of hybrid wetland system in wastewater treatment, or it may not be preferred due to the occupancy of large land area for constructing hybrid wetland system, which is more expensive than the single system. Thus, in order to make hybrid wetland system more popular and people friendly, there is need to conduct regular studies on the treatment of different wastewater at the local level and promote hybrid wetland by using locally available plants and through people's participation in the developing countries like India to curb the growing problem of water scarcity and wastewater treatment.

Mechanism in Constructed Wetlands

Constructed wetland is a complex system comprised of soil, water, plants, microorganism, enzyme activity, etc. So, the mechanism involved in the treatment of wastewater is interrelated to each other. The mechanisms included in it are as follows:

- Suspended particulate matter in wastewater settles in between the filtering bed consists of soil, gravel layers or other adsorbent and root network.
- Removal of chemical compounds through chemical precipitation, when wastewater containing it comes in contact with the substrate, root network and litter.
- Chemical reactions involved by the exchange of ions and adsorption and transform the reactant.
- Decomposition of pollutant by microorganism and enzyme activity into simple inorganic matter.
- Uptake of nutrients by plants and/or transformation of nutrients by microorganisms into inorganic form.
- Pathogens naturally die or reduced by predation.

Advantages of Constructed Wetlands

To treat wastewater, constructed wetlands are affordable and technically sound method for several reasons. The advantages of constructed wetland are as follows:

- Constructed wetlands are easy to build and less expensive than conventional treatment.
- Requires only periodic maintenance than continuous.
- Operation and maintenance cost is significantly less.
- Capable to tolerate fluctuations in influent seasonal variations.
- Facilitates water recycling and reuse for gardening and agriculture purposes.
- Enhances the scenic beauty of the open spaces and hence aesthetically more acceptable.

Limitations of Constructed Wetlands

Despite having a large number of advantages of constructed wetlands, it also possesses few shortcomings. These shortcomings or limitations are as follows:

- It generally requires a larger land area than conventional wastewater treatment systems, which is difficult to arrange in a densely populated country like India.
- Wetland treatment may be economical compared to other options, if the land is available and close to the source of wastewater, which is seldom possible.

- Wetland treatment efficiencies may vary 'seasonally' in response to changing environmental conditions, including rainfall and extremely hot or cold weather.
- Toxic chemical can kill biological components that are sensitive to toxic chemicals, such as ammonia and pesticides.

Design of Constructed Wetlands

Despite various types of research and published studies on constructed wetland, there are not enough conclusions for the standard design of the constructed wetland. It has been found that the efficient constructed wetland system depends on a number of variables such as characterization of the wastewater, type of constructed wetland used, wastewater flow rate, loading of pollutant and retention of wastewater into the system, evapotranspiration, types of plants used, distribution or splitting of wastewater into the bed of the system. Similarly, climatic conditions play an important role in removal efficiency of pollutants as it varies growth of microorganism with the temperature involved in the decomposition of pollutants (Bojcevska and Tonderski 2007; Trang et al. 2010). To maximize the removal efficiency of the pollutants, theoretical and actual hydraulic retention must be considered. Hydraulic retention time (HRT) is the amount of time pollutants retain within the substrate of the system. More the HRT, more pollutant comes in contact with the microorganism attached on the surface of the soil and roots of the plants.

Designing of constructed wetland depends on the removal of pollutant and the hydraulics. The surface area for the horizontal subsurface flow of a constructed wetland can be calculated using the following formula:

$$A = \frac{Q(C_i - C_e)}{k_{BOD}}$$

where A is the surface area; Q is the average daily flow rate of wastewater measured in m^3 per day; C_i is average BOD (biological oxygen demand) of the influent, mg L^{-1}; C_e is the average BOD of the effluent, mg L^{-1}; k_{BOD} is the reaction rate constant, mday^{-1}. For the secondary treatment, the rate constant (k_{BOD}) has been measured as 0.06 and 0.31 for the tertiary treatment of wastewater.

Mostly, decomposition of pollutants is accomplished by hydraulic loading and selection of suitable types of filtering media and vegetation (Kadlec and Wallace 2008). Removal of contamination happens mostly due to the presence of microorganism, and the growth of microorganism increases on the surface of the substrate used in constructed wetlands. Hence, a suitable choice of the substrate is also an important factor in the removal efficacy of nutrients. The root of the plants also used to play an important role in the decomposition of contaminants in constructed wetlands. The roots provide oxygen in the soil that ultimately increases the aerobic condition in the system. The constructed wetlands are designed as attached growth biological reactors. In this regard, sufficient contact with biofilms on substrates such as gravel, plant stems, roots and sediments is important.

In order to get an overview of constructed wetland and wastewater treatment, some of the case studies from India are mentioned below.

Case Studies of Constructed Wetland

Case Study 1: Constructed Wetland for the Treatment of Agricultural Wastewater Constructed Wetland, Indian Agriculture Research Institute (IARI), Pusa, New Delhi

Field study was conducted to reuse treated domestic water for irrigation in the field of institutional area, Pusa, New Delhi, by constructed wetland. The treatment capacity of the system was 2.2 MLD of wastewater with an irrigation potential of 132 Ha annually. Wastewater from the nearby drain was the source in the system. Three horizontal subsurface systems measuring 80 × 40 m in size along with preliminary treatment system comprise of two sewage wells, and grit chamber was constructed. Each system was filled with stratified layer of gravels of different size and planted with *Typha latifolia.* The flow of the water is completely gravity flow to minimize energy consumption. After treatment, it was collected in 80 m × 40 m × 2 m of water tank, from where it is finally pumped of for irrigation in the farms of the institutional agricultural area. It has observed that the treatment capacity of the constructed wetland was efficient in 90–95% nutrient removal, 99% efficient in turbidity, 87% of BOD and 81-91% of heavy metal removal efficiency. This alternative natural method of treatment found much better than the conventional treatment method in both cost (50–65% lower) and energy wise (less than 1%) (Fig. 14.3).

Fig. 14.3 Constructed wetland at Indian Agricultural Research Institute (IARI), Pusa, New Delhi

Fig. 14.4   Constructed wetland at Aravind Eye Hospital, Puducherry

Case Study 2: Constructed Wetland for the Treatment of Domestic
Wastewater from the Aravind Eye Hospital, Puducherry
Constructed wetland of 320 KLD capacity was constructed in the campus of Aravind
Eye Hospital, in order to reduce the demand of freshwater for maintaining 15 acres of
green lawn area and for horticulture. Only domestic wastewater containing both the
grey water and black water collected from the hospital was the source of wastewater.
Grey water and black water were collected in separate settling tanks. The settling
tank containing black water was integrated with the anaerobic baffled reactor. The
partially treated black water and the grey water were collectively passed through
anaerobic baffled reactor and finally to series of horizontally constructed wetland
planted with *Canna indica*. The effluent from the horizontal constructed wetland
was stored in the polishing tank for further reuse. The system was operated since
2003 and observed efficient in removing BOD up to 98%, COD up to 96% and
TDS up to 96%. This project was implemented by the Consortium for DEWATS
Dissemination Society (Fig. 14.4)

Constructed Wetland System: The Policy Implications

The present paper has highlighted the treatment of wastewater by natural methods of
constructing wetlands. Wetland can be constructed for the treatment of wastewater
generated from various sources such as domestic, hospital and industrial. It is one
of the efficient treatment methods for small communities of urban and rural areas.
Apart from water treatment, treated wastewater can also be reused for agriculture
purpose, cooling of towers in industries, toilet-flushing and recharge of water bodies.
The reuse of treated wastewater lessens the burden of municipal corporations' water
treatment plants and increases the effectiveness of the centralized treatment plant.

In order to achieve sustainability in water treatment facility, several steps have to be incorporated such as decentralization of the treatment at the source of wastewater generation for its treatment optimization. Decentralization prevents the pollution of land, surface and groundwater and thus improves the local environmental condition. The policy for the decentralization of the treatment facility needs education and training of all the stakeholders to promote natural treatment at the point of origin before entering into the municipal stream. In India, there is no comprehensive policy on wetland; the existing policy, namely 'Conservation and Management Rule of Wetland, 2010' amended in the year 2017, has specific focus and intent related to the conservation of wetland and its management in the country. Looking into importance of constructed wetland and its utility for the sustainable development and water conservation, there is an urgent need that the government should design and implement the policy mechanism related to constructed wetland for the decentralization of water treatment facility at the national, state and district levels.

Conclusion and Recommendation

Based on the detailed description of wetland system and the analysis of the above case studies, the paper concludes that the constructed wetland can be used as an alternative, easy to maintain, inexpensive and natural reclamation method for sustainable wastewater treatment near the point source for reuse of water for various purposes including agricultural irrigation, pond reclamation and gardening. Hence, constructed wetland not only improves the treatment of wastewater with minimal use of energy but also brings all stakeholders working together to achieve its objectives. So, this paper recommends that the constructed wetland method should be adopted to treat wastewater to reduce the water pollutions of water bodies in urban and rural areas.

Some of the significant recommendations are as follows:

1. Constructed wetlands for wastewater treatment systems will be helpful to treat domestic wastewater (grey and black water) and minimize its pollutants before disposing it into the water bodies.
2. It is also useful to treat industrial, agricultural, storm wastewater, etc. It is a low-cost wastewater treatment system, which reduces energy consumption and requires low maintenance.
3. It shall be adopted as an environmental policy mechanism to reduce the gap between the generation and treatment of wastewater, and support to clean major rivers and water bodies in India.
4. The plants are grown on the constructed wetland, which can be used for making different products and increase livelihood generation among poor sections.
5. Constructed wetland is highly recommended to the rural areas where availability of the land is easy and treated wastewater can be reused in agriculture.

6. Urban areas, where there is an acute deficiency of land, would be appropriate to construct a single vertical flow system and the treated wastewater can be reused for gardening, horticulture and in the industry for cooling of machineries purpose.

Apart from above-mentioned methods, while conceiving the idea of constructed wetland, the issue of climate change should be taken into consideration for effective and sustainable use of constructed wetland system.

References

Ayaz, S. C., Aktaş, Ö., Fındık, N., Akça, L., & Kınacı, C. (2012). Effect of recirculation on nitrogen removal in a hybrid constructed wetland system. *Ecological Engineering, 40,* 1–5.

Bojcevska, H., & Tonderski, K. (2007). Impact of loads, season, and plant species on the performance of a tropical constructed wetland polishing effluent from sugar factory stabilization ponds. *Ecological Engineering, 29*(1), 66–76.

Borner, et al. (1998). In J. Vyamazal, H. Brix, P. F. Cooper, M. B. Green & Habrel (Eds.), *Germany en: Constructed wetland for wastewater treatment in Europe* (pp. 169–190). Leiden, The Netherlands: Backhuys Publishers.

Bulc, T. G. (2006). Long term performance of a constructed wetland for landfill leachate treatment. *Ecological Engineering, 26*(4), 365–374.

CPHEEO, G. (2013). Manual on sewerage and sewage treatment systems.

Juwarkar, A. S., Mamta, V., Meshram, J., Bal, A. S., & Juwarkar, A. (1992). Wasterwater treatment in constructed wetland. *Wetland systems in water pollution control (proceedings) organising committee, sydney* (pp. 35–41).

Kadlec, R. H., & Knight, R. L. (1996). *Treatment wetlands.* Boca Raton: Lewis Publ.

Kadlec, R. H., & Wallace, S. (2008). *Treatment wetlands.* CRC press.

Kadlec, R. H., & Wallace, S. D. (2009). Introduction to treatment wetlands. *Treatment wetlands* (2nd ed., pp. 3–30). Boca Raton: Taylor & Francis Group.

Kurdi, M. S. (2016). Chronic fluorosis: The disease and its anaesthetic implications. *Indian journal of anaesthesia, 60*(3), 157.

Laber, J., Haberl, R., & Shrestha, R. (1999). Two-stage constructed welland for treating hospital wastewater in nepal. *Water Science and Technology, 40*(3), 317–324.

Lienard, A., Boutin, C., & Esser, D. (1990). Domestic wastewater treatment with emergent hydrophyte beds in France. In *Constructed wetlands in water pollution control* (pp. 183–192). Pergamon.

Masi, F., Conte, G., Martinuzzi, N., & Pucci, B. (2002). Winery high organic content wastewaters treated by constructed wetlands in Mediterranean climate. In *Proceedings of the 8th international conference on wetland systems for water pollution control* (pp. 274–282).

Panwar, A. M., & Antil, M. S. (2015). Issues, challenges, and prospects of water supply in Urban India. *Journal of Humanities and Social Sciences, 5,* 68–70.

Ramsar. (2020). The importance of wetland, retrieved from https://www.ramsar.org on 07/10/2020.

Reeb, G., & Werckmann, M. (2005). *First performance data on the use of two pilot-constructed wetlands for highly loaded non-domestic sewage* (pp. 43–51). Leiden, The Netherlands: Backhuys Publishers.

Roy. (2020). Wastewater generation and treatment—Present status in India. Retrieved from https://indianinfrastructure.com on 08/10/2020.

Sahasranaman, M., & Ganguly, A. (2018). Wastewater treatment for water security in India.

Seidel, K. (1965). Neue Wege zur Grundwasseranreicherung in Krefeld, Vol. II. Hydrobotanische Reinigungsmethode. *GWF Wasser/Abwasser, 30,* 831–833.

Serrano, L., De la Varga, D., Ruiz, I., & Soto, M. (2011). Winery wastewater treatment in a hybrid constructed wetland. *Ecological Engineering, 37*(5), 744–753.

Suganya, P. (2020). Status of Swachh Bharat Abhiyan in India. *Purakala with ISSN 0971-2143 is an UGC CARE Journal, 31*(7), 423–433.

Trang, N. T. D., Konnerup, D., Schierup, H. H., Chiem, N. H., & Brix, H. (2010). Kinetics of pollutant removal from domestic wastewater in a tropical horizontal subsurface flow constructed wetland system: effects of hydraulic loading rate. *Ecological Engineering, 36*(4), 527–535.

Tsihrintzis, V. A., Akratos, C. S., Gikas, G. D., Karamouzis, D., & Angelakis, A. N. (2007). Performance and cost comparison of a FWS and a VSF constructed wetland system. *Environmental Technology, 28*(6), 621–628.

Tunçsiper, B. (2009). Nitrogen removal in a combined vertical and horizontal subsurface-flow constructed wetland system. *Desalination, 247*(1–3), 466–475.

United Nations Environment Programme. (2002). *Global Environment Outlook3*, Earthscan, London. Available at http://www.grida.no/geo/geo3/english/pdf.htm.

USEPA (US ENVIRONMENTAL PROTECTION AGENCY). (1999). Constructed wetlands treatments of municipal wastewaters. EPA/625/r-99/010.

Vymazal, J. (2005). Horizontal sub-surface flow and hybrid constructed wetlands systems for wastewater treatment. *Ecological Engineering, 25*(5), 478–490.

Vymazal, J. (2007). Removal of nutrients in various types of constructed wetlands. *Science of the Total Environment, 380*(1–3), 48–65.

Vymazal, J. (2009). The use constructed wetlands with horizontal sub-surface flow for various types of wastewater. *Ecological Engineering, 35*(1), 1–17.

Vymazal, J. (2014). Constructed wetlands for treatment of industrial wastewaters: A review. *Ecological Engineering, 73,* 724–751.

Wintgens, T., Nattorp, A., Elango, L., & Asolekar, S. R. (Eds.). (2016). *Natural water treatment systems for safe and sustainable water supply in the indian context: Saph Pani.* IWA Publishing.

World Bank. (2019). Population, total—India 2019, The World Bank, retrieved from data.worldbank.org on 08/10/2020.

World Health Organization. (2019). *Progress on household drinking water, sanitation and hygiene 2000–2017: Special focus on inequalities.* World Health Organization.

WWAP (World Water Assessment Programme). (2012). *The United Nations WorldWater Development report 4: Managing water under uncertainty and risk.* Paris: UNESCO.

Zhang, D. Q., Jinadasa, K. B. S. N., Gersberg, R. M., Liu, Y., Ng, W. J., & Tan, S. K. (2014). Application of constructed wetlands for wastewater treatment in developing countries—A review of recent developments (2000–2013). *Journal of Environmental Management, 141,* 116–131.

Chapter 15
Biofuels for What? Environmental Sustainability and Discursive Flexibility of Jatropha in India

Rahul Shukla and Sambit Mallick

Abstract Environmental discourses typically represent biofuels as a sustainable and clean source of energy vis-à-vis prevalent fossil fuels. Biofuels are also portrayed as panacea to multiple challenges, viz. climate change mitigation, international trade deficit, energy crisis, unemployment and rural development—especially in developing countries like India. India has been promoting biofuels under the larger ambit of renewable energy sources since late twentieth century. However, various concerns over the adoption of biofuels and its diverse social and political implications have been raised, both in India and the world over. Against such backdrop, this paper attempts to understand the contested meanings arising out of the debates among the scientific communities engaged in adoption, promotion, research and development of biofuels in India. Furthermore, the study examines the legitimizing discourses associated around jatropha (the preferred biofuel crop as per the biofuel policies of the Government of India), and how different conflicting agencies of biofuels embody the crop with several meanings, and in turn draw flexibility to strategically switch over the multiple objectives and challenges which are publicized in the biofuel policies. The study, from science, technology and society (STS) perspectives, argues that jatropha credited with various objectives including improving environmental conditions has not been stabilized yet. Indeed, discursive nature of scientific claim-making in the promotion of jatropha only creates a space to compensate the consequences emanating from the jatropha cultivation in India.

Keywords Sustainability · Renewable energy · Biofuels and environment politics · Jatropha · Discursive flexibility

R. Shukla (✉) · S. Mallick
Department of Humanities and Social Sciences, Indian Institute of Technology Guwahati, Guwahati, India
e-mail: 28rahul@gmail.com

S. Mallick
e-mail: sambit@iitg.ac.in

Introduction

First decade of the twenty-first century marked the surge in crude oil prices owing to the reason of social and political turbulence in one of the major oil-producing regions of the world such as the gulf nations and the northern African countries (IEA 2011). The surge in the prices was also propelled by the increased energy demand in world over. However, the period of the cold war in the 1970s and 1980s has initiated the dynamics of global oil politics and pushed several nations to search for alternative fuels in the backdrop of high dependency of oil over certain regions/nations. The concerns of strategic leverage combined with nation sovereignty and security, arising environmental sensitivities at the global level, and assumed energy crisis opened up the avenues for alternative fuels which could replace socio-technical system of fossil fuels (Kovarik 1998). In another estimate of the World Energy Outlook (IEA 2013), 56% of rise in energy consumption is projected by 2040. By this time (the year 2040[1]), 93% of the total carbon capacity would be in use, leaving a narrow scope for future development of fossil fuels-based economies and allied climate change mitigation objectives (IEA 2018).

The ever-increasing energy requirements as estimated by the reports (e.g. IEA 2013, 2018) and availability of *limited* recourses to meet this demand call for the sources science and technology to cater the projected energy demand. Against such thrust for an alternative fuel, biofuels have been promoted as a sustainable and viable replacement of fossil fuels value chain. The term biofuel refers to 'energy produced from biomass through processes such as solid combustion, gasification or fermentation' (Demirbas 2009). Biofuels have caught the attention of the world as a global solution to energy demand due to manifold possibilities such as 'to reshape livelihoods, patterns of resource consumption, environments and agro-food production systems' (Smith 2010).

Following the global trend for the need of biofuels as an alternative and renewable source of energy, India joined the league and launched the National Mission on Biodiesel (NMB) in 2003, and subsequently, in 2009, India came up with the National Policy on Biofuels (NPB) to accelerate the policies and programmes on biofuels. The NMB was promoted by the Government of India (GOI) as a development instrument with the objectives of achieving energy dependence by reducing oil imports, generating direct and indirect income resources by liking the programme with agriculture sector, and reclaiming 'wasteland' for cultivation purpose (GOI 2003). The NMB envisioned employing new technological implements such as environment-friendly process and products to produce biodiesel from *Jatropha Curcas* (hereafter jatropha)—a shrub of toxic nature that cannot be consumed for dietary purpose. As per the mission document (GOI 2003), stretches of public land along highways, railway tracks and canals would be identified for jatropha cultivation including the existing 'wasteland', 'culturable fallow land', empty land around agricultural fields, joint forestry management (JFM) and common property resources (CPRs) (ibid.). The proponents of jatropha—which includes the factors of not only state policies but corporate influence, farmers' interests, regional aspirations, the technological accessibility—embody jatropha with material flexibility[2] and thereby the flexibility to strategically switch over the multiple objectives and challenges which are publicized in the biofuel policies of India.

Against this backdrop, the study from science and technology studies (STS) perspectives, particularly from actor-network theory (ANT), examines the responses of scientists, government officials from selected scientific institutes, NGOs and state biofuels development boards. ANT provides the understanding of 'sociology of association' (Latour 2005). By sociology of association, Latour refers to identification of multiple 'associations' and 'translations' between various actors—both non-human and human—interacting within certain networks (Latour 2005). In sum, inclusion of non-human actors (and actants) in the associations and translations, and going beyond the social as given are the analytical aspects of ANT. ANT, and STS in general, aims critically addressing policy and governance, attempts to reform the *neutral* perception of science and technology, and in turn develops sophisticated understanding of scientific and technical knowledge. It also attempts to reform technoscience for welfare, equity, justice and environment (Sismondo 2011). This paper attempts to understand the contested meanings arising out of the debates among the scientific communities engaged in adoption, promotion, research and development of biofuels in India. Furthermore, this study examines the legitimizing discourses associated around jatropha, and how different conflicting agencies of biofuels embody the crop with several meanings, and in turn draw flexibility to strategically switch over the multiple objectives and challenges which are publicized in the biofuel policies. The study, from STS perspectives, argues that jatropha credited with various objectives including improving environmental conditions has not been stabilized yet. Indeed, discursive nature of scientific claim-making in the promotion of jatropha only creates a space to compensate the consequences emanating from the jatropha cultivation in India.

The majority of the data are generated in semi-natural settings by interacting/interviewing with the scientists and the government officials at their laboratory or office. The data are collected through semi-structured in-depth interviews with scientists and government officials from the Council of Scientific and Industrial Research (CSIR) and the Indian Council of Agricultural Research (ICAR)-sponsored research institutes, central and state universities, and state biofuel development board. The field sites were selected on the basis that the institutes have been engaged in research and development of renewable energy resources and technologies, especially production of biofuels. Indeed, a few of them are 'Centre of Excellence in Biofuels' and nodal centres entrusted with the cultivation of jatropha, development of the right genotype of the plant and the responsibility of promoting large-scale plantations of jatropha in the country and producing biodiesel subsequently.

Environmental Discourses and Biofuels

Biofuels present one of many enterprising but complicated technologies characterizing what may be broadly considered the 'bioeconomy'—moving from the strictures
of the genetic revolution to wider perspectives of replacing the fossil-based economy
with a bio-based one. Bioeconomy encapsulates a combination of economic activities
inclusive of invention, development and production through biological ingredients,
processes and products which can contribute to socio-economic impacts in developed
and developing countries (OECD 2009). Such interlinkages of idea of development,
environment and technology (here biofuels are representing a technological enterprise) can be traced back from 1950s when the natural world was perceived and
established as a problem, unexploited natural resources by human interventions. By
the 1970s, the environment or nature suddenly appeared to be more fragile (Meadows
et al. 1972). The Club of Rome came up with *The Limits to Growth* in 1970s indicating that natural resources are finite and running out from demands made by human
population—who were apparently exponentially multiplying (ibid.). This has had the
implications for the concepts like economic growth, development, modernization and
technology. The theorist of development took environment as an unsettling exercise
from the hitherto existing notion of untapped natural resources now give way to more
complicated and calculated measures like conservation, preservation and sustainability. This brings an intellectual turn towards reconsidering economic growth with
ecological vulnerability (Pepper 1996). And perhaps, Rachel Carson's *Silent Spring*
(1962) accommodated ecological sensibilities in development initiatives.

In the further development of the interlinkages, international organizations like
IUCN, UNEP and WWF team up and came up with World Conservation Strategy
(WCS) in 1980 imbuing the idea of development with environmental ideas under
the term 'ecodevelopment' coined by Maurice Strong. The WCS report came up as
the *Our Common Future* and often referred as Brundtland Report published in 1987,
and gave the compelling definition of sustainable development.

> Humanity has the ability to make development sustainable to ensure that it meets the needs
> of the present without compromising the ability of future generations to meet their own
> needs. The concept of sustainable development does imply limits—not absolute limits but
> limitations imposed by the present state of technology and social organization on environ-
> mental resources and by the ability of the biosphere to absorb the effects of human activities.
> But technology and social organization can be both managed and improved to make way
> for a new era of economic growth. The Commission believes that widespread poverty is no
> longer inevitable. Poverty is not only an evil in itself, but sustainable development requires
> meeting the basic needs of all and extending to all the opportunity to fulfil their aspirations
> fora better life. A world in which poverty is endemic will always be prone to ecological and
> other catastrophes (WCED 1987: 8).

While the WCS was premised on the belief that the conservation of ecosystems
was crucial to sustaining development, the Brundtland Report reversed the claim by
suggesting instead that the socio-economic betterment of population would result
in better conserved and managed ecologies. Thus, with such a change in emphasis,
the Brundtland Report made a broader appeal to government and policymakers.

But, as Visvanathan (1991: 383) points out, in insisting that development, once again, be put at the centre of addressing environmental concerns, 'deep down, the Brundtland Report still believes that the expert and the World Bank save the world'. Such critiques have obvious implications on the greening, the idea of development for the developing countries (Adams 2008) which was pushed through expert-led policy and good science, rather than no transformative change, no political or radical struggle and challenges to the intellectual shift in environment and development.

Within this broader discourse of environment, development and technology, we attempt to frame biofuels in the global assemblage which had created an enabling environment for the adoption of biofuel production the world over. This assemblage and need for biofuels in India can be traced from the response of a scientist from ICAR-sponsored research institute:

There are various countries who are in advanced stage of biofuel production from corn, sugarcane, palm oil etc. like in Brazil, the USA, the EU countries. But still we went for different approach for biofuels. Why we went ahead with tree borne oilseeds because of food versus fuel debate. From these debates/or from the problems other countries are facing, a lesson learned from others' experiences and we framed our biofuel policy in different line. There are two points which are clearly mentioned in the policy. The first is, we should not go for food crops e.g. soybeans, corn, etc. for biofuel production, and the second is, we cannot target the prime agricultural land. For the purpose degrade land, marginal soil land is to be targeted to avoid the conflict between agricultural land and biofuel feedstock. Earlier biofuel crop plantation got much attention and we had the experience of Eucalyptus plantation from Punjab, Haryana, western UP that farmers gave preference to Eucalyptus than food crops to gain more profit. Therefore, in the biofuel policy it was clearly mentioned that one cannot cultivate biofuel crop in their agricultural land.

However, concerns were being raised by the UN, FAO, OXFAM, local NGOs and researchers about the environment-friendly production process of biofuels, and that arable land was being diverted for biofuels production which was in turn raising food prices (OXFAM 2008). The development of biofuels is involved in debates on monoculture, food versus oil, land grabbing and the not so green production of biofuels (ibid.). Various actors from research organizations and NGOs, farmers and labourers raised issues of concern which were in opposition to the goals and claims of the development initiative. For example, according to one of the respondents from Indian Institute of Science, Bengaluru:

Jatropha alone is not a suitable crop to sustain the biofuels demand. Other countries are not dependent on Jatropha only, they are using some edible oil also, like Malaysia is processing palm oil for biofuels. Though we are not using the edible oil, but surely, we can diversify the feedstock by including other non-edible crops.

As we can observe from those two quotes from the scientists mentioned above, concerns are emerging on the viability and sustainability of biofuels globally, and voices of scepticism are also arising in India on the production and viability of biofuels (Bailis and Baka 2010; Ariza-Montobbio and Lele 2010). A range of authors have focused on the emergence of biofuels (particularly biodiesel) in India; for example, Ariza et al. discuss the role of soil fertility and low yields of Jatropha in Tamil Nadu; Rajagopal speaks about the policy, economic and environmental dimensions of biofuels; Baka reveals the politics of wasteland and land grabbing associated with biofuels in Tamil Nadu; Tompsett centres on the biofuel policy as a development project in Rajasthan; and Shinoj et al. did an economic assessment of a biodiesel value chain in India (Ariza-Montobbio and Lele 2010; Rajagopal 2008; Tompsett 2010). In the following section, we bring up the narratives of renewable, sustainable and clean sources of energy with specific focus on biofuels in India and how these narratives have contributed in facilitating recent jatropha programmes and missions in India even after or in the midst of scepticism of biofuels' trajectory.

Jatropha as Renewable and Sustainable Source of Energy

Biofuels have been presented as an allegedly sustainable and clean replacement to fossil fuels by policy framers and biotech companies (Carolan 2009; White and Dasgupta 2010). Although the concern about potentially detrimental effects of biofuels' large-scale cultivation and consumption in social, economic and environmental facets has been undermined by the policy actors (White and Dasgupta 2010; Shinoj, Raju and Joshi 2011). For example, rising food prices in 2007–2008 is linked with the increased blending of biofuels in the USA, which eventually emerged out as 'food versus fuel' debate in the world over (Paarlberg 2010). In the scenario of expected conflict between food and fuel, the Government of India claims that the Indian approach to biofuels is from international approach as it relies on jatropha—a non-edible feedstock for biofuels. As the NPB states '… it is based solely on non-food feed-stocks to be raised on degraded or wastelands that are not suited to agriculture', thus avoiding a possible conflict of fuel versus food security (GOI 2009: 9).

Generic claims promote jatropha to enhance energy security and climate change mitigation (IPCC 2011). Nonetheless, jatropha is also presented for more specific claims such as it can grow on 'wastelands' and would reduce the carbon debt and eventually it led towards 'low-carbon economy' (Fargione et al. 2008; Walker et al. 2010; Romijn 2011). Such claims on carbon reduction can be traced in an interview with a researcher from University of Hyderabad:

> We are focussing on how a plat fix the atmospheric carbon dioxide and what quality of biomass is produced. We are engaged in such work from almost thirty years. Later we got a project from DST-DBT, they asked us instead of looking other plants, to look this phenomenon in biofuel plants, whether these plants take up the carbon dioxide and convert into oils. So, my lab is interested in basically two aspects: carbon fixation and oil production. We go for trees which can fix lot of carbon dioxide. Carbon dioxide level has increased to 400 ppm, and as the carbon dioxide increases on atmosphere, temperature also increases. What we propose, we should look for those trees which can fix carbon dioxide and at the same time biomass can be treated for biofuels. In this context we proposed jatropha and pongamia. Fortunately, we identified jatropha among many species. It can fix a lot of carbon dioxide in comparison to other trees.

On a similar note, a senior scientist from ICAR-Central Research Institute for Dryland Agriculture, Hyderabad, concurred his opinion on land use for the cultivation of biofuels:

> If we go for large scale plantation jatropha can fix the rising carbon dioxide problem. For this we need marginal lands or waste lands, other agricultural land should not be used. Now the question is, whether jatropha can be grown in marginal lands as it is or not? It cannot be grown on marginal lands without care, but it has to be managed. Jatropha has potential to produce lot of oil and carbon dioxide fixation. If you can provide some care to the plant and wait then it is very useful.

Jatropha supports its promotional claims in terms of carbon balance in comparison with fossil fuels and biofuels produced from other feedstocks as long as its cultivation is restricted to non-farmland and fallow CPRs (Baka and Baka 2013). Moreover, claims such as minimal resource-input requirement and low land use impact have helped jatropha to contend it more sustainable and clean energy source than other sources of biofuels.

Debates surrounding jatropha's benefits and detrimental effects tend to align in two sets. On the one hand, high hopes derived after successful trials of biofuel as an alternative to jet fuel have provided optimism in the aviation sector; on the other, large-scale production raises the question of land diversion and shortage of food (Jha 2008). However, at the local level and domestic use, biofuel from jatropha would provide smoke-free fuel in place of fuelwood for the cooking purpose. It is also hoped that biofuels have potential to provide off-grid energy to the rural community, so they do not have to depend upon organized/grid energy for running small engines for agricultural and minor industrial purposes (Achten et al. 2008; Nielsen et al. 2013).

Environmental discourses centred around jatropha portray it as a tree rather than a shrub or a crop for energy. Therefore, jatropha has been associated with both reforestation programmes and tree-plantation drives not only for deriving biofuel but to earn carbon credits as well (Walker et al. 2010). Additional to restoration and expansion of forest cover, a few claims that jatropha can minimize deforestation owing to its roots capacity to firmly hold soil and pruned branches can provide shelter to other plants and it is helpful in the situation of flood and drought (Achten et al. 2010). A scientist from CSIR-Central Salt and Marine Chemicals Research Institute (CSMCRI), Bhavnagar, mentions about the robust character of jatropha that 'jatropha is a very hardy plant. The Berhampur plantation along the Bay of Bengal has sustained two super cyclones'. Another scientist from the same institute has similar opinion but on different aspect:

> The national agenda for biodiesel has been changed, new scenario is searching for new sources. Our prime minister has talked for solar energy on various occasions. From media what I am getting is solar energy is coming very fast, but other things are also going in parallel. Which one is sustainable for us, I cannot say? What we thought about biodiesel, it did not pick up. If you talk about solar energy it seems very well, but it will not help environment. With biodiesel we can support our environment and make it green.

The Director of Chhattisgarh Biofuel Development Authority comments on the enabling aspect of biofuel production with reference to environment suitability and economics of jatropha:

> After installing the plant, we proposed same to a few private companies. But the government did not come with clear policy. With all the characteristics of environmental friendly like no

Sulphur, no Lead, low viscosity, optimal pour point, safe flash point, and no pollution. We were producing biodiesel from jatropha, and the new government did not allow us to sell the same in market. They said, you cannot sell it directly in the market. Then, to whom we should give? If you supplement environmental economics, which was not considered while calculating its economic viability, with other benefits from jatropha it is quite possible to go ahead with this biodiesel crop. If you add up the environmental economics that how we are decreasing the pollution and what is the cost and implication of the pollution, nobody has done that. If that is being done what would be the net cost.

Promoters of jatropha argue that it does not only improve environment by reforestation, it does enhance local biosphere by controlling soil erosion, enriching soil nutrients, improving water retention and tackling floods. Such 'land improvising' add-ons portray jatropha as one of the prominent crops for 'restoring and reclaiming degraded lands' (Francis et al. 2005). However, there are concerns raised by a few actors who oppose jatropha plantation as a threat to biodiversity which would deteriorate the equilibrium of biosphere (Calvert et al. 2017).

The Scientific Claim-Making: Consensus and Contestation?

In this section, we attempt to understand the conflicting perspectives within an organization, presented as consensus of the organization to the receivers such as apparently the farming community as the receiver of scientific knowledge. This knowledge is presented in coherent and straightforward way to outsider. This selling of dissent (among scientific community) as consent is extended as per technocratic processes and mandates of the organization, viz. mandates of CSIR, ICAR, DBT, etc., on biofuels research. For example, how institutional mandates reflect in unexpected turns and outcomes of research projects, and the same can be observed in the response of a senior scientist from CSIR-National Botanical Research Institute (NBRI), Lucknow:

I am not very sure about policy, but we started working for jatropha in 2005 and very much engaged in it till 2013–14. We were mainly engaged in exploring the various environmental factors which could affect the plant, collection of germplasm, conducting field trial, and accessing the suitability of the plant for various regions especially for draught prone areas. The projects related to jatropha were short lived as compared to other projects we were engaged with. We can say its total failure main reason was economic viability, nothing to do with oil part/content. In research point of view, one cannot expect results in short duration, we should have patient for substantial outcomes. Now the mandate is shifted to biofuels from algal, but personally I feel algal is not viable for India.

Technocratic process embeds the epitomizing of valuing knowledge placed within the rigid cognitive system, excludes many local practices and actually established an 'ideology of extension' (Desai 2006). Knowledge is not understood as 'neutral' but as part of a power field that is continuously transformed and redefined by ongoing discourses and hegemonies. As Latour (1987: 133) writes, we should not forget 'the many people who carry (ideas and technologies) from hand to hand', we should turn towards the network and translations processes that form the arena in which these ideas are moved about. The process of knowledge extension, in spite of obvious hegemonic dynamics, produce pluralities that should not be ignored. In this sense, knowledge should be seen as 'set of practices' (ibid.). Significance of practices and the daily reality of laboratory process can be compared to what Pickering (1995) has called the 'mangle of practices'. With concept of 'mangle', Pickering offers a new approach to science and unpredictability of change that comes within it.

Drawing parallels from Pickering arguments, promotional and oppositional claims about biofuels are polarized. Debates surrounding the potential of biofuels as a substitute for fossil fuels have been vehemently countered by warning arising from unanticipated implications of biofuels. Actors in these debates make reference to scientific base to extend their claims. In such polarized scenario of scientific claim-making, it becomes undecisive for both general public and policymakers to identify and assimilate the scientific claims. In such conundrum, STS attempts to find out the role of science—'the conventional arbiter in disputes about factual matters' (Hansen 2014: 74)—in the controversies surrounding biofuels programmes and policies. Scientific uncertainty related to biofuels debate can be understood by bringing up the unsettled controversies in the domains of renewable energy and bioeconomy such as solar energy, nuclear energy, biotechnology and climate change. Opinion of one of the scientists from CSIR-CSMCRI, Bhavnagar, draws attention on choice of an energy source:

Because vegetation [which is supposed to spread by jatropha plantation] is required for environment, in barren land, and if you plant something over there, it definitely helps microbial flora fauna and also provides shelter to animals of the area. It also helps to reduce environmental pollution and soil erosion. Jatropha is very good plant for wasteland, but there is a myth that jatropha produces a good number of seeds. Though it can grow very luxuriantly in wasteland, rocky land and other different kind of land. We have tasted this in all over the country, it grows very well. But what happens, it also requires a good amount of inputs in the form of water and nutrients for healthy and good production. So, there is a limitation with the plant. If you do not want any production it is good, but if you want good production you have to plant some selected one. It requires elite plants, then only you can get good production.

She further emphasizes on the politics of choice:

The national agenda for biodiesel has been changed, new scenario is searching for new sources. Our prime minister has many times talked for solar energy. From media what I am getting is solar energy is coming very fast, but other things are also going in parallel. Which one is sustainable for us (India) I cannot say? What we thought about biodiesel, it did not pick up. If you talk about solar energy it seems very well, but it will not help environment. With biodiesel we can support our environment and make it green.

On the one hand, Winner (1980) and Beck (1992) bring up the political dimension of technologies that cause the controversies emanating from scientific uncertainties; Winner attaches inherent political attributes to technologies. On the other, Jasanoff (2004) points out the prominence of political competition behind the scientific controversy. Nonetheless, a few scholarships suggest that science and politics are encroaching into each others' boundaries (Nowotny et al. 2001).

The contestation between various scientific claims can be categorized into three conflicting perspectives about biofuels. First is a 'reductionist bio-processing' perspective emanating from biosciences and bioengineering disciplines, here changes at molecular level directed towards achieving divergent application of biofuel sources. Second is 'holistic bioscarcity' perspective emanating from ecology and life-cycle analysis (Hansen 2014). This perspective focuses on detrimental implications of biofuel on ecological cycle. And, the third perspective has apprehensions over adoption of biofuels in either situation. This perspective is in favour of biofuels in line with other biomass energy sources at rural level but not for large-scale production for industrial level.

The bio-processing perspective assumes that it is a generic process technology, by which any biomass can be converted into a plant-based equivalent of crude oil. This suggests that plants through photosynthesis process can capture a huge amount of energy from the sun. Enzymatic degradation of celluloses is one of the scientific advancements in bio-processing technology by which plant-based energy can be extracted. One of the researchers from IISc, Bangalore, has opined the similar approach for biofuels in India:

> We are ranked one in Biosynthesis potential of the land. If we use the Gangetic plain, we can produce enough food for the whole world. The farmers need income other than food. If they produce oil with the remaining potential which is not being used for the food crops, can be diverted for the oil production. We have an issue with market, if the production is more the prices will go down. Even in such cases if they produce both, they can get the price for both the food and the fuel.

Reductionist 'building-block' metaphor is the foundation for actors who subscribe to bio-processing for extending their scientific claims. For them, alteration at the molecular level can be employed to tap energy from organic matter through bio-process technologies (Hansen 2014). They provide a top-down approach to observe potential energy scarcity which is arising owing to suboptimal use of biomass and that can be resolved by regulating biomass policies as these resources are abundant and provide flexibility in use. Such bio-processes claim to address various macro-level issues of society by gradual interventions at molecular level. Thus, we can infer bio-processers couple societal problem with scientific accounts produced at molecular level in two facets. On the one hand, claims have been made for future changes based on technoscience interventions. Such estimates and claims are not established scientific facts rather fetch its authenticity from scientific community. On the other hand, the estimates divert the onus of outcomes/realization of the interventions to policymakers which is beyond the domain of scientific system (ibid.). This also reflects the asymmetry exists in scientific and policy domains, where scientific expertise barges into political validity and that makes it hard for non-scientist policymakers to deny scientific expertise. To tackle this asymmetry emerging from expertise claims, alternative scientific perspective is needed. 'Holistic bioscarcity' perspective is the response to counter such expertise claims.

Scientific accounts in opposition to biofuels promotion and production draw their arguments from holistic perspectives. Researcher from the discipline such as ecology, life-cycle analysis and environmental sciences represents holistic bioscarcity perspective. Scholarship from this perspective counters the claims of bio-processers that present the abundant availability of biomass. Accelerated depletion of fossil fuels has diverted the interests of the world to employ novel technologies to harness the energy from biomass for various purposes, viz. heat and power generation, fuel, lighting, along with the use of biomass for food production. As the development of these novel technologies follows a parallel trajectory, priority becomes the concern for the diversification of biomass for various purposes because it may lead to detrimental implications to land use pattern and exiting biomass application at local level. Therefore, researchers who advance holistic perspective have apprehension over the abundancy claim and argue that biomass would be a limited resource in coming time if its various application could not be rationalized. For example, according to one of the scientists from University of Delhi:

> If you start converting oil from soybeans or mustard you are taking away food in lieu of biofuel. So, biofuel is coming from food which is already in short supply. We import edible oil, for biofuel we need oil which is not edible. If we convert edible oil into fuel, we end up importing more and more edible oil. And edible oil is not cheap. What we look for an economic viable option for fuel, therefore that is another reason edible oil cannot be considered for biofuel, we need cheap biofuel. Biofuel from crops we need land, we cannot go for those land which are already being used for cultivation. What I need a plant which is not competing for land which is used for growing food. So, there are two things: first, edible oil is not an option and is not economic option, and the second, land issue, used for food crop. We will look for those lands which are not used for growing anything, which are wastelands. So, I need a crop which can be grown in wastelands. That is why jatropha was selected for fuel. But again, economics must be worked out. If I grow in road side, I need manpower to pick up the seeds, those seeds now extracted for the oils, and oil needs to be converted into biofuels. All these processes require money. And, even after you end up not getting quality seeds, you are spending more input costs than price of diesel. Jatropha has some stability issues. The biofuel is not very stable. So, in this area I was working on [if I can make it more stable] to make it more stable through increasing mono saturated fatty acids rather than poly unsaturated fatty acids.

Here, we can illustrate biofuels drawn from either first generation or second generation are not fulfilling the expectation. Findlater and Kandlikar (2011) also argue that availability of biomass, even the projections are in theoretical sense, is going to be scarce as the consumer demand for transport market of biofuels is much larger than the demand of global food requirement. And, in turn, that would have implications for land use pattern—diversification of land for biomass/biofuels production.

In the past two decades, technological advancement in liquid biofuels has been on the agenda of many developed and developing economies (IEA 2018). However, from the holistic perspective, it can be noted that transport sector has not been able to utilize biofuel optimally and there is a possibility of creating technological delay and lock-ins when the world is approaching for electric mobility or replacement of combustion engines by hydrogen/hybrid engines. Also, limited availability of biomass in the well-off parts of the globe is possibly supplemented by ecologically and economically sensitive regions of the globe. A senior scientist from CSIR-NBRI, Lucknow, points out the limitations arising from economic viability and environmental implications in the context of India:

Jatropha was taken back because there was no way how to harvest it. The plantation was labour intensive and costly. What inputs we were providing we were not getting back. That is why the government withdrew it. This was a fall, but, still I think jatropha is very good and a potential crop for biodiesel purpose. If you supplement environmental economics, which was not considered while calculating its economic viability, with other benefits from jatropha it is quite possible to go ahead with this biodiesel crop. If you add up the environmental economics that how we are decreasing the pollution and what is the cost and implication of the pollution, nobody has done that. If that is being done what would be the net cost.

The scientific perspective of ecology, environmental sciences and life-cycle analysis thus goes beyond from a macro-perspective of bio-processors to understand biomass as a limited and vulnerable resource of energy, and illustrates technological delay in transition and lock-ins in transport value chain holistically. The life-cycle analysts compare the estimated energy required to cultivate biomass, produce and process feedstock, and refine it for end-use purposes. On the basis of estimates, they depict the suboptimal usage of biofuels in transport sector and counter the claims of transport, agriculture and allied sector who advertise deployment of biofuels under 'green agenda' and climate change mitigation (Sarewitz 2004).

Some scientists have had apprehensions about large-scale jatropha cultivation for biofuel production. They are suspicious about a premature thrust to jatropha plantation. Inadequate understanding of the basic agronomics may result in very unsustainable agriculture. It can be traced in the suggestion of one of the scientists from CSIR-CSMCRI, Bhavnagar:

In genetic engineering front, most important problem we are facing is disease problem like viral infection and fungal problem. Their attack can kill a five-six-year-old mature plant within a span of fifteen days. [Sustainability problem] here, biotechnology is required to develop some disease resistant germplasm. We are also trying to develop genetic markers for conventional breeding and molecular plant breeding.

If we juxtapose bio-processor and holistic perspectives, inference can be drawn that the pluralities in scientific communication among the perspectives are governed by scientific diversity rather than scientific uncertainty over the politicization of biofuel debates. Hansen (2014) suggests that scientific diversity arises in a situation of addressing a scientific enquiry with different, though principally equal, perspectives and when the enquiry is associated with relevant policy concerns those are propagated simultaneously, thereby resulting to varied policy recommendations. These distinctions also germane to disciplinary differentiation of scientific knowledge production. Sarewitz (2004) refers it as 'an "excess of objectivity" where multiple, possibly conflicting, scientific claims are at the disposal of decision-makers simultaneously with no meta-criterion to adjudicate'. Therefore, 'excess of objectivity' in scientific and political enquiry creates a situation for general public and policymakers where complex scientific problem leads to ambiguity and indecisiveness in public understanding of science and policy decisions.

Is Biofuel Green? Case of Environmental Politics in a Village

In the previous section, we delineate the process of scientific claim-making. In this section, we attempt to foreground the social dimension of knowledge vis-à-vis expert knowledge, and further, we try to look into the formation of identities of an individuals or communities in a village of the central India where biofuel has been promoted and produced with the implementation of the National Mission on Biodiesel in 2003.

Jatropha has the capacity to survive in marginal edaphic conditions. Jatropha can thrive up to fifty years with less water, minimal nutrients and fertilizers. These are the reasons which portray jatropha as a hardy, draught-resistant and perennial plant (Fairless 2007). Such attributes make jatropha a prominent choice for biofuel production among various alternatives both in edible and in non-edible oil categories. Jatropha biodiesel can be blended with diesel for large-scale consumption at transport/industrial sector on the one hand, and on the other, at the rural level, it can supplement diesel and kerosene for agricultural purpose and can provide off-grid decentralized set-up for electricity generation. According to one of the respondents, working as an agricultural extension officer:

> We are trying to replicate practices and programmes to involve community for participation on a large scale. We are encouraging the stakeholders to adopt appropriate sustainable technology and path to achieve the goal. The model which we have developed, it encompasses and fulfils aims and objectives of thrust areas of biofuels, extending the activities to all other districts and biofuels a sustainable option. This [biofuel] park has competed five years, and it is successful in achieving the set objectives of conducting research, identifying elite lines, standardisation of protons and methods of trans-esterification of oils, designing oil expellers, decorticators, trans-esterification units, creating awareness among farming community, conducting training programs, setting up nurseries of elite lines, providing planting materials and arranging planting in identified land areas, and so on. Rapid international exploitation of fossil fuels and plateauing of domestic crude oil output have compelled the search for indigenous renewable and viable source of energy, leading to focus us on biodiesel. Our vision is to look forward for a green, healthy, pollution free environment, cistern then he nation, make it self-reliable in fuel industry and to gift the future generation with eco-friendly, replenishable fuel resource. Our mission is to develop decentralised people-oriented model, system and strategies the production of biofuels. To provide knowhow on biofuels from lab to layman's platform and educate them regarding long term advantages of biofuels and, to provide end to end solutions for biofuel programs with appropriate technologies and market network.

Laboratory studies of the 1980–90s (e.g. Latour and Woolgar 1979) have made the shift from common understanding of 'knowledge as system to understanding' to 'knowledge as process'. Scientific knowledge and practical knowledge constantly recreated and upheld because they help the different individuals and institutions to position themselves among their peers and within communities. Below here is the excerpt of a case of the village; here, we try to juxtapose this case and response of the previous respondent (the agricultural extension office—a bearer of scientific knowledge) to understand the politics of environment at local level, although which is not at all isolated from global politics of biofuels.

Now in many families nobody is active in village politics, either in the election of *panchayat* or *sarpanch*. Politics is so bad that it can destroy a simple person. My father is a very simple man, even today ninety percent of the villagers listen to him. If he would contest for *Vidhayiki* (member of the legislative assembly) he can easily win. But some people have damaged his image so badly in the case of beer plant that he is not able to recover from that shock of defamation. Around 5–6 years ago there was a proposal to set up a beer plant. Beer was not going to produce here but from the Indore plant, only bottling and storage were the purpose here in this proposed plant at our village. Husband of the previous Sarpanch, who is also relative to us and a very close friend of my father, opposed so fiercely to my father who was sarpanch at that time, just to create a space for the candidature for his wife in coming election for the post of sarpanch. Because of that protest our family submerged in debt of seven lakh of rupees. The company who was planning to set the plant gave 16 lakhs for the village on behalf of getting the no objection certificate (NOC). Out of that two lakhs were taken by *dalals* (mediators), and rest of the money was distributed among the other 40–50 villagers including *panchs*, so that people of this village would not do any protest for the plant. My father got only 1.5 lakh out of that huge money.

It was expected that this plant would bring a few facilities for the villagers, employment, education, etc. There was an agreement on five points: the company will put effort for development of the village, will open a college, a bigger hospital, construction of dams and plants. But people could not understand this, they were misguided by the person who is my relative and opposed the plant. Similarly, these people also protested the seed plant, in the name of pollution. These protests let village down, otherwise Godhi would have done better than Murmunda (nearby small market place serving the needs of nearby villages). But the political greed for becoming the sarpanch of Godhi destroyed the future of this village. Earlier this village was famous in entire region for its peace, fraternity, and hospitality; now after the protest the village has divided in factions, rivalry, and distrust. During that protest our family was not able to go out of home for three days. Entire village was gathered in front our house with tents and mike. It was a torturous to us, it was illegal also, but nobody came to our rescue. Till date we have not recovered from this trauma.

Other projects were opposed in the name of pollution. The sponge iron factory was fortunate, I think because it was the first one in this region, tough it is more polluting than the other proposed projects like seed plant and beer plant. And our village was less affected by pollution from this iron factory and other village like Kapasdha and Akola, because usually the direction of wind is in the direction of these villages. These village were severely affected by the iron factory. Other projects are also there, people have taken NOC and purchased the land, but after that protest, nobody is daring to start any project in this village.

In the case of Jatropha plantation there was no protest, because it is a government project and it is for the *Paryavaran Sanrakshan* (environmental conservation). Now there is a proposal to set a biofuel expeller plant beside the plantation. The NOC has not been requested by the authorities yet. I do not know what would be the fate of this biofuel plant by looking at the attitude of the village. Here the number of people is more those who do not understand (*Yaha nasamajh log jyada hai*). At least 200–300 people would have got some job by those projects, and unemployment would have vanished from our village, but people could not understand this. Now they are thinking and regretting. The youth of village is jobless, now the protesters are realising.

The above case is carrying the aspect of development through industrialization and therefore arising conflicts for power and resources. As per the Centre for Science and Environment (CSE) study in 1982, 'conflicts in development processes are essentially conflicts for control of these [natural] resources. The protest and struggles make compelling associations or connections between politics, the environment and resource inequalities' (Agrawal and Narain 1996).

Environment movement in India, in fact, can be characterized as a type of 'environmentalism of the poor insofar as they have combined green ecological concerns with the red of class politics' (Guha and Martinez-Alier 2000). The 'environmentalism of the poor' (ibid.) as a typology to explain new social movements in India, however for Baviskar (2008: 171–172), environmentalism, instead—as a form of political struggle—is shaped through a series of connections made between the 'environmentalist representation of social movements, the nature of capital that they oppose and their collaborations with metropolitan audiences and interlocutors'. Here, Baviskar is arguing in sharp contrast to Guha and Martinez-Alier that discourses on environmentalism in India have in actual fact drawn upon a range of sensibilities, ideologies and negotiations between varied groups and interests. In effect, social movements that have been characterized as ecological struggles against development initiatives have not only been 'ideologically hybrid' Baviskar (2008), but the strength of these challenges has also rested meanings that have been deliberately or otherwise kept 'ambiguous and shifting'.

Policy Implications

Globally, biofuels are promoted as environment-friendly substitute for fossil fuels, and developing countries combine this objective with rural employment generation and decreased dependency on imported crude oil. Repercussions of such large-scale adoption of biofuels have been in discussion in public policy debates. For example, increased production of biofuels, led by the overall energy demand, would diversify existing land pattern and include hitherto agricultural land into biofuel plantation. Major implications of changing land use pattern are twofold: on the one hand, it is making agrarian relations more skewed to marginal farmers; on the other, it is one of the reasons of agricultural induced environmental degradation owing to the fact that biofuel plantation requires as much resources as any other food crops, hence making it not carbon-neutral. Nevertheless, an oft-excluded implication of biofuel is emanating itself from policymakers, as biofuels subscribe to science, technology and innovation, and development policy options. Narratives are created and promoted by policymakers and policy-networks, and they often tend to serve the interests of these epistemic communities and in the process reduce the role of indigenous groups by justifying the role of experts and outsiders in the policy framing. These experts marginalize the interests of indigenous groups by labelling and categorizing them and tend to conceive the target groups as passive user of policy. Biofuel policies, unless coupled with public policies that consider the overall demand for energy, changing agrarian relations and significance of indigenous group in knowledge production, alone will eventually become unsustainable.

Conclusion

Although promotional claims, projections and the novel technologies depict an optimistic scenario about jatropha's capacity to serve as a feedstock for biofuels, this optimism is surrounded by the unanticipated implications of large-scale cultivation. One of the major reasons is the wild characteristics of jatropha; the research and development to domesticate jatropha are still in progress. On the one hand, it led to unpredictable yield rate and thereby varied production and projection of biofuel for commercial purpose. On the other, detrimental effects of biofuels on ecology and environment are yet to be fully analysed.

The idea of green development propagated through the Brundtland Report has had 'technocratic element in environmental and ecological thinking' (Adams 2008: 57), which furthermore reflected the 'authoritarian idea that environmental harmony should be sought through central control'. Technology and rational management, therefore, were presented as being politically neutral interventions and that conflicts could thus be literally planned away without any transformative change or political and radical challenge when environmentalism is merged with development and offered as expert policy and good science by the developed world to the developing countries.

In the section of scientific claim-making, we referred Pickering's 'mangle' in the process of resistance and accommodation performed by actants and actors and how they set the locus of any scientific experiments (Pickering 1995). Discursive flexibility associated with jatropha has supported its chances of survival in the case of failure. Through discursive flexibility, jatropha is advanced by the claim that it can fulfil a set of goals depending upon the requirement of market and production chain.

In the last section, we attempted to see the relationship between knowledge (scientific knowledge) and 'social order developed' (Knorr-Cetina 1981) by it in the biofuel promotion and extension. That is, knowledge production culminates in knowledge consumption such as in the case of farming communities engaged in jatropha cultivation. This means that agricultural knowledge extension must consider the historical and social environment it operates in as we saw in the case of village politics over power and resources. Thus, agriculture extension programme for biofuels is a manifestation of relations of the different purposes and agents involved, not only the environment sustainability or assumed energy crisis.

End Notes

1. Carbon capacity indicates the level of carbon dioxide emissions as per the mandated limit without causing significant global warming effects. This implies that there is a limited scope for new fossil fuel projects if climate change mitigation measures are taken into consideration.
2. Borras et al. (2016) put the limitations for a crop to be treated as a flex crop. Crops with multiple uses do not automatically qualify for flexibility. They set three conditions to become a flex crop, namely material basis, technological

possibilities and profit viability. 'If a crop or commodity use can be switched from one specific purpose to another with technical ease and with attractive economic return, then a link between multiple-ness and flexible-ness may have far-reaching political economic implications'.

Acknowledgements We are grateful to our participants who are\were engaged in cultivation of jatropha particularly the farming community for reflecting on their experiences. We are thankful to CBDA, ICAR, CSIR and the universities for allowing us to carry out fieldwork.

References

Achten, W. M. J., et al. (2010). Jatropha: From global hype to local opportunity. *Journal of Arid Environments, 74*(1), 164–165.

Achten, W. M. J., et al. (2008). Jatropha bio-diesel production and use. *Biomass and Bioenergy, 32*(12), 1063–1084.

Adams, B. (2008). *Green development: Environment and sustainability in a developing world.* Routledge.

Agrawal, A., & Narain, S. (1996) [1985]. *The state of India's environment 1984–85.* Centre for Science and Environment.

Ariza-Montobbio, P., & Lele, S. (2010). Jatropha plantations for biodiesel in Tamil Nadu, India: Viability, livelihood trade-offs, and latent conflict. *Ecological Economics, 70*(2), 189–195.

Bailis, R. E., & Baka, J. (2010). Greenhouse gas emissions and land use change from Jatropha curcas-based jet fuel in Brazil. *Environmental Science and Technology, 44*(2), 8684–8691.

Baka, J., & Baka, J. (2013). The political construction of Wasteland: Governmentality, land acquisition and social inequality in South India. *Development and Change, 44*(2), 409–428.

Baviskar, A. (2008). *Contested grounds: Essays on nature, culture, and power.* US: Oxford University Press.

Beck, U. (1992). *Risk society: Towards a new modernity.* Sage.

Borras Jr, et al. (2016). The rise of flex crops and commodities: Implications for research. *The Journal of Peasant Studies, 43*(1), 93–115.

Calvert, K. E., et al. (2017). Geographical perspectives on sociotechnical transitions and emerging bio-economies: Introduction to a special issue. *Technology Analysis & Strategic Management, 29*(5), 477–485.

Carolan, M. S. (2009). Ethanol versus gasoline: The contestation and closure of a socio-technical system in the USA. *Social Studies of Science, 39*(3), 421–448.

Carson, R. (2002) [1962]. *Silent spring.* Houghton Mifflin Harcourt.

Demirbas, A. (2009). Political, economic and environmental impacts of biofuels: A review. *Applied Energy, 86*(1), 108–117.

Desai, B. (2006). Inside out: Rationalizing practices and representations in agricultural development projects. In Development brokers and translators: The ethnography of aid and agencies (pp. 171–195).

Fairless, D. (2007). The little shrub that could: Maybe. *Nature, 449*(October), 652–655.

Fargione, J., et al. (2008). Land clearing and the biofuel carbon debt. *Science, 319,* 1235–1238.

Findlater, K. M., & Kandlikar, M. (2011). Land use and second-generation biofuel feedstocks: The unconsidered impacts of Jatropha biodiesel in Rajasthan, India. *Energy Policy, 39*(6), 3404–3413.

Francis, G., Edinger, R., & Becker, K. (2005). A concept for simultaneous wasteland reclamation, fuelproduction, and socio-economic development in degraded areas in India: Need, potential and perspectives of Jatropha plantations. *Natural Resources Forum, 29,* 12–24.

Government of India (GOI). (2003). *Report of the committee on development of biofuel*. New Delhi.
Government of India (GOI). (2009). *National policy on biofuels*. New Delhi.
Guha, R., & Martinez-Alier, J. (2013) [2000]. *Varieties of environmentalism: Essays north and south*. Routledge.
Hansen, J. (2014). The Danish biofuel debate: Coupling scientific and politico-economic claims. *Science as Culture, 23*(1), 73–97.
International Energy Agency (IEA). (2011). *World energy outlook 2011*. Paris.
International Energy Agency (IEA). (2013). *World energy outlook 2013*. Paris.
International Energy Agency (IEA). (2018). *World energy outlook 2018*. Paris.
IPCC. (2011). IPCC special report on renewable energy sources and climate change mitigation. In O. Edenhofer, R. Pichs-Madruga, Y. Sokona, K. Seyboth, P. Matschoss, S. Kadner, et al. (Eds.), *Prepared by working group III of the intergovernmental panel on climate change*. Cambridge, UK: Cambridge University Press.
Jasanoff, S. (2004). Ordering knowledge, ordering society. In S. Jasanoff (Ed.), *States of knowledge. The co-production of science and social order*. Routledge.
Jha, A. (2008). *Jatropha-fuelled plane touches down after successful test flight* (p. 30). Decemver: The Guardian.
Knorr-Cetina, K. D. (1981). *The micro-sociological challenge of macro-sociology: Towards a reconstruction of social theory and methodology*. Routledge and Kegan Paul.
Kovarik, B. (1998). Henry Ford, Charles F. Kettering and the fuel of the future. *Automotive History Review, 32*, 7–27.
Latour, B. (1987). *Science in action: How to follow scientists and engineers through society*. Harvard university press.
Latour, B. (2005). *Reassembling the social: An introduction to actor-network theory*. Oxford University Press.
Latour, B., & Woolgar, S. (1979). *Laboratory life: The social construction of scientific facts*. Sage.
Meadows, D. H., Meadows, D. H., Randers, J., & Behrens III, W. W. (1972). *The limits to growth: A report to the club of rome*. New York: New American Library.
Nielsen, F., et al. (2013). *Jatropha for local development: After the hype*. Amsterdam: Hivos.
Nowotny, H., Scott, P., & Gibbons, M. (2001). *Re-thinking science*. Cambridge: Polity Press.
Organisation for Economic Co-operation and Development (OECD). (2009). *The bioeconomy to 2030: Designing a policy agenda*. Paris.
Oxfam. (2008). Another inconvenient truth: How biofuel policies are deepening poverty and accelerating climate change. *Oxfam briefing paper* 114. Oxford: Oxford International.
Paarlberg, R. (2010). *Food politics: What everyone needs to know*. New York: Oxford University Press.
Pepper, D. (1996). *Modern environmentalism: An introduction*. London: Routledge.
Pickering, A. (1995). *The mangle of practice: Time, agency, and science*. University of Chicago Press.
Rajagopal, D. (2008). Implications of India's biofuel policies for food, water and the poor. *Water Policy, 10*(1), 95–106.
Romijn, H. A. (2011). Land clearing and greenhouse gas emissions from Jatropha biofuels on African miombo woodlands. *Energy Policy, 39*(10), 5751–5762.
Sarewitz, D. (2004). How science makes environmental controversies worse. *Environmental Science & Policy, 7*(5), 385–403.
Shinoj, P., Raju, S. S., & Joshi, P. K. (2011). India's biofuels production programme: Need for prioritizing the alternative options. *Indian Journal of Agricultural Sciences, 81*(5), 391–397.
Sismondo, S. (2011). *An introduction to science and technology studies*. Second: John Wiley & Sons.
Smith, J. (2010). *Biofuels and the globalisation of risk: The biggest change in North-South relationships since colonialism*. London: Zed Books.

Tompsett, C. (2010). Fuelling development? A critical look at government-centred Jatropha cultivation for biodiesel as promoted by the biofuel policy in Rajasthan, India. Master's thesis, The University of Bergen.

Visvanathan, S. (1991). Mrs. Brundtland's disenchanted cosmos. *Alternatives, 16*(3), 377–384.

Walker, G., et al. (2010). Renewable energy and sociotechnical change: Imagined subjectivities of 'the public' and their implications. *Environment and Planning, 42*(4), 931–947.

WCED (UN World Commission on Environment and Development). (1987). *Our common future.* Oxford University Press.

White, B., & Dasgupta, A. (2010). Agrofuels capitalism: A view from political economy. *The Journal of Peasant Studies, 37*(4), 593–607.

Winner, L. (1980). Do artifacts have politics? *Daedalus, 109*(1), 121–136.

Chapter 16
Alternative Approaches to Measure Sustainability in a Subsistence Economy: Empirical Insight from Shifting Cultivation in Odisha

Amalendu Jyotishi and M. Manjula

Abstract This chapter attempts to critique the existing and popular criteria of measurement of sustainability, especially in the context of subsistence economy. It empirically verifies the alternatives in the context of shifting cultivation in Odisha. Shifting cultivation or swidden has been a widespread form of land use since Neolithic period. Globally, it has been viewed as an unsustainable form of agriculture in the mainstream policy. Therefore, it is an appropriate context to test the alternative/complementary criteria for measuring sustainability. In the process, the chapter suggests three alternative/complementary ways to address sustainability in a subsistence economy based on shifting cultivation. These are, carrying capacity of the land, employment, and consumption needs. The empirical study is based on the shifting cultivation dependent villages in Odisha. The initial study was conducted in the year 2000. Later on, one of the researchers revisited to the region after a gap of 17 years to verify how these three criteria are useful in understanding sustainability of one of the earliest forms of agricultural practices. New policies in this direction also have played critical role in manifesting these three criteria. This chapter shows how policies, awareness and collective action play critical role in reshaping the alternative sustainability goals.

Keywords Shifting cultivation · Alternative approaches to sustainability · Subsistence economy · Cost–benefit analysis · Carrying capacity · Monetization · Odisha

Introduction

Sustainable use of natural resources has gained more importance in economic literature in last few decades. The neoclassical perspective on this is ideologically specific

A. Jyotishi (✉) · M. Manjula
School of Development, Azim Premji University, Bangalore, India
e-mail: amalendu.jyotishi@apu.edu.in

M. Manjula
e-mail: manjula.m@apu.edu.in

© The Author(s), under exclusive license to Springer Nature Singapore Pte Ltd. 2021
M. K. Verma (ed.), *Environment, Development and Sustainability in India: Perspectives, Issues and Alternatives*, https://doi.org/10.1007/978-981-33-6248-2_16

with its focus on market and prices. However, subsistence economic activities are ill-represented in such analysis largely due to the absence of appropriate price-discovery market mechanism on the one hand, and the opportunity cost of the existing livelihood of the population on the other. In addition to these, subsistence economies, especially those dependent on natural resources (as in the case of indigenous communities), are not highly monetized that restricts the economic integration with the mainstream formal economies. In such cases, evaluating sustainability of a subsistence economic system using tools of modern formal economies would not only be inappropriate but also ill-represent the system. Shifting cultivation or swidden system is a case in point which is in existence since Neolithic period and has been largely practised by indigenous communities in almost all parts of the world.

Conventionally, shifting cultivation (also known as Swidden) has been interpreted as an inefficient (economically) and destructive (ecologically) form of agriculture. More recently, shifting cultivation is viewed as an inflexible static system (institutionally) ill-suited for adapting to changes brought about by modernity. This latter view, as illustrated in World Bank study (1992), holds that it slows agricultural production and causes ecological degradation. At the core of these issues is the sustainability of shifting cultivation system. In this chapter, we attempt to verify the criteria of measurement of sustainability in a relatively primitive form of agricultural system and develop some alternative/complementary ways those help understanding sustainability in the natural resources dependent subsistence economy.

Measuring Sustainability: A Critical View on CBA and Exploring Alternative Approaches

The term sustainability has varied meaning in literature. However, in its crudest form, it can mean that no one will be worse off, if not better off, over a long period of time.[1] Often, sustainability of a project or production system is verified through usual cost–benefit analysis (CBA). CBA is a useful tool to understand and evaluate a system of production, however, due to certain inherent flaws and tendency to ignore some important social aspects it has been criticized by many economists and project planners. The issue of incommensurability and problem of monetary valuation and discounting are major criticisms of CBA (Aldred 2006; Ackerman and Heinzerling 2004). The failure of CBA to measure costs and benefits accurately might lead to biased estimates (Hwang 2015). CBA also fails to account for social externalities like changes in social capital and community well-being (Hara and Messner 2000). The ability of CBA to address inter- and intragenerational equity issues is also debatable (Gasparatos et al. 2008). The other methodological drawbacks of CBA are the attempt to value benefits—*which often have dignity and inner worth and cannot be assigned a price*—on the same scale as costs which have price; the assumption of universal trade-offs; the inability to precisely estimate risks and benefits under uncertainty (Ackerman 2008; Gasparatos et al. 2008). In our study, we observed that the essence of understanding the sustainability of shifting cultivation system requires the understanding of factors, which are local in nature. First, it is the capacity of the land to produce for supporting a stream of population over a period of time. In other words, it is important to understand what number of people the system can accommodate and for how many years. Land carrying capacity (LCC) estimates explain how effectively the local land resources support lives and livelihoods. Ecological footprint analysis (EFA) and index system method (ISM) are used to measure land carrying capacity (Quian et al. 2015). EFA is observed to be suitable in resource-output regions, while ISM is observed to be useful in resource-input regions. In the presence of complex multifactorial determinants of sustainability, ISM is proven to be a better approach (Quian et al. 2015). Kessler (1994) integrated the concept of agricultural sustainability into LCC. Komatsu et al. (2005) used the concept to study the linkage between agricultural sustainability and combating desertification to evaluate the influence of land conversion policies on rural supply-demand balance. The second important factor is not only the capacity of land to support a group of people but also for how many people the system can provide employment opportunity. Often, the population support from land perspective does not match with the population support from employment perspective. Though the carrying capacity of the land becomes a necessary condition for use of land resources, its sustainability depends on the number of people it gives employment opportunity. The third important factor that gains importance in understanding the sustainability of shifting cultivation system is, how far the system is capable of feeding the people, i.e. to what extent the system supports the consumption needs of the population dependent in the practice of this system of agriculture. At the core of the third proposition lies the food security issue. However, it becomes more complicated when we link the production system to the monetary system. Where the highly monetized economy may show integration between production and consumption systems in real and monetary terms, it may not be the case in a low monetized economy. Substantive economies often show the tendency of low degree of monetization. Therefore, consumption aspects need further probing into the degree of monetization and its implications on employment and food security aspects.

Cost–benefit analysis (CBA) is a part of the t neoclassical tradition and is currently the dominant approach to decision-making in public domain. This approach seeks an identification of all kinds of impact connected with the alternative courses of action considered and a systematic comparison of these impacts, using money as the 'common denominator'. The analyst aims at one value for each alternative, for instance a net present value, suggesting the relative 'efficiency' of that alternative. The idea behind cost–benefit analysis is that all impacts can be traded against each other in monetary terms, and also the analyst can refer to the 'correct' price of each impact for purposes of societal evaluation. When there is no actual market, reference is made to an imagined or 'shadow market'. Use of CBA acknowledges a number of limits. The criticisms offered at this level in turn would be a starting position for the elaboration of alternative approaches to societal decision making. It is argued that the limits to CBA become accentuated in relation to environmental issues. Soderbaum (1998) identifies a number of reasons for which CBA would not be a good choice, specifically for environmental problems which are multidimensional, multifactorial and complex. First, to make an attempt to deal with multidimensional impacts in one-dimensional terms may not be the best strategy. It is true that analysis invariably involves some simplifications, but higher degree of simplification may imply a loss in relevance. Second, if information about impact is uncertain and fragmentary, then an attempt to express all aspects in one-dimensional terms becomes even more daunting. Third, in the cases of irreversible environmental impacts, the references to present consumers and their monetary valuation seem strange, since the impact concerns future generation as well. Assuming that CBA framework can suggest specific prices in these cases, the relevance of such prices and estimated present values to citizens and decision-makers becomes nebulous. And fourth, assuming that the analyst is able to suggest a price in monetary terms for each impact, according to the rules of CBA, these rules will then necessarily represent a specific ideology or, more precisely, market ideology. What is perceived as public issues are reduced to private issues of willingness to pay for real and imagined commodities. A person, therefore, can essentially be reduced to a consumer of each impact that can be traded against other impacts, in monetary terms.

However, the criteria followed for measuring sustainability in a typical cost–benefit framework has not gone beyond internalizing the externality of environmental degradation in monetary terms. Cost–benefit analysis (CBA) is held as the method of land use choice, since past several decades (Taylor 2001). Despite considerable reforms in recent years towards understanding CBA, the main criticism against this method still upholds because of its limitations of market-oriented approach, which has not developed in a very comprehensive manner specifically in developing countries. In CBA approach, we measure the net income/benefit streams discounted from the future and tend to conclude that discounting the future will take care of the ecological aspects of the resource. Here, there are two important points, which draw our attention while evaluating the land use choice, specifically in a developing country. First, a specific type of land use is chosen, because it ensures certain parts of labour use in the economy. This should be understood in the backdrop of opportunity cost of labour, specifically in a developing country, where opportunity cost of labour is nearly zero. Second, in an economy where people are poor, land use choice is also guided by the factors related to the consumption needs. Therefore, any evaluation of land use choice only with CBA and without considering the employment and consumption factors would be partial, and hence, may not be sustainable. Traditional CBA gives no indication of who benefits from the economic activity and how the consumption needs are derived. Developing regions are subject to high degree of income inequality and capital flight (Todaro 1994). Thus, it is quite possible that significant amount of the monetary benefits of a development project will not accrue to a broad base of regional habitants. On the other hand, in a similar subsistence economy, where product and factor markets are distorted,[2] the direct consumption would ensure standard of living in a better way than indirect consumption via a distorted market. A point which holds for both employment-based analysis (EBA) and consumption-based analysis (CoBA) is on population support. This also prevents improvement of some subgroup (in terms of rewards) at the cost of declines in welfare in others.

Carrying Capacity of Land

One of the important points on sustainability of shifting cultivation system is the *carrying capacity*. Carrying capacity is generally defined as the human–land balance,[3] which is maintained by the native populations practising simple food-producing methods. In other words, it refers to the number of individuals that can be supported in a given area, the level of consumption they are to be supported and the time the area is to be capable of providing the support.

Some human carrying capacity estimation techniques determine when carrying capacity has been exceeded by some behavioural change in the population. Such behavioural changes indicate that the rate of production is unsatisfactory by the population's own culturally defined standards. In this study, some broad indications can be deduced related to sustainable carrying capacity. To the extent, say for example, the ten-year minimum fallow period traditionally in use in the sparsely populated areas appears to be sustainable, whereas the two-year fallow in the densely populated areas results in visible environmental degradation.

The information provided by carrying capacity estimates, when coupled with information from other studies concerning changes in soils, yields and vegetation under different fallowing regimes, can lead to useful conclusions about sustainable population levels, with appropriate assumptions about technology and consumption. However, this case may occur if the technology of production is stagnant. If the swiddeners are adaptable to the technological mix, then there is all possibility that they will go for different types of land usage to enhance the carrying capacity of the land without any visible impact on the environment. Therefore, it is essential to observe the dynamics of land use management, instead of observing shifting cultivation in isolation.

The basic definition of sustainable carrying capacity is patterned after Allan's (1949) pioneering work on estimating carrying capacity for shifting cultivators in Zambia. He defines carrying capacity as *the maximum numbers of persons that can be supported in perpetuity on an area, with a given technology and set of consumptive habits, without causing environmental degradation.* Later, many other researchers tried to formulate carrying capacity (Brush 1975; Carneiro 1960; Conkline 1959; Fearnside 1972, 1986; Gourou 1966; Hayden 1975; Marten and Saltman 1990; Verneer 1970). These formulae for calculating carrying capacity under systems of shifting cultivation can be reduced algebraically to a common form (Faechem 1973). Faechem further reduces the results into an expression indicating that the ratio of what he calls 'theoretical population' to the current population is equal to the ratio of the land available to the land in use.

Employment-Based Approach

EBA evaluates economic activity on the basis of employment generated or population supported by each activity or development path, using a given amount of underlying resources (Taylor 2001). A comparison of swidden with modern logging Dove (cited in Banerjee 1995) concludes that not only are net returns higher than commercial logging, but shifting cultivation also supports more than three times as many people. For a developing economy, larger subsistence support would be a more important criterion for choice of land use. Hence, the basic notion of EBA is to count the number of jobs that an economic activity provides or the number of people it supports over a given period of time. In other words, while quantifying the level of employment, the objective of EBA is to provide numerical measures of *sustainable employment*.[4]

Activities like mining that may provide high 'immediate employment'. However, such activities if based on unsustainable practices would decline within a foreseeable time. Therefore, time horizon of labour engagement is an important criterion of EBA from a sustainability perspective. Based on this, activities can be identified not only as short-term or long-term, but also various other dimensions including seasonality or periodicity of recurrence of the activities. Based on the nature of engagement of labour, it is also possible to identify if there are other specificities including special-ization, gender division of activities, control and ownership. Given these nuances associated with EBA, using discounting future to arrive at a present value would be problematic. As an over-simplistic comparison process, one can compute number of job-days created or proportion of population supported by the activity. Avail-ability of multiple data points across space and over the time may provide relia-bility to such data. Similarly, based on the availability of other resources (natural and physical), space, ownership and user right over the space would also deter-mine the possibilities of labour engagement in perpetuity. The main reason behind ignoring employment-based approach in measuring sustainability, especially by the mainstream economists, is due to its none-flexibility in computation, especially in temporal analysis, in spite of labour engagement or employment being one of the most direct and important measures of welfare. As jobs or employment cannot easily be translated into future purchasing power, bankability, deferred payment, reinvest-ment, transfer, etc., are widely used in temporal analysis of mainstream economics. As Taylor (2001) says, it would serve little purpose to try to transfer jobs, say, from the current generation to future generations in the interest of sustainability. For an activity that targets a given group of population at a given location, discounting and/or transferring jobs over a timeline or across generations is contrary to the general notion and purpose of sustainable development. Furthermore, the intergenerational equity that is inherent in sustainability is better captured without the use of discounting.[5]

Consumption-Based Approach

Consumption-based approach (CoBA) evaluates economic activity on the basis of consumption needs that can be generated from a given land use. Like employment, consumption-based approach provides information regarding how much self-sufficient the communities are in terms of what they produce. In other words, consumption-based analysis can be a good indicator of sustainability, if the communities are able to feed themselves adequately without depending much on external sources. In fact, local availability of resources with least dependency on outside world is one of the components of sustainability (Ramakrishnan 1992). CoBA takes into consideration the production and consumption at the local level. Once it is understood that markets are distorted in the region (specifically for product markets), consumption via market will lead to weaker bargaining power for the population depending on shifting cultivation and associated forms of agriculture. In such cases, directly consumption-based analysis provides information regarding the adequacy of food supply, which is not determined by price. In other words, consumption-based approach enervates the role of price in economic behaviour of decision-making. This assumption would, however, be more realistic in a situation where the dependency on market is less and hence the degree of monetization. On the other hand, market-based analysis depends on a price-making market structure, which largely ignores the factor that how the price is determined or who determines the price. In one of the earlier papers (Jyotishi and Manjula 2017), we have identified that various types of markets operate (specifically for product market) in our study area. Among these, monopsony type of market is dominant and covers most of the commodities. Therefore, deciding land use activity, which depends largely on market (specifically where market is distorted or socially embedded and not competitive in neoclassical terms), would not be appropriate. Consumption-based approach is based on locally produced consumption and not consumption via a price-making market system. Consumption-based approach takes two components into account. First, if the degree of monetization is unfavourable towards understanding an economic phenomenon through a market-based study, consumption-based analysis provides a better alternative and insightful information on a land use activity. Second, it also takes care of the food security dimension of the local population.

Another important advantage of consumption-based approach over the cost–benefit approach is that it gives emphasis to present generation over future generation in terms of sustainable food production and its local availability. If this generation's land use activities substantially do not take care of the food needs, then this may result in shift in the activity, which may induce land use and ecological change. Secondly, the consumption-based land use activities are always need-based and hence, do not enter into the activities which are inflated (or distorted) by market mechanism. Here, one thing is clear that function of market is comprehensive and formal when it can operate in a competitive environemt what Karl Polanyi terms as "exchange at a bargained rate". Substantive economies like that of shifting cultivation operate in the forms that are informal and socially embedded. Formal and comprehensive exchange is just a fraction of total economic activity in the shifting cultivation ssystem. Consumption-based approach is sustainable by design as it emphasizes on land-use practices that emphasizes direct consumption needs in perpetuity.

Consumption-based approach also has edge over the cost–benefit analysis as it is more equitable. Cost–benefit analysis generalizes the benefit aspect and hence neglects the equity share of the benefit flows. CoBA, on the other hand, looking into the consumption needs of the local community, takes care of the equity aspect to a large extent.

Degree of monetization provides more explanation on the nature of the economy, though it is an under-explored area of economic explanation.[6] An economy may be designated as a *Non-Monetized Economy* where the exchanges through a market form are absent or negligible. Such economies are largely based on self-consumption of produce using the family means or organized through the systems of reciprocity and redistribution more than the formal market-dependent medium of exchange. In other words, the dominant aim of production remains one of the self-consumptions with a minimum of commodity production. On the other hand, a *Monetized Economy* or credit economy is one where commodity production takes place through the presence of market (whatever may be the form of market), thereby suggesting the exchange of goods and services, and the associated division of labour in production.

To characterize an economy as non-monetized economy sounds unscientific, as the classification is purely on the basis of a mode of exchange, while scientific classification is based on the mode of production. Non-monetized economy by its very nature belongs to a pre-capitalist period. However, the peculiarity of mode of exchange in the economy of swidden agricultural system on the one hand identifies the correspondence between natural economy and pre-capitalism and on the other hand shows the interface of capitalism through a peculiar form of exchange relation.

Monetization is a conventional index of showing the extent of monetary transactions (largely through cash) made in an economy out of the total transactions of the economy, giving it as a ratio of one to the other. The agricultural sector specifically comprises of both the components of *Non-Monetized Economy* and *Monetized Economy,* whereas non-agricultural sector belongs to only monetized economy. The growth of monetized sector therefore should be (i) mere transactions between the non-agricultural sector and (ii) growing transactions between the non-agricultural sector and agricultural sector.

Methods and Materials for the Empirical Context

Having discussed the alternative/complementary methods of measuring sustainability, we intend to empirically verify it in a specific context. To retreat, following are the objectives of the study:

(a) Understanding the issue of sustainability from the point of view of carrying capacity of land, where land use intensity as a factor explains the carrying capacity.

(b) Employment is a desirable indicator of sustainability, specifically in agrarian sectors of developing economies where off-farm employment opportunities are very less. This, otherwise, also means that in a traditional agricultural system where opportunity cost of labour is nearly zero, employment could be a better indicator of sustainability. Therefore, our objective is to understand the employment pattern among the shifting cultivators.

(c) Sustainability can also be understood from the consumption requirement and how much the present system of shifting cultivation is capable of feeding the population. In addition, it is important to understand the extent to which the economy is dependent on transactional monetized system to fulfil the consumption requirments. Therefore, another imperative objective of the study is to analyse the consumption pattern and factors determining that.

The issues pertaining to shifting cultivation being of ecological, economic and institutional importance, the criteria for selecting field area have to be guided by a few conditions. We identified varying shifting cycles and communities practising shifting cultivation system, ecological and agro-climatic zones along with different types of land use practices in Odisha. Secondly, we decided to study a complete village or hamlet domain for a better understanding of the functioning and linkages of micro-institutions as well as the ecological aspects of land use management and crop management in its totality. Based on the study by Anthropological Survey of India (Bose 1991), all the blocks under shifting cultivation were identified. Accordingly, shifting cultivation in Odisha was divided into three zones, namely north, central and south zones. Among these zones, central zone has the highest extent as well as diverse conditions of topography. Therefore, this zone was preferred for the study, keeping the diversity of shifting cultivation in mind. However, we revisited other regions in later period of time (Jyotishi and Manjula 2017). Within the central zone, data related to slope, altitude, rainfall, accessibility to road, population density and communities practising shifting cultivation were available. We were also careful about selecting different communities and diversity of agricultural practices associated with shifting cultivation in order to select the blocks. Accordingly, four blocks were chosen for the study, namely Kashipur, Muniguda, Bissam Cuttack and Raigada. Five villages from these blocks were considered for primary survey. Interestingly, these blocks also fall into three ecologically important zones, namely Bafalamali, Niyamgiri and Mahendragiri; the first two regions were from Rayagada District and the second one from Gajapati District of Odisha. One village (Brhamarjodi) was chosen from the Bafalamali hill range, whereas two each from Niyamgiri (Sakota and Gandli) and Mahendragiri (Badamasingh and Kalinga) were chosen. The villages chosen have varying shifting cycles ranging from 6 years to 12 years with different cropping patterns, institutional set-ups, different indigenous communities, mode of production, topography, proximity and linkages with market with differing complementary land use systems like horticulture, plantation, terrace cultivation, root crops and fruit-growing farms. Topographically, also the blocks, under which the villages fall, have different slopes and altitudes and different ranges of average rainfall. Besides, the population density as well as accessibility to roads (which can be considered as a proxy for accessibility to market) of the regions is also different. While these villages were considered for the study in the year 2000, we revisited a set of another 6 villages in central and southern region of Odisha in 2015 (Jyotishi and Manjula 2017) where we discussed the institutional changes over those 15 years. Subsequently, we visited northern region of Odisha in 2016 to reflect on the institutional issues as well as sustainability dimensions. A postscript in this chapter reflects on these revisits founded on the comprehensive work done in the year 2000.

For the empirical analysis of this study, qualitative and quantitative data were collected from the five villages considered for our study. Altogether, there were 125 households in these villages that are our unit of study in most of the cases. The five villages chosen for the study have the population depending on shifting agriculture as their main source of livelihood. Besides they engage themselves in other forms of agricultural practice and also collection of various non-timber forest produces. It is also evident from Table 16.1 that the availability of plain land in all the five villages is very low. Therefore, most of the agricultural practice depends on the higher slopes.

Table 16.1 Village-wise indicators

Village	Brhamarjodi	Sakota	Gandli	Badamasingh	Kalinga
1. Community	Paroja	*Dongria Kondh*	*Dongria Kondh*	*Saura*	*Saura*
2. Total HH	29	21	26	32	17
3. Population	136	84	116	148	77
4. Total geographical area (in acres)	420.84	560.07	866.49	358.275	234.41
5. Plain land (in acres)	58.5	66.92	15.06	11.4425	20.155
6. Wasteland (in acres)	333.75	470.07	825.01	268.2425	107
7. 5 as percent of 4	13.90	11.95	1.74	3.19	8.60
8. 6 as percent of 4	79.31	83.93	95.21	74.87	45.65
9. Shifting cycle (in years)	8	10	7	6	6
10. Land use intensity 'R'	25.00	20.00	28.57	33.33	33.33

Note 'R' value is the land use intensity as described by Ruthenberg, i.e. $R = (t'/t'') \times 100$, where 't' is the years in the cropping-and-fallow cycle, while 't = 1' is the year of initial clearing and first year of cropping, where t' is the final year of cropping and t'' is the final year of fallow of the crop-fallow cycle (Ruthenberg 1971)

Source Based on the land records of each village from the revenue department

Shifting cultivation is performed on land that is classified as wasteland in the village revenue records. It was evident during the fieldwork that there is no land right issued by the state for cultivating these lands. However, the communities have claimed de facto right over these lands and they continue their agricultural practice in this terrain. The term wasteland is an oversimplification, in the respective revenue records of the villages. However, on the village site, one would identify various types of vegetation covers. These include swidden land under current cultivation, fallow land at different stages of re-growth, plantations like cashew, plantain, pineapple, fruit-bearing trees like jackfruit, tamarind, mango, etc., and bushy forest as well as forest with substantial wood growth. These villages are surrounded by forests, often under reserve forest category.

Sustainable Carrying Capacity

Sustainability of shifting cultivation can be analysed using *carrying capacity* formulae. There may be odds against carrying capacity analysis; however, with the given data and forecasting possibility of land use and technological change, it gives the values necessary from policy point of view. Looking at the actual and the carrying capacity figures, one can emphasize the need for technological enhancement for betterment of land use or provision of more and better off-farm employment to ease the pressure on land. The carrying capacities were calculated for all the five studied villages using Faechem's (1973) formula given below, and the results are given in Table 16.2.

$$W = a/(CL)$$

where

W carrying capacity = maximum theoretical population.
a cultivable area of land (ha).

Table 16.2 Carrying capacity, actual population and population growth of the studied villages

Villages	a	$C*L$	W	Actual	~W & actual	g
Brhamarjodi	420.84	2.36	178	135	43	0.02
Gandli	866.49	1.76	492	116	376	0.03
Badamasingh	358.28	0.76	472	149	323	0.07
Sakota	560.07	1.37	408	84	324	−0.04
Kalinga	234.41	1.97	119	77	42	−0.01

Note Actual = Actual population of each village in the year 2000. g = growth rate of population from 1991 to 2000. 1991 data are collected from census data
Source The data are collected by the researcher through primary survey

C cultivation factor = number of garden areas required to complete a cycle of cultivation and regeneration = (fallow time + cultivation time)/cultivation time.

L mean area presently cultivated per capita (ha/capita)

The results given in Table 16.2 show that except for the two villages, namely Brhamarjodi and Kalinga, the actual population is substantially lower than the carrying capacity population. In this case, however, we assume that the land use will continue in the present form. If due to technological changes or introduction of more intensive forms of land use, the present area of village can hold even more population, though, sustainability of such changes may be questionable. In two villages, namely Sakota and Kalinga, the population has declined since 1991. Since, the population data have been collected at only two points of time such decline may be considered as temporary. However, the present level of population in two villages as mentioned above is crucial. Of the two, however, the situation in Brhamarjodi is a matter of concern as the village has adopted all possible ways of land use practice in all possible topography of the village area. This village has wet rice cultivation in the plains, home garden and fallow system in the moderately sloped land and swidden in the higher slopes. Therefore, further increase in population will have additional pressure on land, and as a result, it will lead to reduction in the shifting cycle, if off-farm employment opportunities or out-migration does not occur. However, in case of Kalinga, the land use is largely primitive, and hence, there is a possibility of further intensification of land use in selective topography, which can withstand further pressure on the shifting cycle.

Sustainable Employment-Based Approach (EBA)

Traditional shifting cultivation centres around diverse production system on a rotational basis. Fields are usually used for a period of 2–3 years and then allowed to regrow into secondary forests. It is said that, historically, the rotation period regarding the shifting cultivation system was 16–25 years, leading to sufficient sustainable productivity. However, it is claimed that recent population increases combined with conversion of large areas for permanent cultivation as the primary factors for shortening of rotation period to 4–8 years, leading to decreased productivity, increased labour requirement and poor nutrition of the shifting cultivation-dependent families. However, these claims are also questionable as older growth forests are labour intensive in terms of clearing that disincentivizes the shifting cultivators to clear such forest unless there is a critical need to do that. Old growth forests can be used for shifting cultivation if it precedes another activity like logging or other such usages. Population support (as opposed to jobs) as a unit of measurement can be arrived by considering average family size, number of families in a micro-unit (say a village or settlement), with the requirement of swidden land per year, rotational cycles (currently practised or the ideal rotation period whichever is appropriate), land requirement for other agricultural and non-agricultural purposes. Using the total cultivable area of the micro-unit, we can arrive at the figure of number of people supported per unit area.

Estimating the number of jobs a given activity or path will provide over a long period explicitly incorporates sustainability by measuring labour engagement, division of labour within the family or by gender and possibility of support of families into the foreseeable future. The degree of dependency on the activity (shifting cultivation in our case) would determine the degree of support to the family that can be equated to *non-declining welfare* an important measure of sustainability. As Taylor (2001) argued, it is assumed that a job only counts if it does not involve a decline in the current acceptable standard of living. A similar point holds for population support. EBA by not giving differential weights across different types of activities prevents differentiation and specialization. This, however, also prevents discrimination and sustains the welfare generated from an activity, similar to non-declining utility.

There are additional reasons beyond the need to incorporate sustainability into activity analysis for considering EBA as a complement if not substitute for CBA. Employment itself matters significantly in development process, and measuring employment as a part of activity evaluation in developing countries also can be seen as a means of addressing equity issue. EBA is also consistent with ideas and propositions forwarded by some of the harsher critics of standard neoclassical development theory who emphasize on growth and other monetary measurements over employment. Appropriate and ecologically sustainable development requires, among other things, a focus on local economic self-sufficiency and independence from global economic forces. The use of EBA does not guarantee such a focus, but the emphasis on employment over income is more consistent and it emphasizes the equity aspect more than the simple income growth.

Externalities can pose a challenge for EBA. Shifting cultivation-based land use practices faces several such externalities including contesting claim of ownership on land by the state or forest department, limits to territory from an inter-generational perspective, degradation in the ecosystem or even weather variation and extreme weather events. In recent years, availability of employment in several other activities through MGNREGA programme is also a competing use of labour. However, by emphasizing employment, equity and sustainability over the maximization of monetary net benefits, EBA can be a part of a paradigm shift in understanding sustainability (Taylor 2001).

From Table 16.3, we have identified the current number of days employment in all feasible ways. From the figures, it is obvious that a substantial part of employment comes through shifting cultivation. About four to seven months, job is generated by shifting cultivation for each family, whereas it is about two months per adult[7] of a family. One interesting observation that is visible from Table 16.3 is that employment per acre of swidden land is less where diversified land use activities are taken (e.g. Brhamarjodi and Gandli).

Table 16.3 Employment characteristics of the villages

HH and village characteristics	Kalinga	Sakota	Badamasingh	Gandli	Brhamarjodi	Combined
No. of HH	17	21	32	26	17	113
Total population	77	84	149	116	135	561
Avg. family size	4.53	4	4.66	4.46	7.94	4,96
Annual system	21.5	15.45	51.88	48.37	17.6	154.8
Swidden (annually)	20.5	20	20.4	44.5	77	182.4
Shifting cycle	6	10	6	7	9	7
Per HH swidden land required	1.21	0.95	0.64	1.71	4.53	1.61
Employment of total days from swidden land	3688	2611	5243	3675	3583	18,800
Employment per person	47.90	31.08	35.19	31.68	26.54	33.51
Employment per family	216.94	124.33	163.84	134.65	210.76	166.37
Employment per acre of swidden	179.90	130.55	257.01	82.58	46.53	103.01
Employment per acre of swidden and fallow	31.12	18.20	43.73	24.19	14.11	23.86
Employment per adult	73.76			76.56	61.78	66.43
Carrying capacity family size	26	102	101	110	22	–
~ Between carrying capacity and actual family size	9	81	69	84	5	–

Sustainable Consumption-Based Approach (CoBA)

To understand the process of consumption and the interrelation between the local community and the wider society among the shifting cultivator communities, we sought explanation from the field. The results found give interesting explanation of the economic operation in general and the production–consumption relationship specifically. Table 16.4 provides the production details of each study village. The produce is divided into four categories, namely cereals, pulses, oilseeds and others. The other category includes produces like ginger, turmeric, plantain, pineapple, jack-fruits, etc. Oilseeds and pulses to a large extent are produced for the market, whereas cereals are mostly used for self-consumption and other produces are specifically for

Table 16.4 Production characteristics of the villages

Villages	Cereal production	Pulse production	Oilseed production	Others
Brhamarjodi	24772	4646	2201	–
Per household	1457.18	273.29	129.47	
Per person	183.50	34.42	16.30	
Gandli	7330	1544	1707	57,950[a]
Per household	281.92	59.39	65.65	2228.85
Per person	63.19	13.31	14.72	499.57
Badamasingh	31,255	388	–	4840
Per household	1008.23	12.52		156.13
Per person	209.77	2.60		32.48
Sakota	3364	182	290	3850[a]
Per household	186.89	10.11	16.11	213.89
Per person	40.05	2.17	3.45	45.83
Kalinga	9335	948	–	–
Per household	549.12	55.76		
Per person	121.23	12.21		

Note All the figures are in kg except the [a]marked ones. [a]Figures are in rupees

the market only. From the data given in Table 16.4, we also find that insufficient cereal production is substituted by more production of pulses, oilseeds, etc. This process is facilitated then by market system to fulfil the cereal deficit. By this, the producer enters into a circular relation of exchange economy. However, one should bear in mind that a substantial part of cereal need is satisfied by Public Distribution System (PDS), where the people below poverty line get 10 kg of rice for rupees 2 and another 10 kg of rice at rupees 4 per month[8] in the year 2000, which has drastically enhanced subsequently through the *Antyodaya Anna Yojana* (AAY) *program*[9] covering 25 million households. Therefore, a substantial part of cereal need comes through a subsidised scheme of the state, specifically for the villages like Gandli and Sakota.

Production of cereals (largely produced for self-consumption) in two villages, namely Gandli and Sakota, is very less. This implies they depend on other means for their sustenance. However, both the villages have other means of marketable products to compensate their consumption requirements. In case of Gandli, they produce horticultural products like jackfruit, pineapple and plantains and spices like ginger and turmeric, whereas Sakota's dependency is largely on firewood collection for selling in the local market that compensates their subsistence needs. Though cereal production in Kalinga is better than the above-mentioned two villages, the absence of any compensatory marketable product puts them in a vulnerable situation. In such case, they depend largely on off-farm activities like construction and road-making works as and when available. However, due to the presence of diversified agricultural practice in both Brhamarjodi and Badamasingh, the production level is matched with the requirements.

Table 16.5 Consumption characteristics of the villages (*in Rupees*)

Villages	Consumption expenditure	Monetized expenditure	Degree of monetization (in percentage)
Brhamarjodi	469,791.5	69,084.5	14.7
Per household	26,734.8	4063.8	
Per person	3479.9	511.7	
Gandli	311,071.25	81,895	26.3
Per household	11,964.3	3149.8	
Per person	2681.6	706	
Badamasingh	431,521.5	128,955	29.88
Per household	13,485	4029.8	
Per person	2896.1	865.5	
Sakota	269,278.75	70,175	26.06
Per household	12,822.8	3341.7	
Per person	3205.7	835.4	
Kalinga	209,418.75	60,215	28.75
Per household	12,318.8	3542.1	
Per person	2719.7	782	

While observing the dependency on market for consumption need as seen in Table 16.5, we find that the degree of monetization for consumption needs is very low. A few factors explain such low degree of monetization. First, this degree of monetization we carried out was only for consumption purpose. Therefore, it does not explain the extent of relationship with the market through other processes. Since, the food habit of these communities is such that dependency on market for food is insignificant except for a few items including grocery [10] and dried fish. Second, a substantial part of food requirement is fulfilled by PDS at a subsidized price. And third, a number of items including roots, fruits and several forest products are not considered here for calculating the consumption requirements. Therefore, monetization for consumption requirement is often due to shortfall in production of food items. This is also obvious from Table 16.5, where we find that for the village Brhamarjodi degree of monetization is very low as compared to other villages. The land utilization pattern in this village is nearly optimal and most of the households produce sufficiently for their food requirement. This may explain the low degree of monetization in Brhamarjodi. One can observe that the swidden economy in the study villages has not entered into the fold of commodification of products and hence, the resultant change in the division of labour and specialization in producing certain commodities. Rather, the economy is more of substantive in nature where they produce whatever is feasible ecologically, and to their needs.

Self-sufficient production may be a necessary condition for sustainability. However, if the economy is less monetized, there are other possible vulnerabilities to the livelihood. Non-monetized economies also exhibit inflexibility in terms of division of labour, adoption of modern methods and practices and mobility of labour from one occupation to another. On the other hand, highly monetized economies may exhibit another level of vulnerability where the economy shifts from a production–consumption relation to production–market relation often transforming the producer to a money-wage labourer. It may also expose the individuals to the variations in values in money economy with respect to the real economy.

To seek further explanation on the aspect of factors influencing the monetised expenditure (the dependent variable) in these swidden economies, we ran a simple linear regression considering family size, total yearly consumption and total land holding as independent variables. The results obtained give interesting explanations. Descriptive statistics of each variable are given in Table 16.6, and the regression results are given in Table 16.7.

The model has a $\overline{R}^2$ value of 0.77, and family size and total consumption have positive β coefficient significant at one-percent level of significance. Size of holding negatively influences the monetized expenditure. This implies that, as the size of holding increases, monetized expenditure for consumption declines. This is obvious, as with the increase in land size production of food item, which in return leads to less dependency on the market. This also implies that the economies dependent on swidden agricultural system have tendency more towards the non-monetized economy. In other words, this shows the predominance of pre-capitalist nature of economy over the monetized economy. However, increase in total consumption leads to higher consumption in monetary units. Since total consumption includes non-food items,

Table 16.6 Descriptive statistics of the variables used in the model

Variables	Mean	Standard deviation
Monetized expenditure (in rupees)	3631.19	1709.71
Family size	4.96	3.04
Total consumption (in rupees)	14,965.32	10,932.23
Size of holding (in acres)	8.89	6.55

Note Number of observations = 113
Monetized expenditure Consumption expenditure processed through money economy

Table 16.7 Factors determining monetized expenditure on consumption

	Unstandardized coefficients		Standardized coefficients	t-values
	β	Standard error of β		
Constant	1714.7	164.78		10.41[a]
Family size	233.61	79.78	0.416	2.93[a]
Total consumption (in rupees)	0.09	0.02	0.60	4.13[a]
Size of holding (in acres)	−72.02	14.37	−0.28	−5.01[a]

Note Dependent variable: Monetized expenditure per annum
Adjusted R^2 = 0.765; [a] = significant at 1 percent level

an increase in total consumption largely leads to consumption of non-food items, in turn, leading to more consumption in monetary units. Besides, family size also plays a significant role in the process of monetization. The larger is the family size, the higher is the consumption demands, and therefore, there is more dependency on market for consumption needs.

Availability and utilization of resources at the local level are other important factors to determine the livelihood sustainability in an economic system. Therefore, the consumption analysis explains that swidden agricultural systems would be able to carry the present production pattern to the extent land and labour availability does not pose any constraint. Another important point that emerges out of the above analysis is that any changes towards a market-based (monetized) economic system will be unsustainable, if the institutions both external to the production system (i.e. market and easy availability of inputs) and internal to the system (tendency and attitudes towards a money economy) are not corrected. Therefore, any policy towards changing the agricultural system among the swiddeners requires a change in the institutional structures, which can enhance the money requirements for commodity production.

Policy Implications

This paper critiques the cost–benefit analysis (CBA) method and its shortcomings, especially in the context of indigenous community-dominated subsistence economy. In the process, the authors discuss three alternative/complementary methods which explain livelihood security and sustainability. These are carrying capacity of land, labour utilization and consumption-based approaches. We believe that these alternative methods are more useful in a less monetized subsistence economy as compared to CBA where monetary valuation is used as a common denominator. Therefore, this paper has a strong policy guidance and implications while measuring values and sustainability of various alternative projects and activities in a subsistence economy. A large part of tribal-dominated areas in India or elsewhere is also victim of various so-called development projects those lead to livelihood and ecological destruction in the regions. These alternative/complementary methods would provide better understanding of the ecological and livelihood aspects as compared to CBA and hence, could be used for measuring the multidimensional approach to valuation instead of solely depending on monetary valuation based on either real, imaginary or shadow pricing mechanism.

Discussion and Conclusion

Use of three different criteria of measuring sustainability gives a clearer view of the state of art in each of the five villages under study. Land-based analysis tells that Kalinga and Brhamarjodi are at the threshhold point of optimum possible land utilization, whereas employment-based analysis tells that employment generation is shrinking in case of Brhamarjodi in spite of diversified land use activities. Consumption-based analysis on the other hand not only clarifies the self-sufficiency question in production and consumption, but also suggests the nature and direction of the economies concerned. Gandli and Sakota are two villages where cereal production is low but in Kalinga the situation looks vulnerable due to lack of compensatory income generation activities. Degree of monetization further elaborates on the nature of the economies. The low degree of monetization in all the five villages implies the dominance of pre-capitalist mode of production. However, the interesting result found is that the degree of monetization declines with the increasing size of holding. In other words, the economy does not only have low degree of monetization, it has a tendency towards pre-capitalist mode of production with increasing land use activities.

Considering these results in the backdrop of technological changes undergone, the size and growth of population and nature of economy (degree of monetization), we can broadly conclude that different policy thrust is required for each village. In Kalinga, development of land (specifically the land where fallow system is followed) can increase their productivity in a sustainable manner. Besides, off-farm activities in Gandli and Brhamarjodi will allow the people to continue their land activities in the present form for a longer time period. This case also is true for Kalinga. Therefore, from overall sustainability point of view, Kalinga looks more vulnerable, whereas Gandli and Brhamarjodi will fall into this stage if proper institutional reforms are not taken care of. Badamasingh and Sakota are presently in more sustainable form and will continue to be so. But, in the case of Badamasingh, technological changes have taken place in terms of land use activities, Sakota still remains primitive. Grossly the conclusion applicable to all the villages is that if distortion in the market is removed with adequate institutional reforms, all the villages will be better off and in true sense the development can be considered as sustainable development.

Post Script

The field study conducted in the year 2000 provided an insight into the nature and characteristics of dependency on shifting cultivation and its implications on sustainability. 20 years have gone since then. Hence, given the time lapse in the interim period, it is important to understand the reevance of these alternative/complementary measures in recent years. There might have been characteristic changes in land use, employment scenario and consumption pattern in the study region due to policy and other changes in the interim period. We revisited the region in 2015 and 2016 to understand the institutional changes. The revisit in 2015 was elaborated elsewhere (see Jyotishi and Manjula 2017). These recent visits draw our attention to a few aspects. Though land use intensity remains broadly unchanged, intensification in some forms of land use other than shifting cultivation is visible. There has been a significant change in the employment characteristics due to the Mahatma Gandhi National Employment Guarantee Act (MGNREGA)-based programme that provides about 100 days of employment guarantee in the rural areas. This programme has bene-fited these villagers not only in terms of employment but also in terms of monetiza-tion bringing in significant integration with the market economy. Higher disposable and monetized income is also reflected in terms better integration with the market system which was earlier more distorted due to inter-locking of product market with credit market. Similarly, the *antodaya* programme is playing a critical role in supplementing the consumption needs of the households. These changes, to a large extent, have reduced the pressure on land use intensity, increased the employment opportunities and supplemented the consumption needs. However, the sustainability issues have largely shifted away from swidden agricultural systems towards larger development interventions in terms of infrastructure, mining and similar activities.

End Notes

1. Sustainable development as "non-declining utility" is elegantly investigated by John Pezzey in *Economic Analysis of Sustainable Growth and Sustainable Development,* Environment Department Working Paper No. 15, World Bank, 1989. Peezy is critical of other approaches to sustainability because he feels they are non-operational and non-measurable. The relevance of a non-measurable, non-comprehensive definition of sustainable development for policy purposes is thus very questionable. But this does not detract from the powerful insights into the concept of sustainable development that can be derived by adopting the 'utility' approach. Economists who advocate the 'non-declining utility' defi-nition of sustainable development do this by making environmental quality a factor in the 'utility function'. That is, utility, or well-being, depends on the consumption of goods and services and on environmental quality.
   Annex of 'Blueprint for a Green Economy' by Pearce et al. (1997) compiled about twenty-five definitions of sustainable development from different litera-ture. A few of those definitions are as follows:

   (a) The next generation should inherit a stock of wealth, comprising man-made assets and environmental assets, no less than the stock inherited by the previous generation (Pearce et al. 1997).
   (b) Lasting satisfaction of human needs and improvement of quality of life (Allen 1980).
   (c) Lasting and secure livelihood that minimise resource depletion, envi-ronmental degradation, cultural disruption and social instability (Barbier 1987).
   (d) Development that meets the needs of present without compromising the ability of the future generations to meet their own needs (World Commission on Environment and Development 1986).
   (e) Maximising the net benefits of economic development, subject to main-taining the services and quality of natural resources (Barbier 1989).
   (f) Agricultural sustainability is defined as the ability to maintain the produc-tivity, whether of a field or farm or nation, in the face of stress or shock (Conway and Barbier 1988).

(g) Sustainable society is one that lives within the self-perpetuating limits of its environment. That society is not a 'no-growth' society. It is, rather, a society that recognises the limit of growth and looks for alternative ways of growing (Coomer 1979).

(h) A primary goal of sustainable development is to achieve a reasonable and equitably distributed level of economic well-being that can be perpetuated continually for many human generations (Goodland and Ledoc 1987).

(i) …activities should be considered that would be aimed at maintaining over time a constant effective natural resource base. The concept was proposed by Page (1977) and implies not an unchanging resource base but a set of resource reserves, technologies and policy controls that maintain or expand the production possibilities of future generations (Howe 1979).

(j) Sustainability might be redefined in terms of a requirement that the use of resources today should not reduce real incomes in the future (Markandya and Pearce 1988).

(k) Norgaard (1988) gives five comprehensive definitions, which address the sustainability of changing interactions between people and their environment over time. First, whether a region's agricultural and industrial practices can continue indefinitely. Second, whether the region is dependent upon non-renewable inputs, both energy and materials, from beyond its boundaries or beyond its boundaries which are not being managed in a sustainable manner. Third, whether the region is in some sense culturally sustainable, whether it contributing as much to the knowledge and institutional bases of other regions as it is culturally dependent on others. Fourth, the extent to which the region is contributing to global climatic change, forcing other regions to change their behaviour, as well as whether it has options available to adapt to the climatic change imposed upon it by others. Fifth, the cultural stability of all the regions in combination evolving along mutually compatible paths or not.

(l) World commission on Environment and Development (1987) emphasizes on two points while defining sustainability. First, even the narrow notion of physical sustainability implies a concern for social equity between two generations, a concern that must logically be extended to equity within each generation. Second, living standards that go beyond the basic minimum are sustainable only if consumption standards everywhere have regard for long-term sustainability.

2. Market distortion in developing economies is widely discussed in many literatures as well as in Jyotishi (2003, 2005). I have talked about the market distortion existing in the present study area.

3. This can be extended beyond land to the water bodies for fisheries and related activities. Therefore, the usage of term of 'land' to that extent can be considered as a generic term.

4. The main competing type of definition of sustainability centres on the maintenance of human and natural capital stock. Employment, which ensures income

flow and consumption and in the process, ensures standard of leaving of human capital stock, is hence important indicator of sustainability.

5. As without discounting, we are giving equal preference to both present and future as against giving more emphasis to present. Therefore, the notion of sustainability is inherent and more strong in case of EBA. Employment or population support counts the same whether it is at the beginning or the end of the time period. The analysis of benefits is thus blind to which generation is receiving the benefits.

6. This part of the discussion heavily relies on R. S. Rao's work on 'A note on an Aspect of the Indian Economy' (1995).

7. We have considered anybody within the age group of 15–50. This means we have excluded the old and young ones who also work in the field for activities like weeding and collections.

8. This price has been changed by the present government of Odisha in the year 2001. However, the data collected was in the year 2000, and therefore, we refer to that price only.

9. This 'Antyodaya Anna Yojana' (AAY) program implemented in the year 2000 and amended subsequently is targeted at about 25 million households living below poverty line where each household is entitled to get 35 kg of rice or wheat at rupees 3 and 2 respectively per kg.

10. Since the food habit among these communities does not require much oil or spices, grocery items form an insignificant part of total consumption requirement.

Acknowledgements This chapter is a revised version of the earlier work by the first author produced as a working paper (No. 163) of Gujarat Institute of Development Research. The authors would like to thank ICSSR for the funding support to the project for which this paper is a prelude to further investigation. The authors would also thank Foundation for Ecological Security in general and Eastern Regional Office, Bhubaneswar, and Field Office in Keonjhar for facilitating later part of field visit and follow-up research. Support of Pradeep Maharana in facilitating the fieldwork in Keonjhar and engaging in discussion is much appreciated. Also, we acknowledge the support of Mr. Benudhara Suchen in providing information from several villages in Southern Odisha.

References

Ackerman, F., & Heinzerling, L. (2004). *Priceless: On knowing the price of everything and the value of nothing*. New York: New Press.

Ackerman, F. (2008). Critique of cost-benefit analysis, and alternative approaches to decision-making—A report to friends of the Earth England, Wales and Northern Ireland.

Aldred, J. (2006). Incommensurability and monetary valuation. *Land Economics, 82*(2), 141–161.

Allan, W. (1949). Studies in African land usage in Northern Rhodesia. *Rhodes Livingstone Papers, No. 15.*

Allen, R. (1980). *How to save the world*. London: Kogan Page.

Banerjee, A., K. (1995). Rehabilitation of degraded forests in Asia, World Bank Technical paper Number 270, World Bank, Washington, DC.

Barbier, E. (1987). The concept of sustainable economic development. *Environmental Conservation, 14*(2), 101–110.

Barbier, E. (1989). *Economics, natural resources, scarcity and development*. London: Earthscan.

Bose, S. (1991). Shifting Cultivation in India, Anthropological Survey of India, Ministry of Human Resource Development, Department of Culture, GoI, Calcutta.

Brush, S. B. (1975). The concept of carrying capacity for systems of shifting cultivation. *American Anthropologist, 77,* 799–811.

Carneiro, R L. (1960). Slash-and-burn agriculture: A closer look at its implications for settlement patterns. In F. C. Wallace (Ed.), *Men and cultures: Selected papers of the fifth international congress of anthropological and ethnological sciences, september 1956* (pp. 229–234). Philadelphia: University of Pennsylvania Press.

Conkline, H. C. (1959). Population-land balance under systems of tropical forest agriculture. *Proceedings of the Ninth pacific science congress* (Bangkok, 1956).

Conway, G., & Barbier, E. (1988). After green revolution: Sustainable and equitable agricultural development. *Futures, 20*(6).

Coomer, J. (1979). The nature of the quest for a sustainable society. In J. Coomer (Ed.), *Quest for a sustainable society*. Lordon: Oxford.

Pearce, D., Markandya, A., & Barbier, E. B. (1997). *Blueprint for a green economy*. London: Earthscan Publications.

Faechem, R. (1973). A clarification of carrying capacity formulae. *Australian Geographical Studies, 11,* 234–236.

Fearnside, P. M. (1972). An estimate of carrying capacity of the Osa Peninsula for human population supported on shifting agriculture technology. In Organisation for Tropical Studies (OTS), *Report of research activities undertaken during the summer of 1972* (pp. 486–552). San Jose, Costa Rica: OTS.

Fearnside, P. M. (1986). *Human carrying capacity of the Brazilian rainforest*. New York: Columbia University Press.

Gasparatos, A., El-Haram, M., & Horner, M. (2008). A critical review of reductionist approach for assessing the progress towards sustainability. *Environmental Impact Assessment Review, 28,* 286–311.

Goodland, R., & Ledoc, G. (1987). Neoclassical economics and principles of sustainable development. *Ecological Modelling, 38.*

Gourou, P. (1966). *The tropical world: Its social and economic conditions and its future status* (4th edn.) (S. H. Beaver & E. D. Tabunde, Trans.). New York: Longman.

Government of India, Census Data, www.nic.in.

Hara, O. S., & Messner, S. (2000). In K. Puttaswamaiah (Ed.), *The limits of economic rationality: Social and environmental impacts of recreational land use, in cost benefit analysis: Environmental and ecological perspectives*. Transaction Publishers.

Hayden, B. (1975). The carrying capacity dilemma: An alternative approach. In A. C. Swedlund (Ed.), *Population studies in archaeology and biological anthropology: A symposium* (pp. 11–21). Washington, D.C.: Society for American Archaeology, Memoir 30.

Howe, C. (1979). *Natural resource economics*. New York: Wiley.

Hwang, K. (2015). Cost-benefit analysis: its usage and critiques. *Journal of Public Affairs, 16*(1), 75–80 (2016). https://doi.org/10.1002/pa.1565.

Jyotishi, A. (2003). Ecological, Economic and Institutional Aspects of Shifting Agriculture: A Study in Orissa, Unpublished Ph.D. Thesis Submitted to Bangalore University, Bangalore.

Jyotishi, A. (2005). Transcending Sustainability Beyond CBA: Conceptual Insights from Empirical Study on Shifting Cultivation in Orissa, GIDR Working Paper No. 163.

Jyotishi, A., & Manjula, M. (2017) Revisiting Statutory Laws And Customary Norms Governing Swidden Agricultural Systems.

Kessler, J. J. (1994). Usefulness of human carrying capacity concept in assessing ecological sustainability of land-use in semi-arid regions. *Agriculture, Ecosystems & Environment, 48,* 273–284.

Komatsu, Y., Tsunekawa, A., & Juc, H. (2005). Evaluation of agricultural sustainability based on human carrying capacity in drylands–a case study in rural villages in Inner Mongolia, China. *Agriculture, Ecosystems & Environment, 108,* 29–43.

Markandya, A., & Pearce, D. (1988). Natural environments and the social rate of discount. *Project Appraisal, 3*(1).

Marten, G. G., & Saltman, D. M. (1990). The human ecology perspective. In G. G. Marten (Ed.), *Traditional agriculture in Southeast Asia: A human ecology perspective.* London: Westview Press.

Norgaard, R. (1988). Sustainable development: A co-evolutionary view. *Futures, 20*(6).

Qian, Y., et al. (2015). A comparative analysis on assessment of land carrying capacity with ecological footprint analysis and index system method. *PLOS ONE, 10*(6), e0130315. https://doi.org/10.1371/journal.pone.0130315. Accessed February 23, 2019.

Ramakrishnan, P. S. (1992). *Shifting agriculture and sustainable development* (Vol. 10). UNESCO, Paris and The Parthenon Publishing Group.

Rao, R. S. (1995). A note on an aspect of the Indian Economy. In R. S. Rao (Ed.), *Understanding semi-feudal semi-colonial society collection of essays.* Hyderabad: Perspective Publishers.

Ruthenberg, H. (1971). Farming Systems in the Tropics, Clarendon Press, Oxford, UK; second edition (1976) London: Oxford University Press.

Soderbaum, P. (1998). The political economics of sustainability. In S. Faucheux, M. O'Connor, & J. van der Straaten (Eds.), *Sustainable development: Concepts, rationalities and strategies.* London: Kluwer Academic Publishers.

Taylor, F. D. (2001). Employment-based analysis: An alternative methodology for project evaluation in developing regions, with an application to agriculture in Yucatan. *Ecological Economics, No. 36,* 249–262.

Todaro, M. (1994). Economic development. New York: Longman, White Plains.

Verneer, D. E. (1970). Population pressure and crop rotational changes among the Tiv of Nigeria. *Annals of the Association of American Geographers, 60,* 299–314.

Chapter 17
Tribal Health and Sustainable Development: Traditional Knowledge Practice and Medicinal Plant

Samita Manna and Aritra Ghosh

Abstract Sustainable livelihood ensures proper human development providing with the basic necessities of everyday life. India is a country of multi-ethnic groups having more than 500 tribal communities along with different religious groups. Among these tribal groups more than 80% people live in different forest environment. Forest environment-based tribal communities fulfil most of their basic needs from the surrounds. In general, these tribal people preserve good notions of health among them as per their perceptions. Overtime they are facing crisis due to non-sustainability of production, consumption and uses of forest goods. Bestowing to the welfare approach adopted since independence for the tribal communities, a special human value loaded attitude is shown to them for shielding their human rights, however, protection of their rights in nature has continuously being ignored. However, sustainable development helps to preserve the natural resources from the environment for its economic growth and social viability from present generation to future generation. To be more precise, the tribals are the sources of the indigenous knowledge of the medicinal plants used for healing and curing diseases for better living and solving problems of day to day life. Therefore, medicinal plants should be protected and be used by them for sustaining their rights as humans. In this article, by emphasizing on the significance of the sustainability of medicinal plants and its importance in tribal life, the authors focus on an alternative development method which can improve the social environment in general and the tribal development in particular.

Keywords Sustainable development · Tribal health · Indigenous knowledge · Alternative development

S. Manna (✉)
Department of Sociology, University of Kalyani, Kalyani, West Bengal, India
e-mail: samite.manna@gmail.com

A. Ghosh
Department of Sociology, Serampore Girls' College, Hooghly, West Bengal, India
e-mail: aritrasocio1987@gmail.com

Introduction

Rio Earth Conference in 1992 observed a real global problem which is alarming before us due to environmental degradation, depletion of natural resources, deforestation, natural catastrophe, etc. As a result, sustainable development is a new paradigm shift of economic development. Sustainable development stands for protecting the needs not only of the present generation population but also it fulfils the needs of future generation. To meet this end, the new strategy of alternative measures is as how to sustain our forest-based environment. But it is really difficult, a long trajectory path has already been travelled, many of our traditional wisdom are also out of reach. Forests being the main supplier of raw material for attaining developmental goals have always remained the centre of focus of planners. Moreover, the extreme poverty of the tribes often forces them to cut down mercilessly the trees from the jungles, even the urban dwellers also encroach the forest for their dwelling and harvesting purpose. As a result, the healthy life of the jungle dwellers is no longer prevailing.

There is no denying fact that the tribal people are enriched with their indigenous knowledge and traditional wisdom. Specially the undivided Midnapur District of West Bengal has been referred as a fertile field of medicinal plants (Pakrashi and Mukherjee 2001). They live in nature, use natural resources and enjoy nature for maintaining their life and livelihood. From time immemorial they know how to protect themselves from different types of diseases and ailments. As a result, they go to their Gunins (medicinal men) who collect and prepare different types of medicines by collecting different types of medicinal plants and its roots, leaves, fruits, etc. These tribal people are often considered as the reservoir of our traditional knowledge and wisdom. Throughout the country, the tribals have their own distinct pristine culture separated from the others. The indigenous knowledge about medicinal plants and its uses help the tribal group to sustain their society, environment and economy (Dasgupta and Sarkar 2005; Joshi 1993).

Sustainable Development and Tribal Health

Human beings are at the centre of concerns for sustainable development. They are entitled to enjoy a healthy and productive life in harmony with nature (Principle of the Rio Declaration Environment and Development). Even though the processes related to sustainable development are not so easy in our present-day society but for better understanding of sustainable development, a balanced and integrated analysis from the point of *Sustainable Development Triangle*: economic, social and environment is considered to become a serious thought for us.

The significance of medicinal plants for sustainable human health cannot be overlooked. These plants have healing/therapeutic properties in one or any of their organs. The use of these plants is increasing worldwide. They are used in several conditions to augment and maintain human health. In sustainable human health management, medicinal plants have played a vital role which has led to the growing interest in alternative therapies and therapeutic use of plants (Akinyemi et al. 2018).

The term 'health' is a part of bipolar conceptualization. It is opposed to 'disease' at the other pole. The term assumes a connotation in the common parlance which refers to the balanced state of body and mind. The World Health Organization (WHO) seeks to define health as 'a state of complete physical, mental, and social wellbeing and not merely the absence of disease or infirmity'.

Good health stands for good society. But country like India has been facing several hydra-headed problems of its people since Independence where health problem is not an exceptional one. The colonial administration purposefully made a clear-cut distinction between tribal and non-tribal. Majority of the tribal groups throughout the country till today are facing different types of problems which are not directly linked with the medical issues but other associated factors like environmental degradation, illiteracy, lower age at marriage, high mortality, malnutrition, conservative outlook in relation with the supernatural power, etc., also make an overall impact on the tribals' body and mind. Article 21 of the Constitution rightly points out that 'no person shall be deprived of his life or personal liberty except according to procedure established by law'. So it denotes that maintaining good health of a person is one of the fundamental rights. As a result, good health not only deals with medical care of an individual but also it depicts an integrated overall development of a society with its cultural, economic, educational, social and political aspects.

The poor tribal people are till today socio-economically lagging from the mainstream of population. They live in different rural areas not far from the jungles (Bhowmick 1991, 1994; Mahapatra 1994; Sachchidananda 1994). The rich biodiversity has enabled them to sustain their life for many years. As a result, the tribes are now in transition (Atal 2015; Manna and Ghosh 2016; Manna and Sarkar 2016). The dualities of tradition and modernity create many social, economic and cultural threats in their life of which climate change is a central one. Biodiversity is people's resource especially for the poor of the third world countries. They have to depend on the natural resources for food, shelter and other essentialities of everyday life.

Climate change (Karnan and James 2009; Pelling 2011; Wapner and Elver 2016) is undoubtedly one of the most transformative issues of the twenty-first century. Unlike previous environmental problems, the effects of climate change are global in scope and cut across many different sectors. As such, climate change is not a singular task that can be left to any one specialized agency. The United Nations Framework Convention on Climate Change (UNFCCC) is just one piece of the puzzle and its efforts to mitigate and adapt to climate change were not intended to substantively address biodiversity concerns (Roberts 2010).

Climate change is not new, and species have traditionally responded to such change over evolutionary time scales. But the key question today is how organisms will respond to the current apparently rapid rate of anthropogenic climate change (Root et al. 2003; Round and Gale 2003).

Member states express their commitment in the 2030 Agenda for Sustainable Development to protect the planet from degradation and take urgent action on climate change. The Agenda also identifies, in its paragraph 14, climate change as—

One of the greatest challenges of our time' and worries about 'its adverse impacts undermine the ability of all countries to achieve sustainable development. Increases in global temperature, sea level rise, ocean acidification and other climate change impacts are seriously affecting coastal areas and low-lying coastal countries, including many least developed countries and Small Island Developing States. The survival of many societies, and of the biological support systems of the planet, is at risk.

Sustainable Development Goal 13 aims to 'take urgent action to combat climate change and its impact', while acknowledging that the United Nations Framework Convention on Climate Change is the primary international, intergovernmental forum for negotiating the global response to climate change.

India is a poor rural-based developing country where majority of the rural people directly depend on climate-sensitive sectors like agriculture, forests and fisheries. These sectors need natural resources like water, biodiversity, mangroves, coastal zones, grasslands for their subsistence and livelihoods. (Further, the adaptive capacity of dryland farmers, forest dwellers, fisher folk and nomadic shepherds is very low). Climate change is likely to impact all the natural ecosystems as well as socio-economic systems as shown by the National Communications Report of India to the UNFCCC.

Munasinghe (1992, 1994) proposed the term *sustainomics* to describe 'a trans-disciplinary, integrative, comprehensive, balanced, heuristic and practical meta-framework for making development more sustainable'. Sustainability and development can be studied through this approach which stands for 'science of sustainable development'. Such a synthesis will need to draw on a wide range of core disciplines from the physical, social and technological sciences. Methods that bridge the economy–society–environment interfaces are especially important. Environmental and resource economics attempt to incorporate environmental considerations into traditional neoclassical economic analysis (Freeman 1993; Teitenberg 1992). Costanza et al. (1997) rightly point out that recently ecological economics deals with environmental problems emphasizing the importance of key concepts like the scale of economic activities. In this context, the present study has focussed on the various uses of the medicinal plants by the tribal people where indigenous knowledge of the tribal people is seen. So it is one form of 'Sustainable Development Triangle' which emphasizes to restore traditional tribal economy strengthening social networking within the natural environment which is a new societal paradigm of the sustainable development.

Sustainomics is also related to recent initiatives on a 'sustainability transition' and 'sustainability science' (Clark 2000; Parris and Kates 2001; Tellus Institute 2001). Newer areas of ecological science such as conservation ecology, ecosystem management and political ecology have birthed alternative approaches to the problems of sustainability, including crucial concepts like system resilience, and integrated analysis of ecosystems and human actors (Holling 1992).

In this context, we may refer neologism which focuses attention explicitly on sustainable development and avoid the implication of any disciplinary bias or hegemony. For example, both biology and sociology can provide important insights into human behaviour which challenge the 'rational actor' assumptions of neoclassical economics (Gintis 2000; Robson 2001).

The substantive trans-disciplinary framework underlying sustainomics leads to the balanced and consistent treatment of the economic, social and environmental dimensions of sustainable development (as well as other relevant disciplines and paradigms). Balance is also needed in the relative emphasis placed on traditional development versus sustainability. Sustainable development itself involves every aspect of human activity, including complex interactions among socio-economic, ecological and physical systems. The scope of analysis needs to extend from the global to the local scale, cover time spans extending to centuries, and deal with problems of uncertainty, irreversibility and non-linearity. The sustainomics framework seeks to establish an overarching design for analysis and policy guidance, while the constituent components provide the 'reductionist' building blocks and foundation.

Sustainable Development: Man–Nature Relationship

Survival of human beings basically depends on environment in which they dwell. Development of mankind is closely associated with the environment. Hence, environment and development are interdependent. Relationship between man and environment has been point of discussion for a long time by the anthropologists who engaged themselves to understand man relationship with environment (Vidyarthi 1963; Mukherjee and Mukherjee 1971; Moghadam et al. 2015).

The symbiotic relationship between man and nature was never constant but has been changing constantly. Man was 'nature homospecy' since its inception. The tribal people have gained a lot of empirical knowledge and experiences to live in forest with its natural resources. The traditional wisdom is based on the intrinsic understanding that man and nature are dependent to each other and their coexistence make the society vibrant and useful. This eco-centric view develops a distinctive attitudes and typical behaviour patterns for accepting plants, animals, lands, water and other natural resources available in their everyday life. The tribals have the wisdoms of vast indigenous knowledge about flora and fauna used by them to meet their basic primary needs of food, shelter and well-being. Their belief systems, social, economic and rituals activities are intricately interwoven around forest and its neighbourhood. In their daily life, they never kill an animal, a bird or cut the tree or plant with which they claim totemic affiliation (Bhargava 2002; Choudari 2007; Guha 1983; Kulkarni 1987; Manna 2000).

The livelihoods of tribes are greatly impacted due to the effects of population explosion and their regular exploitation over natural resources resulting to depletion and deterioration of forest environment.

Indigenous People and Indigenousness

Indigenous knowledge (hereafter 'IK') is rooted in the lived experiences of indigenous peoples; these experiences highlight the philosophies, beliefs and educational processes of tribal communities. Indigenous people come to know things by personal observation and interactions in their daily lives (Berger 1987; Flavier 1995; Warren 1990).

In contemporary discourses 'indigenous peoples' primarily refer to ethnic groups that have historical ties to groups that existed in a territory prior to colonization or formation of a nation-state. These groups of people are sometimes considered as the original inhabitants of a country having a distinct cultural heritage of its own. In many cases, they live a segregated life away from the so-called mainstream of the population.

Traditional herbal medicines are rooted in indigenous knowledge systems. These cognitive systems play a crucial role in decision making with respect to the use of medicinal plants resources and are embedded in the lifestyle of the local community (Astutik and Kimengsi 2019).

Xaxa (1999) opines that the notion of "indigenousness" is a political construction rather than an empirical reality. Following the same line of argument Kujur (2010) analyses:

> The Adivasis' consciousness has come about to promote their rights and privileges because their very survival is at stake. They are the victims of exploitation and alienation at all levels. Hence in the absence of a mechanism or powers to safeguard their interests, a new form of identity or indigeneity is crystallizing among the tribes across India. The people now use the notion of indigenousness to identify and define themselves in differentiation from the non-tribal population.

Current discussions of 'indigenous peoples' take place against the background of three conceptual frameworks that emerged after World War II: Human Rights, collective rights, and the rights of aboriginal peoples (Bowen 2000). According to Bowen (ibid.), current debates within this conceptual framework concern 'the legitimacy of religious or other cultural norms as sources of individual rights or limits on rights claims'. The empirical foundation of this kind of formulations can be found in different kind of customary laws existing across the world.

Now pertinent question that comes to the fore is: 'who are the indigenous peoples of India?' In 1987 Indian Council of Indigenous and Tribal Peoples was established. This prestigious organization is also indirectly linked with the United Nations. After a thorough analysis of the historical backgrounds and socio-cultural peculiarities of the communities concerned, the organization conclusively said that the 'scheduled tribes' of India fall in the said category (Das 2001).

The tribals had developed a distinct and particular way of life in remote forests, hills, deserts—far away from the so-called mainstream of the population—in a word in the zones of less interaction. Many sociologists and anthropologists believe that they were comparatively self-sufficient in the lap of the nature.

Objectives

This paper has tried to focus cn the several issues related to Tribal Health and Sustainable Development. These are as follows:

1. to examine how do the tribal people use different types of medicinal plants to make them free from diseases and ailments.
2. to point out various problems faced by the tribal people due to climate change and environment degradation.
3. to focus on a new societal paradigm relating to sustainable development of the tribal health through using medicinal plants for curing and preventing diseases which is one of the burning issues of the present-day society.

Naturally a development with sustainability of the environment may be one of the visions or strategies of the environmental sociologists in contemporary society.

Study Areas and Methods

Indigenous communities have preserved their traditional knowledge on the uses and management of wild plant resources (Rajbhandari et al. 1995; Coe and Anderson 1999; Manandhar 1994, 1995, 2002). The traditional wisdom regarding the use of medicinal plants is not only useful for conservation of cultural traditions, but also for community health care and deve.opment of some new drugs for the common people (Pei 2001; Gazzano et al. 2005).

With the rapid growth of science and technology tremendous changes have taken place in everyday life of the people throughout the world. India is not an exception. The loss of traditional knowledge and culture of human life is due to loss of plant species as these irreversible changes not only affect their material cultures but non-material cultures have also been affected by these processes of change. Humans are no longer simply members of homogeneous group. They are the integrated part of the complex cultures. Till today most of them are dependent on nature, utilizing their environment and ecology with the help of technology. Even today they combat against their ill health and other crisis situation with their indigenous methods of utilizing their natural resources (Haldar et al. 2008).

In India, there is a diversity of various ethnic groups. It is known that more than 500 different tribes and other ethnic groups (Jain 1991) and their combined population comprise more than 7.5% of the total population of the country. In this context, Paschim Midnapore is enriched for its ethnic population as jungles, hills, forests, etc., are the natural habitat of these people. Five ethnic groups have been selected for the present study. They are the [1]Lodhas (hunters and gatherers), the [2]Santals (mainly agriculturists), the [3]Mundas (agriculturists), the [4]Mahalis (bamboo workers) and the [5]Bhumij (agriculturist and hunting-forest products gatherers). These tribal people are the autochthones group of Paschim Midnapore in West Bengal and they have very close association with nature.

From the age-old times, different plants have been used as sources of medicines by the tribals. They have good faith on the traditional system of herbal medicines and they also rely on it. According to the World Health Organization (WHO), nearly 80% of the people depend upon traditional medicines for primary healthcare need. The tribal people of the Paschim Midnapore have also shown a great belief on the traditional knowledge of medicine (Manna 2003). The medicinal plants are usually collected from the nearby forests of the district. They prepare the medicines from

Table 17.1 Socio-economic characteristics of the sample population

Characteristics		Lodha	Santal	Munda	Mahali	Bhumij
Age (in years)	19–25	13 (26.00)	7 (14.00)	10 (20.00)	10 (20.00)	9 (18.00)
	26–32	36 (72.00)	37 (74.00)	34 (68.00)	35 (70.00)	32 (64.00)
	Above 32	1 (2.00)	6 (12.00)	6 (12.00)	5 (10.00)	9 (18.00)
Total		**50 (100.00)**	**50 (100.00)**	**50 (100.00)**	**50 (100.00)**	**50 (100.00)**
Education	Illiterate	24 (48.00)	18 (36.00)	21 (42.00)	21 (42.00)	23 (46.00)
	I–IV	16 (32.00)	25 (50.00)	24 (48.00)	20 (40.00)	21 (42.00)
	IV+	10 (20.00)	7 (14.00)	5 (10.00)	9 (18.00)	6 (12.00)
Total		**50 (100.00)**	**50 (100.00)**	**50 (100.00)**	**50 (100.00)**	**100.00**
Occupation	Agricultural labour/labour	50 (100.00)	49 (98.00)	49 (98.00)	24 (48.00)	49 (98.00)
	Basket maker	0	0	0	26 (52.00)	0
	Others	0	1 (2.00)	1 (2.00)	0	1 (2.00)
Total		**50 (100.00)**	**50 (100.00)**	**50 (100.00)**	**50 (100.00)**	**50 (100.00)**
Monthly income (Rs.)	2000–5000	49 (98.00)	47 (94.00)	50 (100.00)	49 (98.00)	50 (100.00)
	Above 5000	1 (2.00)	3 (6.00)	0	1 (2.00)	0
Total		**50 (100.00)**	**50 (100.00)**	**50 (100.00)**	**50 (100.00)**	**50 (100.00)**
Types of family	Joint family	11 (22.00)	10 (20.00)	7 (14.00)	11 (22.00)	8 (16.00)
	Nuclear family	39 (78.00)	40 (80.00)	43 (86.00)	39 (78.00)	42 (84.00)
Total		**50 (100.00)**	**50 (100.00)**	**50 (100.00)**	**50 (100.00)**	**50 (100.00)**

Figures in the parentheses indicate percentages
Source Field study conducted by the authors in 2012

several plant species and apply to the common people for remedial purposes. These medicine men belong to five ethnic groups—the Lodha, Munda, Santal, Mahali and Bhumij and they use various plant species for the cure of same type of diseases or different types of diseases. So there is diversity in the method of preparation and the uses of plants as folk medicines. (The present study shows highlight the various uses of herbal medicines prepared by the medicinal men/medicine men (Gunin) from different plant sources with their indigenous knowledge and methods of using these medicines for curing several types of diseases by the five ethnic groups of the district Midnapore, West Bengal, India.)

In this article five tribal communities the Lodha, Santal, Munda, Mahali and Bhumij are taken into consideration for knowing their impact of changing biodiversity on their everyday life specially related to their health. 50 households of each community who are living in nearby [6]Arabari jungle ranges (Belti jungle extends from Hijli to Keshiary—a wide forest range belonging to Paschim Midnapore) are surveyed in the year 2012 in detail.

Table 17.1 depicts the five important socio-economic characteristics of the five sample tribal groups of Paschim Midnapore, West Bengal. Age, Education, Occupation, Income and Family type are considered as the important variables for knowing their socio-economic status. It is really interesting that all five tribal groups show a similar trend of socio-economic life situation of their daily life. Majority of them belong to Nuclear Type of families and most of them are day labourers. Majority of the tribals' monthly income Rs. 2000–5000 and around 40% tribals are illiterate. The similar trends of socio-economic characteristics of the similar tribes of Paschim Midnapore are also observed in different studies done by P. K. Bhowmick 1963, 1994 and S. Manna 2010. Significant changes in their socio-economic life are not observed even in this study done in the year 2012. So these ethnic groups are till today maintaining their traditional ways of life in spite of so many development projects which are carried out by the different governmental agencies since independence.

In Table 17.2, 25 tribal medicine men from 5 tribal communities viz. Lodha, Santal, Munda, Mahali and Bhumij of Midnapore District, West Bengal have been studied. These tribal medicine men live in 7 villages under 4 Gram Panchayats of Midnapore District. The study reveals that these different medicinal men (*Gunins*) from different ethnic groups are enriched with their traditional wisdom or knowledge relating to curing diseases with the help of their age-old traditional-based knowledge. Each ethnic group has developed its own medicinal men (Gunin/Priest) within the community itself. But these medicinal men from various ethnic groups are engaged with curing different diseases not only for their own community members but the other sick men of different caste groups and all ethnic groups of the local villages.

Table 17.2 Village wise distribution of tribal medicine men/medicinal men

Panchayat	Village	No. of the medicine men	Community
Daharpur	Daharpur	02	Lodha
		07	Munda
	Santarangi	04	Munda
		01	Bhumij
		01	Mahali
Makrampur	Borageria	02	Santal
	Makrampur	01	Munda
		01	Bhumij
Makrampur	Pichabani	01	Mahali
		01	Munda
	Joli Padima	02	Munda
		01	Mahali
Dahjuri IV	Dahijhuri	01	Lodha
Total 4	7	25	5

Source Field study conducted by the authors in 2012

Table 17.3 shows that 64.00% sample medicinal men belong to the age group between 50 years and 60 years above. In general, they are selected as village leaders or priests by their own community members due to possessing some charismatic traits. The women in tribal societies are not allowed to act as *Ojhas* (medicinal men); generally they men are assigned to this particular role. Traditionally, in all tribal communities the aged male members are preferred though in some cases younger members are also allowed to serve the community members as medicine men. The age distribution of the traditional medicine men is described in Table 17.3.

Majority of the children in all communities (Table 17.4) are highly affected by Cough, Cold, Fever, Vomiting, Itching, Loose motion, Jaundice, Bleeding and Snake bite. On the other hand, the adult men and women are generally affected by Jaundice, Cough, Cold, Fever, Vomiting, Itching, Loose motion, Bleeding, Snake bite and Arthritis on a regular basis. For their regular and easy measure, they primarily depend on the local medicine men (Gunins) who produce the herbal medicines with their traditional knowledge and wisdom. Their positive awareness is also expressed regarding the acceptance of allopathic treatment by the establishment of rural health centres and the positive role of ICDS. These types of diseases are very common among the tribal people of this region observed by the study conducted Bhowmick (1990) in his Chapter 13 article—Socio-cultural and Environment Factors of Health: A Micro Study in a Similar Ecology.

Table 17.3 Age wise distribution of the tribal medicine men

Community	20–29 (years)	30–39 (years)	40–49 (years)	50–59 (years)	60+ (years)	Total
Munda	2 (8.00)	1 (4.00)	2 (8.00)	5 (20.00)	5 (20.00)	15 (60.00)
Santal	–	1 (4.00)	–	1 (4.00)	–	2 (8.00)
Lodha	–	–	1 (4.00)	–	2 (8.00)	1 (4.00)
Mahali	–	–	1 (4.00)	2 (8.00)	–	3 (12.00)
Bhumij	–	1 (4.00)	–	–	1 (4.00)	2 (8.00)
Total	2 (8.00)	3 (12.00)	4 (16.00)	8 (32.00)	8 (32.00)	25 (100.00)

Figures in the parentheses indicate percentages
Source From the field study by the authors

But gradually there is a declining tendency of the use of medicinal plants by the tribal people of the District Paschim Midnapore, West Bengal, India. It may be due to lack of knowledge about herbal medicine, malpractice of the 'Ojha' and scarcity of medicinal plants due to deforestation and acquisition of lands for industry and human habitation.

Reaction on Climate Change and Degradation of Biodiversity

The tribal people of the sample villages have been facing severe problems in their daily life due to climate change and degradation of biodiversity which affect their socio-economic life. To know it in detail the sample populations (250 Tribal Head of the Households) have been asked to point out their perceptions and reactions on climate change which they have noticed within the last 10 years. Irrespective of their ages and the ethnic variations all of them positively respond their strong feelings towards climate change. All of them have raised their voices regarding changing climate. They have said that temperature is increasing every year due to deforestation, industrialization, urbanization and climate change. Consequently, homeostatic condition of the biodiversity is disturbed. As a result, the children are badly affected with various diseases. Modern medicines are often out of reach among them. Traditional medicines are over time eroding due to climate change impacting on biodiversity.

They have also noticed the seasonal changes which affect their normal rhythmic life. They have pointed out about the late running of the seasons; i.e. a few seasons are merged together. Late appearance of rainy season disturbs the traditional crop pattern that affects the biodiversity in greater extent. Traditional crop productions are discontinued as, for example, the Aman crop (one type of seasonal paddy) production is now totally vanished and, in its place, the high yield variety paddy seeds are supplied by the local government for more production by taking less time. Late

Table 17.4 Diversity in the use of medicinal plants by the different tribal people of the district Paschim Midnapore, W.B., India

Disease	Groups	Name of the tree (root/leaf/flower/bark/fruits)	Scientific name
Cough and cold	All five tribal groups	Basak plant (extract of leaves)	*Adhatoda zeylanica* Medic.
		Tulsi plant (extract of leaves)	*Ocimum tenuiflorum* L.
		Manasa plant (extract of leaves)	*Euphorbia neriifolia* L.
	Specific tribal groups	Mango tree (extract of leaves) (Lodha)	*Mangifera indica* L.
		Karanja tree (extract of seed) (Munda)	*Pongamia pinnata*(L.) Pierre
		Ginger (extract of root) (Santal and Munda)	*Zingiber officinale* Rosc.
Fever	All five tribal groups	Tulsi plant (extract of leaves)	*Ocimum tenuiflorum* L.
		Seuli tree (extract of leaves)	*Nyctanthes arbor-tristis* L.
		Basak plant (extract of leaves)	*Adhatoda zeylanica* Medic.
		Kalmegh plant (extract of leaves)	*Andrographis paniculata* Wall.ex Nees
	Specific tribal groups	Challa plant (only bark) (Santal)	*Holoptelea integrifolia* Planch.
		Valia plant (gum of the fruit) (Santal)	*Semecarpus anacardium* L.*f.*
Fever	Specific tribal groups	Apang tree (extract of root) (Lodha)	*Achyranthes aspera* L.
		Iswarimul (extract of root and leaves) (Munda)	*Aristolochia indica* L.
		Amrul (extract of leaves) (Lodha)	*Oxalis corniculata* L.
Headache	All five tribal groups	Ghrita Kumari (extract of leaves)	*Aloe barbadensis* Mil.
	Specific tribal groups	Karala plant (Munda and Bhumij)	*Momordica charantia* L.
		Mahul plant (extract of bark) (Lodha and Santal)	*Madhuca indica* J.F.Gmel.
		Helencha plant (extract of leaves) (Lodha)	*Enhydra fluctuans* Lour.
		Guava tree (extract of root) (Munda)	*Psidium guajava* L.
		Garlic plant (extract of root) (Santal)	*Allium sativum* L.
		Mustard (extract of seed) (Santal)	*Brassica nigra* (L.) Koch

(continued)

Table 17.4 (continued)

Disease	Groups	Name of the tree (root/leaf/flower/bark/fruits)	Scientific name
		Gulach tree (extract of leaves) (Munda and Santal)	*Plumeria alba* L.
		Marie gold (extract of leaves) (Lodha and Munda)	*Tagetes erecta* L.
Vomiting	All five tribal groups	Arjun tree (extract of bark)	*Terminalia cuneata* Roth
	Specific tribal groups	Lemon tree (extract of fruit) (Munda and Bhumij)	*Citrus limon* (L.)Burm.*f*.
		Kundri tree (extract of leaves) (Lodha)	*Coccinia grandis*(L.) Voigt
		Piplas tree (extract of leaves) (Munda and Santal)	*Litsea glutinosa* (Lour.) C.B.Robins.
		Gaisira (extract of root) (Santal)	*Asparagus racemosus* Willd.
		Jam tree (extract of root) (Lodha)	*Syzygium cuminii*(L.) Skeels
		Iswarimul (extract of root) (Munda)	*Aristolochia indica* L.
		Kasmia (extract of root) (Lodha)	*Lannea coromandelica* (Houtt.)Merr.
		Akanda Katha (extract of bark) (Munda and Santal)	*Stephania hernandifolia* Walp.
		Sal tree (extract of bark) (Lodha and Santal)	*Shorea robusta* Roxb.ex Gaertn.*f*.
Bleeding	All five tribal groups	Tulsi plant (extract of leaves)	*Ocimum tenuiflorum* L.
	Specific tribal groups	Pasukedar plant (extract of root) (Lodha and Santal)	*Curcuma aromatica* Salisb.
		Dudhi Lata (gum of plant) (Lodha and Munda)	*Ichnocarpus frutescens* R.Br.
		Latapata plant (extract of leaves) (Lodha)	*Mikania cordata* (Burm.)B.L.Robinson
		White Akanda (gum of plant) (Lodha)	*Calotropis gigantea*(L.)R.Br.ex.Ait
Bleeding (continue)	Specific tribal groups	Durba Ghas (extract of leaves) (Munda and Mahali)	*Cynodon dactylon*(L.) Pers.
		Bisallakarani (extract of leaves) (Munda)	*Barleria lupulina* Lindl.
Loose motion	Specific tribal groups	Sal tree (extract of bark) (Lodha)	*Shorea robusta* Roxb.ex Gaertn.*f*.

(continued)

Table 17.4 (continued)

Disease	Groups	Name of the tree (root/leaf/flower/bark/fruits)	Scientific name
		Iswarimul (extract of root) (Lodha and Santal)	*Aristolochia indica* L.
		Guava tree (extract of leaves) (Munda)	*Psidium guajava* L.
		Arjun tree (extract of bark) (Munda)	*Terminalia cuneata* Roth
		Tulsi plant (extract of leaves) (Santal)	*Ocimum tenuiflorum* L.
		Gaisira (extract of root) (Munda)	*Asparagus racemosus* Willd.
		Mango tree (extract of leaves) (Lodha)	*Mangifera indica* L.
		Mahua tree (extract of Leaves) (Lodha)	*Madhuca indica* J.F.Gmel
		Black pepper (fruit) (Santal)	*Piper nigrum* L.
		Kundri (extract of Root) (Lodha)	*Coccinia grandis*(L.) Voigt
		Akanda plant (extract of leaves) (Munda)	*Calotropis gigantea* (L.)R.Br.ex.Ait.
		Anantamul (extract of Root) (Santal)	*Hemidesmus indicus*(L.) R.Br.
Itching	All five tribal groups	Nargi plant (only stem)	*Eupatorium odoratum* L.
		Dhulimera plant (only stem)	*Clerodendrum indicum*(L.) O.Ktze.
	Specific tribal groups	Bon Chakunda (extract of leaf) (Santal)	*Cassia alata* L.
		Nengus tree (only root) (Lodha)	*Mucuna prurita* Hook.
		Karanja tree (boiled leaves) (Munda and Santal)	*Pongamia pinnata*(L.) Pierre
		Kasmila (extract of bark) (Lodha and Santal)	*Lannea coromandelica* (Houtt.)Merr.
		Karanja tree (boiled leaves) (Munda, Santal, Bhumij)	*Pongamia pinnata*(L.) Pierre
		Chatina plant (Mucilage of Plant) (Lodha)	*Alstonia scholaris*(L.)R.Br.
Jaundice	Specific tribal groups	Apang tree (extract of fruit) (Lodha and Munda)	*Achyranthes aspera* L.
		Arhar tree (extract of leaves) (Munda and Santal)	*Cajanus cajan*(L.)Millsp.

(continued)

Table 17.4   (continued)

Disease	Groups	Name of the tree (root/leaf/flower/bark/fruits)	Scientific name
		Bon Jamir (extract of fruit) (Santal)	*Citrus limon* (L.)Burm.*f.*
		Tungur (extract of leaves) (Lodha)	*Cajanus cajan*(L.)Millsp.
		Anantamul (extract of root) (Lodha and Santal)	*Hemidesmus indicus*(L.) R.Br.
		Pipul (extract of root) (Munda)	*Ficus religiosa* L.
		Neem (extract of leaves) (Lodha)	*Azadirachta indica* A.Juss.
		Tumeric (extract of root) (Santal)	*Curcuma domestica* Valeton
		Seuli (extract of root) (Santal)	*Nyctanthes arbor-tristis* L.
		Durba Ghas (extract of leaves) (Lodha)	*Cynodon dactylon*(L.) Pers.
Arthritis	All five tribal groups	Satamuli plant (extract of root)	*Asparagus racemosus* Willd.
	Specific tribal groups	Marrie gold (extract of leaves) (Munda and Santal)	*Tagetes erecta* L.
		Apang tree (extract of root) (Lodha and Munda)	*Achyranthes aspera* L.
		Bisallakarani (extract of leaves) (Lodha and Santal)	*Barleria lupulina* Lindl.
		Jai Bahadur (extract of leaf) (Lodha)	*Arisaema tortuosum* Schott
		Gulancha (extract of root) (Munda)	*Tinospora cordifolia* (Willd.) Miers ex Hook.*f.*et Thoms.
Snake bite	All five tribal groups	Patal garu (extract of root)	*Rauvolfia tetraphylla* L.
	Specific tribal groups	Bisallakarani (extract of leaves) (Santal)	*Barleria lupulina* Lindl.
		Pasu Kedar (extract of root) (Munda and Santal)	*Curcuma aromatica* Salisb.
		Shankachura (extract of leaf) (Lodha)	*Sansevieria trifasciata* Prain
		Chandor plant (extract of root) (Munda and Bhumij)	*Rauvolfia tetraphylla* L.
		Sibjata (extract of leaves) (Lodha and Mahali)	*Sansevieria roxburghiana* Schult.*f.*

(continued)

Table 17.4 (continued)

Disease	Groups	Name of the tree (root/leaf/flower/bark/fruits)	Scientific name
		Lal Manasa (extract of root and leaf) (Lodha)	*Synadenium grantii* Hook,f.
		Arum (Mankochu) (extract of root) (Munda)	*Alocasia indica*(Roxb.)Schott
		Neem tree (extract of root) (Santal)	*Azadirachta indica* A.Juss.
		Bichuti Pata (extract of leaves) (Lodha)	*Fleurya interrupta* Gaudich.
		Ginger (extract of root) (Munda)	*Zingiber officinale* Rosc.

Source Field study conducted by the authors in 2012

monsoon and decreasing rainfall, due to global warming, affect the local vegetables Green local vegetables are not adequately found, in addition new hybrid varieties of vegetables are produced. It demands high pesticide and organic fertilizers. Local fishes and water bodies animals like small geol fishes (*Siluriformes*), crabs (infraorder brachyura), mollusks (Mollusca) of various types and different types of local prawns (*Dendrobranchiata*) are not available as earlier,

Biodiversity is not simply a crisis due to disappearance of species but it also threatens the livelihood of millions of people in general and the tribes in particular.

Protection, conservation and regeneration of natural resources are the best solutions to achieve the sustainable environment. Indigenous knowledge systems are also much beneficial for the sustainable livelihood of a local community in the balanced environment situation. The interior tribes still live relatively in isolation of hills and forests.

Policy Implications

Since Independence, the tribal people are not at per developed with the other non-tribals due to their poor socio-economic status and backwardness. Poor infrastructure in economy and society is one of the most important barriers for economic development.

Framing different policies approved by the Government and Non-Government Organizations may save these vulnerable groups of our society. A few suggestions are as follows in relation to sustainable development of the tribal health by restoring natural resource-based medicinal plants.

- The medicinal plants should be restored with the help of the local government and concerned authority. Forests have to be preserved for maintaining healthy life not only for humans but also for the animals and biological species which will help to maintain the homeostatic condition of the nature. This traditional health practice is known as "Living Bio- health Culture" which is very important specially at this juncture (Post-[7]COVID-19) when we are trying to establish 'Self Reliant Bharat'. As it is known to us that at the dawn of civilization human beings have been relying on Herbal Medicine having its strong immunity power.

- Conservation of medicinal plants largely depends upon forest protection and management where tribal medicine men of the local areas are the fittest persons for proper identification of the local medicinal plants. Local Panchayats should identify the local Gunins and record their names. These local Gunins of respective areas should be honoured for their knowledge and Wisdom. Like other folk artists they should be provided a monthly ex gratia for incentive.
- They should be trained on quality preparation of medicine by standardized technique to strengthen their practice and to prove authenticity in their field. With this attempt. The young expertise will come in forward and will able to sustain their economy. Nearly 80% of the rural folk show their intention to accept herbal medicine for their common disease and ailment.
- Tribal people of the respective villages should be encouraged to establish 'Home Herbal Garden' and side by side Forest Department of respective forest areas may open 'Medicinal Plant Nursery'. Recently WHO also involves in this process (2004). National Bureau of Plant Genetic Resources (1995) has already started their identification of the herbal medicines from different forests.
- The integration of the knowledge and patronage at the government level is needed to save them from extinction. Medicinal men should be allowed to sell their medicine in open market after judging their authenticity. For this reason, government should take measures for legal recognition of them as indigenous health care practitioners.
- Forest should be protected from the traders who are meant for marketing and commercialization of those medicinal plants and who have been damaging and killing mercilessly the medicinal plants of the forest.

Conclusion

From this study it is very clear that the tribal people throughout the country are facing many challenges to overcome the problems of their daily life. They are now in transition. Forest-based economy fails to fulfil their needs as they have loosened their right to use the forest of their own. From earlier time to till date they have some affinity with forest. Most of them like to accept medicinal plants for getting rid of diseases. Many plants are very common plants which they have accepted for healthy living. The five tribes from the common ecology have shown more or less the same responses in relation to using the medicinal plants. Variations are also there but the plants are common with their various uses in the same locality.

The study also shows that the tribals are facing tremendous crisis due to climate change and environmental degradation. The tribals have to adjust with the changing climate. Most of the time, they are not ready to get their traditional food items. New environments have already destroyed their healthy flora and fauna. They are now gradually accepting the new ways of life with full of discomfort and dissatisfaction.

Sustainable development is now a global as well as local political agenda. Globally different issues related to climate change are a concerned matter. In general, the poor people of developing countries are the worst sufferer, as many of them have to depend directly on nature or forest products. Industrialization, urbanization, technological development, climate change, etc., make the life of the forest dwellers vulnerable. Their vulnerability is also stated by them through this study.

The tribal people like to maintain a bridge between tradition and modernity. On the one hand, they want to restore their indigenous knowledge through sustainability. On the other hand, government provides them modern medical facilities for the betterment of their quality of life. But they are not in good position to accept it. So the alternate strategy, i.e., the sustainable development through the restoration of medicinal plants is only an alternative development project considered to be effective development model for the poor tribal people. According to them jal (water), jamin (land) and jungle are now adversely used by the so-called development agencies, as a result, climate change is the outcome of it. So they want to save natural resources like water, forest, land. The tribal people throughout the country irrespective of their distribution in various parts of the country, still they are claiming themselves as the original inhabitants of the forest or jungles. They maintain a spontaneous dual relationship with the plants and animals of the forest. In time of crisis, they depend on forest. They collect fruits, foods, fodder and different types of medicinal plants and use them in their day to day life. They use the local herbal medicines prepared by the local Gunins or medicinal men by those collected plants, herbs, roots leaves, fruits, etc., and the medicinal men keep them comfortable creating soothing environment for their nourishment.

Naturally, by protecting their medicinal plants as well as encouraging the *Gunins* (medicinal men) a healthy sustainable relationship between men and nature could be developed which may be referred as alternative development strategy among them.

But with the name of development, the tribal people have to be withdrawn from their original inhabitants and day by day they are becoming the foreigners. The medicinal men are also becoming alienated from the uses of plants and in many cases the uses of medicinal plants are totally prohibited as the middlemen create hindrances by ensuring their rights on those medicinal plants. In this way, body and minds of tribals are gradually alienated from their nature. Developing sustainable economy which can give them support and security and makes them stable and healthy with eco-friendly relationship. Time has come to overview the situation where sustainable development is only the strategy of alternate development. We are hopeful that new rays of light will sweep away the darken phases of tribal life. Gradually with the tune of development with the sustainability, dancing faces of tribal people will be seen as a new societal paradigm shift of modern socio-cultural and political agenda of the contemporary society.

End Notes

1. **The Lodhas**

   The Lodhas of West Bengal were one of the ex-criminal tribes in India. But now they are identified as denotified community. In earlier days the Lodhas were engaged in hunting and food gathering economy. They fully depended on forest and forest products. But that situation did not continue forever. The primitive people were once brought up around the forest which provided them the raw materials of their livelihood as well as the joys and sorrows of their lives. They were automatically tuned with the forest environment but gradually they have been alienating from the forest by losing their free entry into it. Eventually, they have started working as agriculture workers and engaged themselves in cultivation. The Lodhas are strictly endogamous family. Poor socio-economic conditions compel them to live in nuclear families.

2. **The Santals**

   The Santal community belongs to the Austro-Asiatic group. The Santal society is patriarchal and the dominance of the male is found in every affair of their life. They are in general bilingual. They are mainly involved in agricultural work.
   Originally the Santals were nomadic in nature and they roamed about from place to place in search of food and shelter. With the passage of time they gradually settled and established their permanent habitat in different parts of our country.

3. **The Mundas**

   The Mundas hold a unique position in the tribal map of India. The name 'Munda' was coined by their ancient Hindu neighbours. The Munda call themselves as *Horoko* (men). The Munda tribe consists of an elder and younger branch and the Munderi-speaking people are called Kolarians. Generally the Mundas live in villages and most of them are agriculturists. Their society is patriarchal and patrilineal. The Mundas have belief in the potent evil powers of witch.

4. **The Mahali**

   The word 'Mahali' was originated from the two words: 'Mah' means bamboo and Ali means specialized. Etymological the word 'Mahali' indicates those people who are specialised in bamboo work. They are also known as traditional bamboo workers. They have many tribal attributes of their own. Mahalis have greater tribal affinity to that of the Santals and they have also originated from the big Santal or Hor race.
   At present they are cultivators. They participate in agriculture in different ways as owner—cultivator, share-cropper, contract-cultivator and also agricultural labourer. Traditionally, they mainly prepared bamboo baskets which were used by the betel leaf dealers in large quantities to pack the leaves for disposal.

5. **The Bhumij**

   The Bhumij are one of the non-Aryan Hinduised tribes found in Manbhum, Singhbhum districts of Bihar and Midnapore and Bankura districts of West Bengal. They are the original inhabitants of Dhalbhum, Barabhum and bagmundi estates of Bihar. They are chiefly located in the area between the Kasai and Subarnarekha rivers of Paschim Midnapore.

   The etymological meaning of the term 'Bhumij' or 'Bhumija' means 'born of the soil'. During that time perhaps the immigrant Hindus might give the name 'Bhumija' as they were the early settlers of the land. The Bhumij of the Jungle Mahals were the terror of the surrounding districts and were under the nick name 'Chuar'. Their outbreaks were called 'Chaaris'. In most of the cases they were under the Chiefs of local area.

6. **Arabari** is the name of a forest range of West Midnapore, West Bengal. It is bordered with the Dalma range of East Singhbhum and Jharkhand. This forest range covers Midnapore area of West Bengal—Jangalmahal, Lodhasuli, Salboni, Godapiasal, Gurguripal, Hoomgarh, Goaltore, Gohaldanga, Tetulmuri, Bhadutala, Ranjha, Lalgarh, Chandrakona, Khasjangal, Dhamkura, Joypur, etc. The main flora of the forest range is the Sal trees. The participating villagers including local tribes are given exclusive rights to use all minor forest products of the Sal trees, kendu leaves, dry twigs, seeds, etc. for their daily use.

7. **COVID 19** is a life-threatening dangerous virus appeared from the province of Wuhan, China, at the end of December 2019. Gradually the people of whole world are badly affected through it as it is pandemic in nature. 7 June 2020 morning news display that the number of total deaths in the world is more than 4 Lac. But surprisingly, we would like to say that on first week of May 2020, a few villages Kukai, Keshiary, Daharpur, Bakhrabad, etc. of Paschim Midnapore have been observed during the time of providing relief, it has been found not a single tribal family is affected with the virus in spite of villagers poor living. They are totally abstained from repeated hand washing, wearing masks, maintaining social distancing, etc., which are literally impossible for them. A further detailed study is needed but the present observation justifies that these population have some distinct cultural traits and strong immunity to fight and protect them from the deadly poisonous virus COVID 19.

References

Akinyemi, O., Oyewole, S. O., & Jimoh, K. A. (2018). Medicinal plants and sustainable human health: A review. *Horticulture International Journal, 2*(4), 194–195.

Astutik, P. J., & Kimengsi, J. N. (2019). Asian medicinal plants' production and utilization potentials: A review. *Sustainability, 11*(5483), 1–33.

Atal, Y. (2015). *Indian tribes in transition: The need for reorientation.* New Delhi: Routledge India.

Berger, J. (1987). *Report from the frontier: The state of the world's indigenous peoples.* London: Zed Books LTD.

Bhargava, M. (2002). Forest, people and state. *Economic and Political Weekly, 37*(43), 4440–4446.

Bhowmick, P. K. (1990). *Applied-action-development anthropology.* Calcutta: Institute of Social Research and Applied Anthropology.

Bhowmick, P. K. (1991). *The Chenchus of the forests and plateaux: A hunting-gathering tribe in transition.* Calcutta: Institute of Social Research and Applied Anthropology.

Bhowmick, P. K. (1994). *The Lodhas of West Bengal: A socio-economic study.* West Bengal: Rarh Samskriti Sangrahalaya.

Bowen, J. R. (2000). Should we have a universal concept of 'indigenous peoples' Rights'? Ethnicity and essentialism in the 21st century. *Anthropology Today, 16*(4), 12–16.

Choudari, B. (2007). Forest and Tribals: A history review of forest policy. In C. Kumar Paty (Ed.), *Forest government and tribe* (pp. 1–17). New Delhi: Concept Publishing Company.

Clark, W. C. (2000). Visions of the 21st century: Conventional wisdom and other surprises in the global interactions of population, technology and environment. In K. Newton, T. Schweitzer & J. P. Voyer (Eds.), *Perspective 2000: Proceedings of a conference.* Ottawa: Economic Council of Canada.

Coe, F. G., & Anderson, G. J. (1999). Ethnobotany of the Sumu of Southern Nicaragua and comparison with Miskitu plant lore. *Economic Botany, 53*(4), 363–386.

Costanza, R., et al. (1997). *An introduction to ecological economics.* USA: St. Lucia's Press.

Das, J. K. (2001). *Human rights and indigenous peoples.* New Delhi: APH Publishing Corporation.

Dasgupta, S., & Sarkar, A. (2005). *Reflection of ethno science: Study on the Abujh Maria.* Delhi: Mittal Publication.

Flavier, J. M. (1995). The regional programme for the promotion of indigenous knowledge in Asia. In D. M. Warren, L. J. Slikkerveer & D. Brokensha (Eds.), *The cultural dimensions of development: Indigenous knowledge systems.* London: Intermediate Technology Publications.

Freeman, A. M. (1993). *The measurement of environmental and resource values: Theory and methods.* Washington DC: Resources for the future and RFF Press.

Gazzano, L. R. S, Lucena, R F. P., & Albuquerque, U. P. (2005). Knowledge and use of medicinal plants by local specialists in a region of Atlantic forest in the state of Pernambuco. *Journal of Ethnobotanical Ethno-medicine, 1*(9).

Gintis, H. (2000). Beyond homo economicus: Evidence from experimental economics. *Ecological Economics, 35*(3), 311–323.

Guha, R. (1983). Forest in British and Post British India: A historical analysis. *Economic and Political Weekly, 18*(44), 1940–1947.

Haldar, A. K., Singh, P. K., & Coomar, P. C. (2008). *Tribal ethnomedicine—Problem and prospect.* West Bengal: The Institute of Social Research & Applied Anthropology.

Holling, C. S. (1992). Cross scale morphology, geometry and dynamics of ecosystems. *Ecological Monographs, 62,* 447–502.

Jain, S. K. (1991). *Dictionary of Indian folk medicine and ethnobotany.* New Delhi: Deep Publications.

Joshi, P. C. (1993). Culture, health and illness: Aspects of ethno medicine in Jaunsar Bawr. In S. K. Biswas (Ed.), *Central Himalayan Panorama* (Vol. 1). Calcutta: Institute of SocialR Research and Applied anthropology.

Kannan, R., & James, D. (2009). Effects of climate change on global biodiversity: a review of key literature. *Journal for the International Society for Tropical Ecology, 50*(1), 31–39.

Kujur, J. M. (2010). Cultural rights of adivasis. *Social Action, 60*(4), 323–338.

Kulkarni, S. (1987). Forest legislation and tribals: Comments on forest policy resolution. *Economic and Political Weekly, 22*(50), 2143–2148.

Mahapatra, L. K. (1994). Concept of health among the Tribal population groups of India and its socio-economic and socio-cultural correlates. In S. Basu (Ed.), *Tribal health in India.* New Delhi: Manak Publications Pvt Limited.

Manandhar, N. P. (1994). The ethnobotanical survey of herbal drugs of Kaski district. *Fitoterapia, 65*(1), 1–13.

Manandhar, N. P. (1995). Ethnobotanical notes on unexploited wild food plants of Nepal. *Ethnobotany, 7*(1–2), 95–101.

Manandhar, N. P. (2002). *Plants and people of Nepal.* Oregon, USA: Timber Press Portland.

Manna, S. (2000). *The fair sex in Tribal cultures.* New Delhi: Gyan publishing House India.

Manna, S. (2003). Role of medicine men in the traditional Tribal heritage. *Man in India, 83*(3 & 4), 407–418.

Manna, S., & Ghosh, A. (2016). Souria Paharia are now in transition: Beliefs, rituals and practices—A micro study in Godda District, Jharkhand. In P. K. Bhowmick, P. K. Singh & M. P. Rajak (Eds.), *Changing Tribal society in India.* New Delhi: Abhijeet Publications.

Manna, S., & Sarkar, R. (2016). 'Birhors' ways of life in 21st century: A micro study in a few villages of Hazaribag District, Jharkhand. In P. K. Bhowmick, P. K. Singh & M. P. Rajak (Eds.), *Changing Tribal society in India.* New Delhi: Abhijeet Publications.

Moghadam, D. M., Singh, H. J., & Yahya, W. R. W. (2015). A brief discussion on human/nature relationship. *International Journal of Humanities and Social Science, 5*(6), 90–93.

Mukherjee, M., & Mukherjee, D. (1971). *Indian Tribes.* Calcutta: Saraswat Library.

Munasinghe, M. (1992). Environmental economics and sustainable development. Paper presented at the UN Earth Summit, Rio de Janeiro and reproduced as Environment paper 3. USA: World Bank.

Munasinghe, M. (1994). *Sustainomics: A transdisciplinary framework for sustainable development, keynote paper 50th anniversary sessions of the Sri Lanka Association.* Colombo: Advance of Science.

Pakrashi, S. C., & Mukherjee, S. (Eds.). (2001). Medicinal and aromatic plants of West Bengal Midnapore. West Bengal Academy of Science and Technology and Development of Science and Technology and NES, Kolkata.

Parris, T. M., & Kates, R. W. (2001). *Characterizing a sustainability transition: The international consensus. Research and assessment systems for sustainability discussion paper.* Cambridge: Harvard University Press.

Pei, S. J. (2001). Ethnobotanical approaches of traditional medicine studies: Some experiences from Asia. *Pharmaceutical Biology, 39*(1), 74–79.

Pelling, M. (2011). *Adaptation to climate change—From resilience to transformation.* New York: Rouledge.

Rajbhandari, T. K., et al. (1995). *Medicinal plants of Nepal for ayurvedic drugs.* Natural Products Development Division, Kathmandu, Nepal: HMGN.

Roberts, J. (2010). *Linking climate change with biodiversity related multilateral environmental agreements.* Italy: International Development Law organization and the Centre for International Sustainable Development Law.

Robson, A. J. (2001). The biological basis of human behaviour. *Journal of Economic Literature, XXXIX*, 11–33.

Root, T. L., et al. (2003). Finger prints of global warming on wild animals and planta. *Nature, 42*(1), 57–60.

Round, P. D., & Gale, G. A. (2008). Changes in the status of Lophura in Khao Yai National Park, Thailand: A response to warming climate? *Biotropica, 40*, 225–230.

Sachchidananda, (1994). Socio-cultural dimension of Tribal health. In Salil Basu (Ed.), *Tribal health in India.* New Delhi: Manak Publications Pvt Limited.

Teitenberg, T. (1992). *Environmental and natural resource economics.* New York: Harper Collins Publication.

Tellus Institute. (2001). *Halfway to the future reflections on the global condition.* Boston: Tellus Institute Press.

Vidyarthi, L. P. (1963). *The Maler: A study in nature-man-spirit complex of a Hill Tribe.* Calcutta: Bookland Private Limited.

Wapner, P., & Elver, H. (2016). *Reimaging climate change.* London: Rouledge.

Warren, D. M. (1990). *Indigenous knowledge systems and development. Background paper for seminar series on sociology and natural resource management.* Washington, D.C.: The World Bank.

Xaxa, V. (1999). Tribes as indigenous people of India. *Economic & Political Weekly, 34*(51), 3589–3595.

Chapter 18
Indigenous Knowledge of Women: Alternative Methods of Water Conservation in Rajasthan

Ritu Sharma

Water scarcity is directly proportional to socio-economic challenges faced by even young girls drop-out in schools due to harassment in carrying water. The intervention is therefore required to solve the global problems of access, equity and sustainable use by active inclusion of women's knowledge.
—Ban Ki-moon[1]

Abstract Alternative methods integral in locating gender dynamics in natural resource intervention to value water as a resource. Women are neglected in water-policy, when in fact, they mitigate acute shortage by collecting water from farthest places to restore its multiple use. Their training in water conservation evolves around fetching from water-bodies to be passed on to generations via folklores and local narratives. The social fact of patriarchy exerts constraints on women and water-burden falls upon them as 'cultural labour' reiterating inequal right (collection and storage) as compared to male counterpart. Women's role has been recognised globally, yet their inclusion in governmental planning and water resource management is dismal. This highlights the feminist paradigm of 'knowledge as power' detecting alternative methods disguised in cultural and religious discourses of water conservation. Based on vast in-depth literature of identifying women as 'sustainable-community' becomes central to ecological consciousness of preserving water. Thus, the aim is to encourage gender mainstreaming by accommodating indigenous knowledge of water conservation as an alternative in Rajasthan. This paper is unique in two ways; firstly, recommending alternative methods of water conservation encompassing local needs as relevant within both the theory and practise intersecting sustainability and justice and secondly, to have all-inclusive approach of gender mainstreaming to natural resource governance. Hence, the paper addresses disproportionate relationship with water as embedded in culture both in scarcity and abundance due to its run-offs, wastage and problems of in terms of gender-equity and

R. Sharma (✉)
Kamala Nehru College, University of Delhi, New Delhi, India
e-mail: senoritaritu@yahoo.com

M. K. Verma (ed.), *Environment, Development and Sustainability in India: Perspectives, Issues and Alternatives*, https://doi.org/10.1007/978-981-33-6248-2_18

governance. Therefore, interrogating the water conservation techniques and alternative paradigms of its preservation by women intertwined as primary-handlers cum leader of sustainability.

Keywords Gender · Indigenous knowledge · Community · Sustainability · Alternative methods of water

Introduction

India would be facing its worst water crises in the availability and access with no drinking water for 40% by 2030 (CWMI Neeti Aayog 2019). This is an (inter)national emergency appeal by respective government bodies and public at large. To save 135 crore of population with water scarcity is dealt in multiple ways in theory and practice largely handled by women in household and agrarian reforms. As per water Index report, annually calculating the water fetching capacity is more than 14,000 km of women in rural areas with almost 75% households have no availability to water (ibid.). Then be it the relationship with the water body (wells, ponds, lakes) or piped water (storing in earthen pots, sieving it by cotton cloth), it is primarily women's traditional wisdom that is being utilized for conservation in times of crises. Significantly, it is, low costs and friendly budget that suits the localized conditions of geopolitical and climatic conditions of uneven terrains of water stress regions. This challenging situation needs to be explored by dynamic approach of gender (UN 2006); tapping into their 'indigenous knowledge as alternative methods' for water management. Gender is social construction with subjective interpretation and variations. Be it, developing (South Asia) or developed (USA and Canada) regions, water contestations obstruct socio-political participation of capacity building needed for gender equality. Surprisingly enough, it is well-known fact, that more than one marriage is solemnized owing to long distance of availability of water that require labour which resulted in decency as 'water-wives' (Denganmal, Maharashtra)[2]. More so, groundwater reserve is replenished via recharging traditional check dams by 'water-mother' (Gopalpura, Rajasthan); or 'water council' (Gujarat) to resolve conflicts related to water dispute of distribution and provision. Hence, the worst affected are women and children as direct bearing of water collection falls upon them as 'primary collectors'. Concerning this, the paper investigates the importance of both men and women on water governance within the western and southern parts of Rajasthan, India. The hypothetical finding of this study reflects water as a developmental paradigm within ample and scarce ecological resource pushing women as leaders in equity and justice. Their engagement in control, allotment for sustainability of water for domestic purposes underrepresented across the world (OECD 2008). To augment an understanding of these issues, the study develops as co-production of knowledge subject to women and water, challenges low levels of participations and possibility of inclusion in government policies. The intriguing question raised here is that should (wo)men be the formal partners in alternate methods engrained in their cultural framework? Or is the grass-roots understanding of valuing water as a resource vis-à-vis inclusion in development paradigm is enough? As often, gendered division of labour is seen as political implication in natural resource management. The crux is to invent traditional harvesting methods by women that should be utilized by government policy makers as alternative methods for water scarcity.

Theoretical Analysis and Reflection

The intriguing interlinkage of women and water stems from sustainable development goals to third world perspective affecting all aspects of human lives. The contradicting possibilities of water stress regions in drought and ample rainfall seemingly carry a universal approach for natural resource intervention. With increasing pressure on booming population the most devastated prediction is expected that more than a dozen nations would be water scarce by 2025. Scarcity is socially constructed critically emanated from capitalist greed and Anthropocene approach which disbalances demand and supply dedicated to theme 'Coping with water scarcity' stemming from climate change (United Nations Water Scarcity Factsheet 2013). Thus, water has been gendered and so does its exploitation stemming from early education to job-market and all the other components of healthy life as deprivation of human rights to women (South Asia Inequality Report 2019). In the above context, the relationship has a direct bearing of MDG on (Millennium Development Goal and Beyond 2015: Goal 4; 5) gender-related implications which affects the social and economic capital of women in terms of leadership, earnings and networking opportunities. Subsequently, can divert their attention on their children well-being and livelihood by reliable knowledge intervention of water conservation techniques. Water and development are inevitable for wholistic analysis of Rajasthan. Also, highly populated with largest area of erratic rainfall demands management of water resource bodies. However, material transformation and gendered relationship with water scarcity can disbalance the two. Gender dynamics is contested within this paradigm by the socio-economic, political, cultural and ecological consciousness. Women in drought-prone areas are more vulnerable to the climatic and topographical atrocities reproducing distinct norms and folkways of traditional society. This entails political economy of feminist discourse central to identity and power struggles of water storage and its sustainability.

Keeping the above background of ecological deterioration under the patriarchal domain of subordination (has been passed on to subsequent generations keeping the social structures intact) in socially constructed meanings to (wo)man to exploit the mutual interdependence (Beck 1992: 67). We need to look into the human constructed activities as the causal relationship for drought and water scarcity by which certain environmental conditions are defined as unacceptably risky, and thus contribute to 'state of crises' (Hannigan 1995: 16–35). So far, the patriarchal disdain governed the authoritarian hegemonic control over earth's resources further complicates debate. This adaptation brings about the theoretical premise of Marxian base (infrastructure and built environment) affecting adversely the social structure (relations and organization) guided by the superstructure (ideas, belief and value formation). However, it is not clear as human's would distinctly follow anthropocentric approach owing to 'metabolic rift' of capitalism (Marx Capital II 1984). Hence, ecological analysis devoid of socio-political analysis is incomplete to interpret the crises. The deep association treating earth as live matter needs to be ignited in indigenous India. Therefore, researching discourses relating to women and water conservation; socio-cultural discourse and women in political participation discourse are some key areas to identify challenges and opportunities from the lessons learned from this experience.

In the case cited above, there are innumerable definitions of sustainable development to be taken into account, prima facie; of intergenerational equity by elevating responsibilities of sustenance to reverse depletion for future generation (Report Environment and Development 1987). Alongside, women are the wageless care workers looking after the old and the young; engaged in reproduction cycle are deprived of being beneficiaries-sharing direct partnership with government and its agencies, whereby women's worth of fetching water is estimated to be 150 million rupees annually would amount to 10 billion rupees annually, if put into the GDP as national exchequer (GOI Neeti Aayog report 2018). Thus, conservation practise invariably resolves poverty and economic conditions by reversing roles of (wo)man. Under a development model, the gender subjectivity has to be churned through woman inclusion in environmental policies. To accommodate women in participatory governance at a more productive level of actively reformulating policies and planning at micro-levels of production. The central crux is to unravel the intertwined relationship of women as experts on indigenous knowledge subject to their direct relationship with the water use for drinking vis-à-vis domestic purposes. Despite their role as water collectors and handlers, their participation in governance is restricted to traditional distinction of subordinate status conferred by the male heads in villages of Rajasthan. From this vantage point, the study entails utilization of their work at forefront by exploring their cultural socialization at well to 'not waste single drop of water' critical in meaningful participation transforming water handlers to the level of water leaders.

Women are the focal point of my empirical study largely interrogating their participation as stakeholder and shareholder combined to natural resource management viz. water. By and large, women are worst affected as legitimate power locates discrimination and misogyny at the behest of compromising sustainable measures and development. Women stand for all domestic responsibilities (cleaning, cooking, care economy and sanitation, and health) bereft of wages. Having said this, cultural-feminism theory since 1970s-80s considers nature and women best suited for liberal and socialist framework. This encompasses the complexity of interaction of water use within the socio-cultural paradigm with miniscule participation in conflict governance. Ample studies have focused on their productive role in agriculture but domestication subjugates them in double burden at home and expecting resilience in the face of natural disasters as well. In the 1980s 'ecofeminists' claimed that as nature and nurture are intertwined and so does the dualism of women and conservation binding them into unit, thereby, this proximity reiterates women as close to natural resources intervention and sustainable use of water (Shiva 1998: 9–12). 'I saw my mother and other family members struggle to get drinking water in my childhood. Now, I see my wife and my daughters-in-law go through the same trouble. Nothing has changed', as stated by old villager Jethu Bhai of Ramsar block. The socio-cultural norm of fetching water from few kilometres (4–5) away is the old practice due to subsequent drought in seasons (Drought in Rajasthan First Post March: 26). So far, water conservation is decided largely by the experiences and ideas of patriarchy.

Therefore, the need is to address the women's opportunities vis-a-vis designing the water conservation models by recognizing differential needs and priorities of (wo)men. The hypothetical finding indicate the conscious role of women as primary providers and handlers, subsequently to become water leaders in many projects launched for water conservation. Hence, the need to have all women inclusive approach to elevate their status as natural managers for water.

Sequencing of Argument

1. Environment does not differentiate on its own as drought and floods are universal catastrophe. Nonetheless, affected by gendered participation explores constraints for proper channel of co-ordination and correlation at a global paradoxical situation of scarcity and abundance.
2. With existing knowledge (population pressure) and valuing water as a scarce resource can address the cultural shift of gender inequality; by equal provision and share of responsibilities in water allocation, as this encourages women otherwise devoid governance-giving priority to family and children.
3. The study includes recommendations for filling the above knowledge gap and initiating community approach of 'cultural consciousness for water conservation' by restoring and recharging the traditional knowledge on water by women as leaders in community outreach for sustainability and alternate methods.
4. Hence, exploring women as resilient in resolving the water scarcity by participation in water management (Baolis: small lakes; Beris: shallow storage areas near lakes and Taankas: deep underground storage tanks well covered) would reverse gender disparity. They mitigate rainwater harvesting by proper management and all-inclusive approach in policy and planning measures.

Alternative Methods and Water Management

Water is collective right and so would be its equity and distribution a social right. Then be it, the comparison of (Dungarpur and Jaisalmer) or differential rainfall deprivation affects equally across the nation. The inhabitants of Rajasthan are anxious to solve water problem directly related with development. Thus, there has been consistent brainstorming to resolve the water crises by peasant's unions VMKS (*Vagad Mazdoor*

Kisan Sangathan); working towards tribal rights of the indigenous people over forests and its 'tribal self-governance' as direct stakeholders of its (ab)use and produce in terms of initiatives and action towards government pro-poor policies (MNREGA) and its implementation in Dungarpur, Rajasthan. In follow up, the maintenance of water bodies(ponds), uplifting of groundwater table and inclusion of local tribal population in decision making of governance is prerequisite. Also, because the depletion of water is calculated as one metre annually (Hindi Water Portal 2018). This direct relationship of paradigm with water as rarity take days to find as the only source here is groundwater (wells, aquifers and ponds). The ironical yet scientific rational behind recharging depends on the amount of meagre rainfall in Thar desert. The Aakar charitable trust did a public-private partnership (40–60%); build 60 check dams supplying water to 200 wells. This is small embankments of two feet high square (Chauka) filled as nine inch deep. These mud walls slow the rainwater giving enough time to percolate and excess water flow through channels to recharge groundwater (Clean Water report July 9 2019). However, the theoretical premise of scarce rainfall in only seasonal further problematizes the freshwater to be filled in wells. Other than these, three major recharge bodies viz. taanka; beri; and johad are used in these dry to semi-arid regions whose images are displayed below.

Rajasthan is an admixture of contradictory possibilities from (Jaisalmer) drying water tables to excessive run-offs in arid stony terrains (Dungarpur) unable to cope water conservation. Women are traditional managers to store and provide water within their domestic sphere and hence bear the responsibility of collection from far off lands. This leaves no time to invest in productive job of earning and impacts adversely to little children and their education at large. Both Jaisalmer and Dungarpur are poles apart and thus are critical areas in the sense of deprivation and marginalization to the access of water. 'Thar desert' is an arid prone area of extreme temperature (51–21 degrees) compared to quite dry and milder than most cities 'Dungarpur' as the southern most part of *Rajasthan*. As there is no perennial source of running water (rivers) the state is dependent on surface water. This bears a direct correlation with the failure of massive state funded schemes highly dependent on ground water which has depleted further with 'two dozen of districts out of the total of 33' declared as severe drought/famine areas (Rajasthan Water Crises DNA).

Another component is community participation involves lesser costs and solves water scarcity with high benefits as depicted below. Significantly, construction of 'community ponds' entails low costs and higher benefits seeking public ownership within the alternative approach of sustainability. This perspective bears a 50:50 ratio partnership in construction, maintenance and budgeting of water bodies coordinated by local gram level panchayats and their collective efforts (India Water Portal 2019). The supportive argument is the conservation of water in drought or water wastage with scientific mechanism to storage and conservation by women. Both districts are hit by drought, former relies on rivers and groundwater management and the other is fluctuating rains and low groundwater availability. Hence, correlation needs to be drawn in paradoxical situation of development arid and semi-arid remote areas of underdevelopment. As per the census of India population of *Rajasthan* in 2018 was 7.83 crore and 3604 cities. And geographical area is as big as largest in India

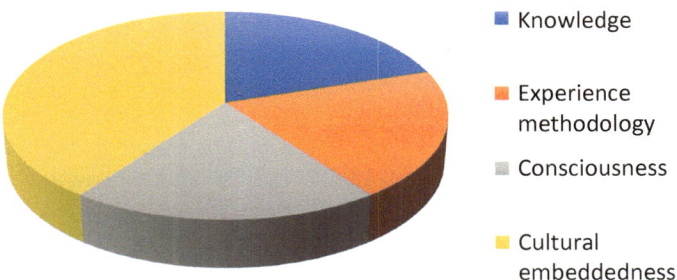

Fig. 18.1 Community approach for water management. *Source* Author

(area-wise—342,000 km). The point is to draw a trajectory indicates disbalance of population versus resources and vulnerability of third world state. Alongside, women at large, are neglected and hence are poorly representative in government proposed water policies and conservation measures directed by the state (Fig. 18.1).

As Hoekstra (2015) once said water just like oil is a geopolitical commodity, where people or nations would do anything to get full control and access of water, thereof Chennai crises appeal to anticipating wars affecting the political economy. The water footprint 'is an analogue to the ecological and carbon footprint' talks of use and consumption (Hoekstra 2015: 36). Moreover, rather than recognizing, its dualistic epistemology, the emphasis is on dialectics from prehistoric 'little' traditions of worshipping nature as 'mother goddess' to institutionalizing structural variations of gender inequality owes to water governance. The two diverse areas of *Rajasthan* (*Jaisalmer-Dungarpur*) are chosen due to their extreme similar conditions as they have significant dependence on the forest and/or natural resources available in their neighbourhood that depend on borewells as deep as 150–300 ft. Due to its stony nature the capacity to retain moisture in *Dungarpur* is quite weak. Alongside the arid dry deserts of western *Rajasthan*, almost rainless as bad as 120 mm, the rest of the time water is stored in wells and ponds. Mostly, little earthen embankments are also constructed to store rainwater alongside to groundwater recharge covered by thorns to keep away from animals. Therefore, mixed efforts are being observed in southern and western parts, in reality the indigenous knowledge of preserving and using every drop of water by the local women for most of its use is a socio-cultural construct embedded in Rajasthan. This solves the dual purpose of groundwater recharging vis-à-vis rainwater harvesting.

A large tank (Gadisar) is been built in Jaisalmer by the Garhwal kings, presumed to be the largest tank, wherein a large barrage was built to trap the silt to facilitate the clean flow of water to be stored directly for drinking water (Saxena 2017: 92). Any act of water pollution (bathing, cleaning or animal) may be the reason to vanish under the silt. As water is available only for 5–6 months under good rain. The poor were usually unable to obtain water from distant places and often resorted to drinking mixed water with yogurt. Taanka is an age-old traditional practice to augment water in the Thar desert, an initiative taken by villagers to dug underground and finish it with cement to preserve rainwater as seeped through these circular holes to be used in crises. Similarly, *Par* is a seepage area near flowing water to restore sweet water in shallow well, and if dug deep become a 'Beri' a catchment area to restore seepage water but deep as 3–4 m underground and can refill water overnight. Even today, water is recycled and cycled for washing, cleaning or animal consumption, tanks are cleaned, water is filtered by voile cloth, and cleaned further by dissolving *fitkari* (alum), and segregated for drinking and other purposes solely to be managed by women as water leaders. There is also the concept of modern-day 'water budgeting' by the traditional name of 'Anga' (unit) or 'paytan' (catchment), wherein the tanks are cleaned (desalination) before the onset of rains by the egalitarian distribution of work and water among themselves (community). The distribution is based upon counting humans as per house and animals as per head. Water harvesting, distribution and its use were thus regulated by retaining water near beris, tanks and Kuis and Baoli's (small wells), and tanka, regulated by the community as displayed below. Then there are step wells (baolis or jhalaras), generally are philanthropic as water has inherited a divine intervention of sustenance. People usually give colloquial terms such as 'sagar ka kua' (wide well as ocean) or 'seer ka kua' as natural seepage supplying to wells is comparatively small (Saxena 2017: 91) (Figs. 18.2, 18.3 and 18.4).

As depicted above uneven rainfall caters to multiple ponds and talaabs and hand-pumps in comparison with Jaisalmer with few beris and taankas, most of which have been revived recently. Therefore, both in Dungarpur and Jaisalmer women have been

Fig. 18.2 Depiction of Taanka. *Source* Taanka (Wikipedia)

Fig. 18.3 Depiction of traditional well. *Source* Wells (down to earth)

Fig. 18.4 Depiction of Beri (near ponds to recharge water). *Source* (Beri) Hindustan Times

proactive in collection and preservation of water. In one statement, 'When we moved to the village as teenaged brides, we used to travel 20 km to fetch water', recalls Nakata. This is the way their lives changed as recollected by women from their early experiences, that how they desilted and excavated beris again merged under the sand by building walls around it so as to able to supply almost twenty-five hamlets for the whole year. There is a difference in language and dialect of these two diverse cultural regions of the state. 'Nature is not a public distribution system', hence its solutions have to be invented with the emerging crises stated by a villager who works with NGO (Thar Integrated Social Development Society). 'It has its own way of distributing resources and we can't do anything about it except working on our habits. I was 11 years old when, for the first and the last time, I used two buckets of water for bathing. My grandfather slapped me and said that I was a sinner in the eyes of God', recalled by Jethu Singhji. 'That's how our habits and approaches were moulded' (India Water Portal 2019).

Moreover, societal structures/cultural milieu contradicts the bargaining negotiations of labour (collection, storage and provision) within male-headed households. Also as the water collection is a gendered labour, women emerge as natural mangers for the culture of sustainability and can be thus termed as innovators and caretakers of the water conservation (Joshi et al. 2005). This would be discussed at length in the discourses noted below. Notwithstanding, compels a gendered analysis to address needs, interests and constraints to water resource management. In fact, judicious use, conservation and restoration demand planning based upon participatory democratic partnership between rural and urban sectors. It should be implemented on four levels on village levels, district level, block level and city levels. This should include the alternative methodology adapted by women indigenous knowledge based upon spatial understanding of geo satellite mapping, land-use, and governmental and non-governmental documentation of policy and project.

Water is thus a significant component of socio-cultural and political economy of survival for women, at large whether recognized or not seems to be withhold the empowerment discourse of village women intact. They are the pivot of water management from collecting to handling to water storage in scarcity. In all the global forums from Brundtland's commission to MDG Goals, the relationship of water and gender has been referentially established. The study on the role of woman in water conservation is further divided in three broad themes. Firstly, to explore gender mainstreaming with water discourse; secondly, drawing a cultural metanarrative discourse on the indigenous methodology; and finally, to incorporate the gendered dynamics in policy initiatives and planning discourse is a critical perspective in sustainable development paradigm.

Gender Mainstreaming and Water Discourse

Women and water are intertwined and therefore gender inclusion policies should be targeted to perpetuate equality and inclusion for the larger goals of representation at all levels (UNDP Report 2006). The primary threat for ecological reserves is when the power is centralized in the hands of few controlling the system as fetching determines inequality. The massive time spent on domestic works (cooking, cleaning, washing) is a negative correlation to income or health of home managers. 'It is estimated that, every day women need to make six trips fetching water, often walking 10–12 km on an average, carrying up to fifteen litres each trip. As they load jars or buckets on their heads to carry water, the pressure creates various physical disorders like back, feet and posture problems. Many a young girls are forced to drop out of education in order to help their mothers and other ladies in the household performing similar duties' (South Asia Inequality Report 2019). Similar on the lines of national exchequer status of India, or sub-saharan Africa and other undeveloped countries total worth of women's spend in collection and procurement of food with water would be almost the GDP of European nations at large (France). Thus, the above context is drawn to build a theoretical perspective can change the inequal relationship mobilizing gender equality to social justice. Inadvertently, the time spent can be utilized by the young girls in attaining education and preparing for higher education to build their economic standing directly with no compensation required (UNDP Report 2006).

Secondly, series of discourses via political intervention of agency and structure needs to be disseminated filling the feminist discourse of material struggle over control and management of water. In this process, actor-network theory of resource intervention vis-à-vis shaping of meaningful definitions of woman and man is constructed. The indigenous methods of boiling or earthen pot treatment to filter water by women is an age-old budget friendly practise. Despite establishing a relationship of empowerment and governance, the vacuum of women leadership is missing in promoting the water cause. Hence, this aspect needs to be institutionalized in framing training module to awaken consciousness. It is, therefore, imperative to encourage women to contest elections and be a part of project design to elevate status in Rajasthan (Fig. 18.5).

Fig. 18.5 Women travelling in Jaisalmer to fetch water long distance. *Source* (Women in desert) Smithsonic

Thirdly, critical interventions are restoration by proactive traditional methods of water storage is the counter argument in deconstructing gender roles by positioning 'women' as primary caretaker of 'water'. As has been discussed above, seen within valuable inputs from the localized spatial cultural landscape mere advantaged as natural heir of the rights and privilege of care economy to dominate all the spheres of life. Furthered by the ritualistic bondage of caste, class and ethnicity compels them to be validated in the patriarchal norms and social system. This unjust discrimination exists between the gendered roles subject to private and public sphere. The capitalist domain of patriarchy exploits the women on equitable human rights as compared to men in terms of protection and safeguarding and control over decision making. (Shiva 2010: 45).

Alongside, the conditions in South Asia are vulnerable to external factors reimposing the traditional oppression in newly emerged ways of development. As the wageless care economy of women contributing towards the private sphere of household in exempted from measurement of economy and GDP. According to the ILO (Report on Care Work and Care Jobs for the Future of Decent Work 2018), 'globally, women perform 76.2% of total hours of unpaid care work, more than three times as much as men. In Asia and the Pacific, this rises to 80%'. Thereby, continuing work after discontinuity is the main barrier of progressing and managing promotion in the waged workplaces due to reproductive labour and domestic responsibilities.

Lastly, feminist environmentalism, victimization due to environmental degradation is seen in gendered binaries. This implies, despite being the victims and actors of environmental movements, their roles and implications are assessed in contradiction to equality disjunct from policy and planning (Agarwal 1992). At the same time, there is a contradiction in the ecofeminist discourse and it's interrelated material sources of domination which impinge critically on the political levels of ideological construct merely to conceptually link the symbolic construction of culture with nature, i.e. women with men. 'For women…death of prakriti, simultaneously a beginning of their marginalization, devaluation, displacement and ultimate dispensability, the ecological crisis is at its root the death of feminine principle' (ibid.). Thus, women are elevated as symbols of power but are pulled downwards by its glorification restricted to 'parochial loyalties' within families having to meet its everyday demands of work by exploiting feminity. This was the material sustenance of Shiva's argument in relation to 'staying alive' of women within the feminist discourse of being powerful only when connected to natural resources. The comparison that ignited the research has been the disconnect of nurture from nature or water from women in collective ways of categorizing traditional as disempowered from the obvious roles in their daily lives (O'Reilly 2006: 958–972). Though the studies are not enough in establishing a relationship of women and water, this paper argues to mobilize their political participation by recognizing their intersecting knowledge of prakriti-nature and shakti-nurture (Shiva 2010), other than embedded empowerment and modernity.

Cultural Metanarrative and Water Discourse

Women with water is embedded within the cultural framework for its divine intervention in daily lives. Hence, worshipping its natural resource of precipitation (megh) by worshipping Hindu deity 'Shiva' celebrating it with 'Dal, Baati and Choorma' (traditional whole wheat and pulse preparation) offered to all those who offer their prayers to 'shiva-linga' bathing it with charna-amrut mixture (yogurt, milk, ghee, honey and Basil-leaf). This entire ritualistic system is called as 'sahastra Dhara' (thousand flow of water) to please the rain god. Another interesting incident is from the village in *Barmer (Rajasthan)* theatrical performance of child enacting as a frog strolling around the village only to gain 'wheat-grains' singing along—'aako bhayo ghar ghar dhoondhaiyo' (seeking water strolling in houses to bestow good luck), finally to be cooked with jaggery only to be eaten as a feast by villagers. This again is symbolic of the frog seeking to appeal the rain god to bestow rainfall in dry areas. Participation is ensured across religion and caste in this procession to be ended near the dry well, where the sweet dish is prepared to feast by all (God's sweet). Then another interesting folktale depicting animals (dears) strolling and sacrificing lives for each other in dearth of water for two. They both died wanting the other to survive.

Neh Ghano-jal thodo......to pee,...tu pee... kahveta ... dono tajo sharer

(Less water but immense love sacrifice mortal body for each other)

There are thousands of adage attached in this region where water as natural resource is literally worshipped. The point is to determine water as society and culture as consciousness in *Rajasthan*. The point is to determine 'water as society' and 'culture as consciousness' in innumerable sources of water resources such as lakes, tanks and wells (samanad and sarovar, nadi, johad, bandha, sagar) in Rajasthan. The statement holds true of the socio-cultural life of the desert people in the sense, access, availability and accountability of conservation becomes cultural matrix for sustainability. As stated and scientifically proven, the state receives scanty rainfall (120 mm) and thus is declared drought stricken demands a contextual paradigm of ideological development to material reality of the analysis (Dunlap and Catton 1994: 65). This asserts the fact, locating the water conservation methods knitted upon the distribution, geopolitical mapping with cultural variation; socio-political mobilization reflects the discourse of gender inequality. There can be variation in relation to cultural ethos shaping 'social facts' of the cause, nonetheless remains a problem. 'Paani to ekattho karneo padi; paani bina kee koni', 'lugaii ka kee koni kaam chulha chowka jamanoe hai'—(water has to be conserved as the duty of living life; 'there's nothing without water'—the ultimate responsibility of conservation is on women solely). Is it, then not a cultural study or a challenge to combine the intertwined conflict disseminating from the traditional understanding of woman and their roles in water problem? It is time to interrogate their asymmetrical power relations emanated from varied cultural associations with nature and revive their alternative techniques of cultural conservation with water.

Thus, water conservation is a cultural application factored upon explicit experience of being a women that can aid in rainwater harvesting and policy development. Role of women is crucial in facilitating water management for sustainable use in drought prone Rajasthan with scanty short-term rainfall. These are few of the illustration of how the (in unilocal dialect) base of the pot leaves a permanent mark on the borders of the wells when water is fetched by women. Resultantly, if are disadvantaged then changing this situation requires an action plan for multiple benefits such as political; economic and social welfare policies. (Wo)men taking a lead in forefront would liberate them from hesitation and mobilize them for decision making. On the other side, this way women would save time which would be utilized for income generation and training themselves for skill development for extra income. Other than this, focus on children can bring better results (as there would be no drop out of girls from school and they can concentrate on education rather than fetching water saves their times spent). But the very fact that somebody is there because of the labour put in by women is a fact undenied. Women being the water providers can facilitate its management and governance via cultural consciousness of saving water. The mode of the my secondary research includes working closely through local NGO's (non-governmental organization) documentation and reports; data/statistics of government and interviews after participation in their cultural programmes reflexive of their commitment to save natural resource. The local dialect and narratives are a storehouse to explain the struggles over revival and regenerating a lost natural resource. As the syntax of gendered discourse is well elaborated above, the disadvantaged position to share wageless household jobs stands unequal with respect to water-related activities. The nuances of knitting across the cultural yardstick of conservation of natural resource make women as progeny to find alternative methods in lieu of the global crises of water. However, the question stems in relation to their participation in policy making and planning programmes to delve local level initiatives to resolve the ground water depletion in dry arid zones. The *anganwadis* and the traditional people of villages can combat by pushing forward the women of the house and by sharing their inequal-work and encouraging their inclusion to work with government aided and partial aided agencies to come with construction and rejuvenating water bodies of these areas.

Development and (Wo)Men Discourse

Though there are multiple factors to determine the sustenance of natural resources viz. climatic conditions, land use, geographical mobility yet proactive roles in government policies have meaningful interventions. Gender is utilized as an analytical tool to understand roles and expectations in natural dependent relationship with water. More importantly, the role of women is precarious towards a gender discourse to facilitate access and control on water via resource management. The strategy required above cannot be universally applied to the heterogeneous conditions, yet gender equitable approach can be encouraged disguised in a traditional society of intersectionality of caste, class and ethnicity. However, focusing on negotiating identities as water saviours can get away with underlying questions of gender dynamics of inequality in household activities from collection of water to cooking food, child care and other activities. This shapes their relationship with the other units of economic, religious and political units of social life. In this relation, the discourse is to locate women among the structural constraints and possibilities for women at central to the structures of water conservation methods. Traditionally, women are assigned for struggles for managing water in domestic responsibilities and outside too. Therefore, the term 'water providers' are invariably being used for the capitalist enterprise whereby the water tanks charge much higher prices for the water.

In this regard, it becomes imperative to note the primary role of gender as was recognized in the sustainability of water; meaning gender needs should be part of the overall policy framework (Dublin-Rio Principal in Global Water Partnership 1992 Principle 3). In the sense, major responsibility of pulling them to policies for conservation was suggested with specific needs and empowerment of women (second World Water Forum in 2000), by more participation in management. After this a gender-equal approach came as a highlight (Human Development Report 2006) on freshwater with equal rights of (wo)man. Thereafter, gender sensitivity came with (Millennium Development Goal) women and sanitation to connect women empowerment (SDG: 6; 5); relationship of 'water and women' well elaborated in sanitation report synthesis. Though facilitates maximum socio-economic returns by engaging women at more ground levels, yet unclear on execution on individual levels. Within these landmarks framing a connect beyond rhetoric explore the alternative approach of storage and conservation (Fig. 18.6).

Since this paper is blend of a comparison of the cultural underpinning vis-à-vis indigenous knowledge of women in water scarce region of Rajasthan. The attempt is made to locate 'gender role' and its outreach for 'community development' in 'water-use' in Rajasthan. As the contribution of (wo)man is inconsistent and mono-dimensional in designing the project based as per localized needs. This study can effectively bring SDG 5 (UNDP 2005) by partnering and integrative approach women and water into implementation of MDG 2015 by active participation and recommendation. Nevertheless, certain constraints and possibilities emerge in terms of capacity building of women in (informal) sectors subject to, storage, recharge, management and resource control.

Policy Implication

The need to address the constraints and possibilities represents empowerment by inculcating women into management, planning and decision-making process for various government and non-governmental water projects. Discourses such as linking water as a natural resource with women as responsible for its (ab)use and storage make them more competent in collection and sustenance. These interpretations meanings take on a distinctly gendered character. This has long been advocated by UN

Fig. 18.6 Rajasthani women drawing water from well (Jaisalmer). *Source* Agefotostock

agencies from 1992 to 2015, yet the practical approach seems to be only partially fulfilled. Due to which, intervention of academic discourse and theoretical underpinning is required for strongly recommending gender inclusion and education in this direction via training modules and workshops of SHG's (self-help groups). The paper would reverse the paradigm shift of gendered control by utilizing women's traditional knowledge to mitigate the water scarcity by meaningful interpretation. Hence, the approach benefits mutually 'water and shortage' the implications as a direct repercussion of all-inclusive approach as formal partners. Thus, its use and abuse have to be closely monitored by community management in villages and hinterland. Women have longitudinal view of their experiences, beliefs and ideas about their roles and responsibilities of emerging as water leaders from primary handlers. They can act as a catalyst to conserve water resources due to their groundwater knowledge of land and soils. By including them in local level bodies to panchayat samities can monitor water use in household to irrigation and agricultural use. Deciding upon their work efficiency, training to adopt water-intensive techniques in drought-prone areas can be transmitted in reviving and restoring traditional water bodies (wells, ponds, lakes, catchment areas) to restore rainwater harvesting system. The dependency of rainfall in recharging groundwater is the basis of legal framework of justifying water right. Besides, the rational pragmatic policy (UN 2005–15), the political and administrative representation are required in water recycling, desalination, fluoride-free and water pollution. Therefore, strongly advocates the equality of opportunity in governance would bring a change in income generation, care of the children and their future as overall development of women at large.

Conclusion

To conclude, the crux of this paper is to use traditional water harvesting system as an alternative method to solve the problem of water scarcity. This needs to be done by systematic analysis of women's invariable nature-nurture relationship integral in drought-prone areas. As it is, with almost disappearing water resources and ground water depletion, surface water is not enough to meet the demands of the villages and the hamlets. Women's participatory role has been discussed and rationalized in international platform, but lacks action-oriented approach in developing country. They have been victims of patriarchal division of labour vis-à-vis lacks control in monitoring of water, revival of water bodies and rainwater harvesting. In this way, ecofeminist position recognizes the lacuna found in dry states despite their recognition worldwide-brings forth the issues of revision and refinement of a direct inclusive policy both in theory and practise. Ironically, their ingrained traditional understanding reflects close associations with households and agricultural usage and distribution are the key to develop a model on alternative methods for water conservation. This is how indigenous knowledge within localized conditions of sustainability can be utilized to bring forward their sense of revival of traditional water reservoirs. By developing various discourses of women with cultural metanarrative, develop to manage on alternative source to preserve water conservation techniques. This can happen, by reiterating the gender dynamics of exploitation and providing combat by their participation in policy and planning, which in turn, will reverse gender roles and more equitable division of labour. By and large, their meaningful purpose can guide the community to bring a change in policy reformulation as per the needs and requirement in desert water crises management.

End Notes

1. There has been alarming studies across the globe in relation to understand water crises as social crises. And as it is a collective issue has to be dealt with collective efforts of the people across caste, class and ethnicity. In this regard, the developments from the sustainability to the message from the UN secretary general Ban Ki-moon on 22 March 2011-on world water day are descriptive to bring the stark reality and correlation of the vulnerable conditions of the women and young girls average of spending time in collecting and fetching water is contributing to their adverse effects in education, health and violence in the society. Therefore, the need to pro-governmental policies and programmes is required and a global effort to bring the inclusion and equality for all in terms of water for all with no deprivation in good health should be the target. http://www.unis.unvienna.org/unis/en/pressrels/2011/unissgsm247.html.

2. The condition is miserable in drought-stricken desert areas, then be it Africa or India, women walk more than 2 miles carrying and balancing 2–3 pots approximately 15–20 l of water every day is critical areas of development and human rights. https://edition.cnn.com/2015/07/16/asia/india-water-wives/index.html.

References

Agarwal, B. (1992). *The gender and environment debate: Lessons from India* (Vol. 18(1), pp. 119–158). USA: Feminist Studies. Inc. https://doi.org/10.2307/3178217. https://www.jstor.org/stable/3178217. Accessed on June 10, 2020.

Agefotostock. https://www.agefotostock.com/age/en/Stock-Images/Rights-Managed/DPA-AMA-89210.

Beck, U. (1992). *Risk society: Toward a new modernity.* London: Sage.

Beri. https://www.downtoearth.org.in/coverage/beris-to-the-rescue-40065/Beri.

Clean Water Report on Amla Ruia: India's 'Water Mother' Who Helped Provide Water to Over 300 villages in Arid Region. http://clean-water.co.in/2019/07/09/amla-ruia-indias-water-mother/. Accessed on June 10, 2020.

Dunlap, R., & Catton, W. (1994). Struggling with Human Exemptionalism: The Rise, Decline and Revitalization of Environmental Sociology. *The American Sociologist, 25*(1), 5–30. https://www.jstor.org/stable/27698675 (Accessed on December 2019).

Drought in Rajasthan: Over Rs 7,000 crore spent on projects, but not much water has flow through western region. First Post. https://www.firstpost.com/india/drought-in-rajasthan-over-rs-7000-crore-spent-on-projects-but-not-much-water-has-flown-through-western-region-6331911.html March 26, 2019. Accessed on June 4, 2020.

Dublin-Rio Principal in Global Water Partnership. (1992). Principle 3. https://www.gwp.org/contentassets/05190d0c938f47d1b254d6606ec6bb04/dublin-rio-principles.pdf. Accessed on July 5, 2020.

Hannigan, J. A. (1995). *Environmental sociology: A social constructionist perspective* (Chapters 1 & 2). London: Routledge.

Hoekstra, Y. A. (2015). The water footprint: The relation between human consumption and water use. In M. Antonelli & F. Greco (Eds.), *The water we eat.* Switzerland: Springer International Publishing.

How Local democracy is solving water issues in Southern Rajasthan. https://www.indiawaterportal.org/articles/how-local-democracy-solving-water-issues-southern-rajasthan. Accessed on May 17, 2020.

Human Development Report. (2006). Beyond scarcity: Power, poverty and the global water crisis. http://hdr.undp.org/sites/default/files/reports/267/hdr06-complete.pdf. Accessed on November 30, 2019.

International Decade for "Action for Life" 2005–15. https://www.un.org/waterforlifedecade/gender.shtml. Accessed on January 30, 2020.

Joshi, D., Lloyd, M., & Fawcett, B. (2005). *Women 2000 and beyond published top promote Beijing declaration and platform for action.* https://www.un.org/womenwatch/daw/public/Feb05.pdf. Accessed on January 25, 2019.

Marx, K. (1984). *Capital II: A critique of political economy* (Vol. II). London: Lawrence & Wishart.

O'Reilly, K. (2006). Traditional women modern water: Linking gender and commodification in Rajasthan. *India Geoforum, 37,* 958–972. http://jcsites.juniata.edu/faculty/pelkey/sdarticle.pdf. Accessed on June 1, 2020.

OECD. Annual Report on Sustainable Development. Work in the OECD 2008. https://www.oecd.org/greengrowth/42177377.pdf. Accessed on May 17, 2020.

Rajasthan water Crises in 19 districts, nearly 17,000 villages face acute shortage. https://www.dnaindia.com/india/report-rajasthan-water-crisis-in-19-districts-nearly-17000-villages-face-acute-shortage-2203359. Accessed on June 8, 2020.

Report of ILO. Report on Care Work and Care Jobs for the Future of Decent Work 2018, Care work and Care jobs for the future of decent worth. https://www.ilo.org/wcmsp5/groups/public/—dgreports/—dcomm/—publ/documents/publication/wcms_633135.pdf. Accessed on March 25, 2020.

Saxena, D. (2017). Water conservation: Traditional rain water harvesting systems in Rajasthan. *International Journal of Engineering Trends and Technology (IJETT), Seventh Sense Research Group, 52*(2). (Open Access Journal). https://www.researchgate.net/publication/322051596_Water_Con

servation_Traditional_Rain_Water_Harvesting_Systems_in_Rajasthan. Accessed on May 25, 2020.

Shiva, V. (1988). 2010. *Women in nature. In staying alive: Women, ecology and development* (Ch. 3; pp. 38–54). London: Zed Books.

Shiva, V. (1998). Women water rights. *Waterlines, 17*(1), 9–12. https://www.jstor.org/action/do. Accessed on September 20 2019.

Smithsonianmag. https://photocontest.smithsonianmag.com/photocontest/detail/young-women-carrying-water-from-distance-in-thar-desert-of-rajasthan-nikon.

Taanka. https://en.wikipedia.org/wiki/Taanka.

The challenge of water. India's ability to manage and govern water will determine its future. https://timesofindia.indiatimes.com/blogs/toi-edit-page/the-challenge-of-water-indias-ability-to-manage-and-govern-water-will-determine-its-future/. Accessed Jan 4, 2020.

Traditional Wells. https://www.hindustantimes.com/jaipur/traditional-wells-to-be-revived-in-desert-areas-to-fight-water-crisis/story-qsZrSWcOhNS3TICmtquT5K.html/ traditional wells.

UN Millennium Project task force on water and sanitation Project lead authors Robert lenton, coordinator Albert o Wright coordinator Kristian Lewis. https://www.unoosa.org/documents/pdf/psa/activities/2005/graz/watercomplete-lowres.pdf. Accessed on June 9, 2020.

UN. Sustainable Development Goal 6 Synthesis Report 2018 on Water and Sanitation. https://sustainabledevelopment.un.org/content/documents/19901SDG6_SR2018_web_3.pdf. Accessed on February 7, 2020.

UN Sustainable Development Goals https://www.un.org/sustainabledevelopment/gender-equality/. Accessed on January 7, 2020.

UNDP Annual Report (2005). https://www.undp.org/content/undp/en/home/librarypage/corporate/undp_in_action_2005.html (Accessed on February 2020).

UNDP. (2006). Mainstreaming gender in water. United Nations Development Programme (UNDP), 2006. IWRMGenderResourceGuide-English-200610.pdf. Accessed Jan 20, 2020.

United Nations. Water Scarcity Factsheet. 2013. Available online: http://www.un.org/waterforlife decade/scarcity.shtml. Accessed on April 15, 2020.

Water Crises in Rajasthan Hindi water portal. https://hindi.indiawaterportal.org/content/water-crisis-rajasthan/content-type-page/53102. Accessed on May 25, 2020.

Water Crises? India won't be at Sea-report by Neeti Aayog. https://economictimes.indiatimes.com/news/politics-and-nation/water-crisis-india-wont-be-at-sea/articleshow/70136203.cms?from=mdr July 9, 2019. Accessed on June 1, 2020.

Water and Women: Navdanya report by A research foundation for Science and technology for national Commission of women. https://www.navdanya.org/attachments/Water_Democracy1.pdf. Accessed on January 15, 2020.

We can end Poverty, Millennium Development Goals and 2015. https://www.un.org/millenniumgoals/. Accessed on June 6, 2020.